THE CAMBRIDGE DICTIONARY OF SPACE TECHNOLOGY

The Cambridge Dictionary of Space Technology is a comprehensive source of reference on the most important aspects of this fast-developing field, from basic concepts to advanced applications. With some 2300 entries, it lists fundamental terms that will remain in common usage for the foreseeable future and includes a selection of historical and highly specific entries adding context and depth.

The unprecedented breadth of coverage ensures that there are entries on all major subject areas. While the emphasis is on defining the meaning of a word or phrase, entries have been written with the intention of enhancing the understanding of the subject, both for the practising specialist and the interested layman. To assist the reader in research on a given topic, related entries are highlighted in the text and other important entries are cross-referenced.

The Cambridge Dictionary of Space Technology will be indispensable to anyone with an interest in space activity.

MARK WILLIAMSON is an independent Space Technology Consultant advising the space industry and the space insurance community. A Chartered Physicist and Chartered Engineer, he has more than 20 years' experience in satellite communications engineering, technical management and space consultancy.

As a technical author, he has written some 250 articles on various aspects of space technology, and several dozen papers on subjects as diverse as space law, space education and space history. He is also the author of a student text, *The Communications Satellite*, and the editor of an international space industry magazine.

He enjoys sharing his interest in space technology with audiences at all levels, both at home and abroad, and is proud to have co-presented the 1983/84 IEE Faraday Lecture series, entitled 'Let's Build a Satellite'.

To my son, who could say 'star', 'moon' and 'space' before his second birthday. Here's another couple of thousand words for you, Matthew!

THE CAMBRIDGE DICTIONARY OF

SPACE TECHNOLOGY

MARK WILLIAMSON
Space Technology Consultant

CAMBRIDGE
UNIVERSITY PRESS

PUBLISHED BY THE PRESS SYNDICATE OF THE UNIVERSITY OF CAMBRIDGE
The Pitt Building, Trumpington Street, Cambridge, United Kingdom

CAMBRIDGE UNIVERSITY PRESS
The Edinburgh Building, Cambridge CB2 2RU, UK
40 West 20th Street, New York, NY 10011–4211, USA
10 Stamford Road, Oakleigh, Melbourne 3166, Australia
Ruiz de Alarcón 13, 28014 Madrid, Spain
Dock House, The Waterfront, Cape Town 8001, South Africa

http://www.cambridge.org

First published 2001

Printed in the United Kingdom at the University Press, Cambridge

Typeset in Swift Regular 9.5/14pt, in QuarkXPress™   [SE]

A catalogue record for this book is available from the British Library

Library of Congress Cataloguing in Publication data

Williamson, Mark.
The Cambridge dictionary of space technology / Mark Williamson.
    p.   cm.
ISBN 0 521 66077 7
1. Astronautics – Dictionaries.   2. Aerospace engineering – Dictionaries.   3.
Astronomy – Dictionaries.   I. Title.
TL788.W54 2001
629.4′03–dc21   00–059884

ISBN 0 521 66077 7 hardback

PREFACE AND USERS' GUIDE

Space technology covers a great many aspects of science and technology, and since its birth in the late 1950s, the subject has changed almost beyond recognition. This dictionary has been compiled with the aim of providing a convenient source of reference to some of the most important aspects of this developing technology.

One of the most difficult aspects of writing such a book is deciding what to put in and what to leave out. The basic premise was that it should be a dictionary rather than an encyclopedia. In other words, the emphasis is on defining the meaning of a word or phrase, and in some cases giving notes on its derivation and usage, rather than providing large amounts of background information. The entries have been written with the intention of enhancing the understanding of the subject, both for the practising specialist and the interested layman. To this end, the dictionary includes terms at all levels of sophistication: from the academic to the trivial; from the precision of engineering terminology to the frivolity of jargon.

The entries in this dictionary are arranged in alphabetical order. To assist the reader in research on a given topic, related entries are highlighted in the text and other important entries are included as a 'see also...'. Although every effort has been made to make the cross-references self-consistent, there may be cases where useful references have been omitted. I hope that they are few and far between.

In an attempt to cover the fundamentals of space technology, the dictionary includes material which can be placed under the following headings:

- Spacecraft Technology
- Communications Technology
- Propulsion Technology
- Launch Vehicle Technology
- Space Shuttle
- Manned Spaceflight
- Unmanned Spacecraft
- Materials
- Propellants
- Orbits
- Physics and Astronomy
- Space Centres and Organisations
- Miscellaneous

To facilitate the use of the dictionary, a classified list of all the dictionary entries under these headings is included at the end of the book. Owing to limitations on space and the intended nature of the book, the number of entries on specific spacecraft and space missions has been kept to a minimum.

A degree of judgement is implicit in the compilation of such a book, not only in what to include but also in the relative importance of the topics. It is inevitable that some readers will disagree with some of the entries, especially where preferred usage is under discussion. Where terms mean different things to different people, I have attempted to make this plain, but, for the sake of the novice, have indicated the most common usage.

This is a completely revised and expanded edition of a dictionary that was first published by Institute of Physics Publishing in 1990, since when the space community has witnessed many exciting new developments. The introduction of new terminology to accompany these developments is a continual process and, unfortunately, no book can hope to cover it all. For this reason, *The Cambridge Dictionary of Space Technology* concentrates on the fundamental terms expected to remain in common usage for the foreseeable future, adding a selection of historical and highly specific entries to add context and depth to the subject. Readers with any suggestions for additions or improvements are invited to contact the author through Cambridge University Press.

Mark Williamson
Kirkby Thore, Cumbria
April 2000

NOTE ON ACRONYMS AND ABBREVIATIONS

Many technical subjects are replete with acronyms and abbreviations, and space technology is one of the worst offenders. For simplicity, most people in the industry refer to all groupings of initial letters as 'acronyms', but this is not strictly correct. As a practical guide to usage of these terms, particularly regarding pronunciation, the following distinction has been made wherever possible in this dictionary:

an **acronym** is a grouping of initial letters, or similar contraction of words, that can be pronounced as a word: e.g. ESA ('eesa'), EGSE ('egsey') and ECLSS ('eecliss');

an **abbreviation** is a similar grouping which is usually pronounced as separate individual letters: e.g. EBU, EHT and EVA.

There are a few which can be pronounced in both ways (e.g. TWTA/'tweeta') and, where relevant, this is mentioned in the text.

ACKNOWLEDGEMENTS

Space technology draws heavily on a number of classical scientific disciplines including physics, chemistry, astronomy and, with the inclusion of manned spaceflight, biology. It is, however, the applications of these fundamental subjects that have expanded the field of space technology to the multidisciplinary subject it is today. Thus it includes aspects as diverse as orbital dynamics, communications, propellants and materials technology.

For this reason, in compiling the first edition of this work, I obtained the assistance of a number of friends and colleagues who specialised in one or other of these fields and asked them to review sections of the original manuscript covering their speciality. I am indebted to them for the time they gave so freely and would like to acknowledge them here. Although their contributions have improved the book, they are not of course accountable for the text as it appears – I have assumed editorial control and any errors are entirely my own responsibility!

I would particularly like to thank Alan Hutchinson, Ron Jones, Ron Cooper, Jehangir Pocha, Karen Burt, Geoff Statham and Dr W. Berry whose contributions were invaluable. I am also grateful to Tom Keates, Anthony Giles, Ian White and Paul Brooks for their helpful comments. Many thanks are also due to Claude Bonnet, Richard Barnett and Clive Simpson . . . and to Alan Burkitt for suggesting I write a dictionary of space technology in the first place.

Finally, I would like to thank my wife, Rita, for her helpful comments on the manuscript and her considerable efforts in putting the first edition into alphabetical order. The wonders of word processing software have made this second edition so much easier to compile and update, but I shall never forget the reams of paper involved in the production of the first edition.

MW

A

A-4 The original engineering designation for the **V-2**. See **V-2**.

ablation The erosion of a surface, usually of a spacecraft **heat shield** on **re-entry**, due to friction with the molecular constituents of the **atmosphere**. A surface designed to ablate in such circumstances is known as an 'ablative surface' or 'ablative coating'.

ABM An abbreviation for:

  (i) **apogee boost motor** – see **apogee kick motor (AKM)**.

  (ii) anti-ballistic missile, a missile designed to destroy another ballistic missile in flight and the subject of the 1972 ABM Treaty, designed to limit the development and deployment of ballistic missiles.

abort The termination of a space **flight** or **mission** (used both as a verb and a noun). As an illustration, the **Space Shuttle** has four abort alternatives which can be used during the launch phase in the event of a **Space Shuttle main engine** failure:

  (i) Return To Launch Site (RTLS). Available from separation of the **solid rocket boosters** (**SRBs**) to the time when the next alternative (AOA) becomes available. The Shuttle flies on to burn the remaining **propellant**, turns back towards the **launch site**, jettisons the **external tank** and glides back to the launch site runway for a landing.

  (ii) Abort Once Around (AOA). Available from about 2 minutes after SRB separation to when the next alternative (ATO) becomes available. The vehicle attains a **sub-orbital** trajectory, circles the Earth once and returns to the launch site.

  (iii) Abort To Orbit (ATO). Available when the Shuttle has passed the AOA-point and can attain orbit. Although the orbit attained may be lower than originally intended, the **de-orbit**, **re-entry** and landing would be similar to a normal mission.

  (iv) Transatlantic Landings (TAL). An additional abort alternative which overlaps the latter part of the RTLS option, added when emergency runways became available on the eastern side of the Atlantic.

absolute temperature Temperature measured on the 'absolute scale'. See **kelvin (K)**.

absolute zero The lowest temperature theoretically attainable; the temperature at which all molecular motion ceases (-273.15 °C). See **kelvin (K)**.

AC power A source of electrical power supplied with alternating current. See **power**. The great majority of spacecraft power systems use direct current (DC), AC being used only for special applications in scientific satellites and spacecraft payloads. The **Hubble Space Telescope**, for instance, uses a 20 kHz AC supply.

[See also **regulated power supply**]

acceleration due to gravity – see **gravity**

accelerometer A device for measuring acceleration.

[See also **inertial platform**]

access arm A projection from a launch vehicle **service structure** which can be rotated towards the vehicle for access, either for general maintenance or the loading of a **crew**. If the arm is swung away from the vehicle at the moment of **launch**, usually when it carries **umbilicals** for **propellant** or electrical services, it is also known as a **swing-arm**.

[See also **white room**]

access panel – see **closure panel**

access tower – see **service structure**

acoustic test chamber A ground-based test facility which simulates the acoustic environment experienced by a spacecraft during **launch**.

[See also **thermal-vacuum chamber, vibration facility, anechoic chamber**]

activated charcoal canister – see **environmental control and life support system (ECLSS)**

active A term applied to any device or system involved in mechanical or electrical action, or capable of a productive reaction to external stimuli; the opposite of **passive**. For example, an **amplifier** is an active device in a **communications system**, since it makes an active contribution to the input **signal**; a connector is a passive component. Similarly, a **heater** is an active device in a **thermal control subsystem**, while **thermal insulation** provides passive thermal control. While most spacecraft **sensors** or remote sensing **payloads** are passive, the **synthetic aperture radar (SAR)** is an active device.

active satellite An archaic term for a satellite with an active payload, typically a **communications payload** (see, for comparison, **passive communications satellite**). The first satellite to carry an active radio repeater was Courier 1B – see **repeater**.

actuator Any device which produces a mechanical action or motion; a servomechanism that supplies the energy for the operation of other mechanisms.

Spacecraft actuators forming part of a spacecraft **attitude and orbital control system** include **reaction control thrusters**, **reaction wheels** and **momentum wheels**.

[See also **nutation damper, solar sailing, orbital control, attitude control**]

ADCS – see **attitude determination and control system**

aerial – see **antenna**

aerobraking Aerodynamic braking, an orbital **injection** technique which uses frictional forces generated within a planetary **atmosphere** to decelerate a **spacecraft**, thereby reducing the amount of **propellant** required. Aerobraking typically begins in the higher levels of the atmosphere and lowers the **apoapsis** in a gradual process (e.g. as performed by the Mars Global Surveyor spacecraft in 1997/98); the spacecraft may attain an **orbit** or conduct a **re-entry**. Contrast: **aerocapture**.

[See also **atmospheric drag**]

aerocapture Aerodynamic capture, an orbital **injection** technique which uses frictional forces within a planetary **atmosphere** to decelerate a **spacecraft** in a single pass (as opposed to **aerobraking**, which is a gradual process). In aerocapture, the spacecraft penetrates more deeply into the atmosphere and must be fitted with a **heat shield** to dissipate the higher levels of **aerodynamic heating**.

[See also **re-entry corridor, atmospheric drag**]

aerodynamic heating An increase in the **skin temperature** of a vehicle due to air friction, particularly at supersonic or hypersonic speeds [see **Mach number**]. Sometimes called 'kinetic heating', although this can be caused by other forms of friction due to motion.

[See also **re-entry, aerocapture**]

aerodynamic stress A generic term for the forces to which a **launch vehicle** or **spacecraft**, etc., is subjected during its passage through an **atmosphere** (during **launch, re-entry**, etc.). See **aerodynamics**.

aerodynamics The study of air flow over a body and the resultant aerodynamic forces. See **lift, drag, thrust**.

[See also **aerospace vehicle, lifting body, lifting surface, fairing, dynamic pressure**]

aerospace A modifier describing a relevance to both air and **space**: e.g. aerospace industry, **aerospace vehicle**.

aerospace vehicle A term used for a space vehicle which, as part of its **launch** or **re-entry** phase, is also capable of flight in the **atmosphere** using a **lifting surface** (namely a wing) as well as conventional rocket propulsion.
[See also **spacecraft**, **single stage to orbit**, **reusable launch vehicle (RLV)**, **Hotol**]

aerospike A system of shock-waves (an 'aerodynamic spike') formed in the stream of exhaust gases from a **rocket engine** using a **plug nozzle**, as opposed to a bell-shaped **exit cone**. An abbreviated name for the engine itself: see **aerospike engine**.

aerospike engine An alternative to the bell-shaped **exit cone** used in the **nozzle**s of contemporary **rocket engine**s, whereby a number of small **combustion chamber**s, with accompanying **nozzle** outlets, discharge combustion gases

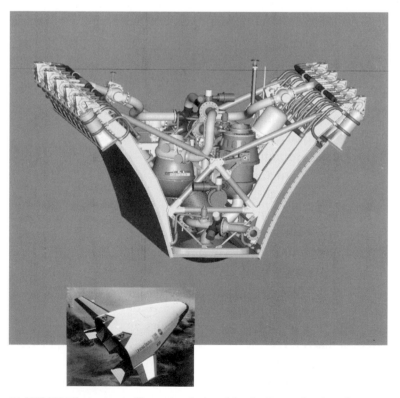

A1: XRS-2200 linear **aerospike engine** designed for the **X-33** technology demonstrator.
[Boeing]

against the outer surface of a truncated wedge or cone [see figure A1]. The system of shock-waves formed below the engine is known as an **aerospike**. In the case of the wedge (or ramp) design, the nozzle-outlets are arranged in a line and the device is known as a linear aerospike engine; in the conical design, where the nozzles are arranged in a ring, it is termed a plug nozzle or annular nozzle.

The aerospike engine is more efficient than the conventional bell nozzle in that it offers an automatic adjustment to the variation in atmospheric pressure between ground level and the upper **atmosphere** [see **plug nozzle**]. The linear engine design formed the basis for **NASA**'s **X-33** programme to develop the technology for a **reusable launch vehicle (RLV)**.

aerozine-50 A **liquid propellant** comprising 50% **hydrazine** and 50% **UDMH** (unsymmetrical dimethylhydrazine).

[See also **liquid propellant**]

Agena A rocket **stage** used with the **Thor**, **Titan** and Atlas **launch vehicles**. See **Atlas**.

AI An abbreviation for **artificial intelligence**.

airborne support equipment (ASE) Equipment flown on a **spacecraft** or **launch vehicle** to support a **payload** (physically or in terms of electrical supplies, etc.); for example, a **cradle** (and its associated systems) designed to support a spacecraft in the **payload bay** of the **Space Shuttle**, or a Spacelab **pallet**.

[See also **flight hardware, ground support equipment (GSE)**]

air-breathing rocket – see **combined cycle engine**

airframe The supporting structure and aerodynamic components of a **launch vehicle** or **aerospace vehicle**. The term, borrowed from aviation technology, tends to be applied only to space vehicles which have some contact with the Earth's **atmosphere** during their launch phase, and not to **satellites** and other similar **spacecraft**.

[See also **thrust structure, inter-tank structure, inter-stage, fairing, fin, skin, longeron, stringer, ogive, skirt, shroud, SYLDA, SPELDA, SPELTRA**]

airlock An airtight chamber which allows **astronauts** and/or equipment to leave and/or enter a **spacecraft** without depressurising the entire vehicle. Most early space **capsules** were far too small to include a separate airlock so any **extra-vehicular activity (EVA)** required all crewmembers to don their **spacesuits** before the capsule was depressurised. However, the Space Shuttle **orbiter** has a removable airlock which can be installed in one of three different positions dependent on the mission: inside the crew compartment, allowing maximum use of the **payload bay**; inside the payload bay attached to the aft cabin bulkhead; or on top of the pressurised 'tunnel adapter' which links a **Spacelab** payload to the orbiter **cabin**. Two spacesuits are stored in the airlock and, during EVA, it can supply oxygen, cooling water, electrical power and communications services to the suited astronauts.

AIT – see **assembly, integration & test**

AKM – see **apogee kick motor**

albedo The ratio of the intensity of light reflected from a body to that received from the Sun (in the 'visible **spectrum**' unless otherwise specified).

The fraction of the incident solar radiation returned to space by reflection from a planetary surface (solid or gaseous) is called 'planetary albedo', the average value of which for Earth, for example, is 0.34. In contrast, the thermal energy re-radiated by the Earth is known as 'earthshine'. Although important to the thermal design of spacecraft in **low Earth orbits**, albedo and earthshine are only significant for geostationary spacecraft carrying devices at **cryogenic** temperatures.

[See also **thermal control subsystem, geostationary orbit**]

Alcantara The location of a Brazilian **launch site** (at approximately 2° S, 44° W), used mainly for its domestic small satellite launcher, **VLS** (Veiculo Lancador de Satellites).

ALH84001 The designation of a **meteorite** found in the Allan Hills region of Antarctica but believed to originate from Mars. In 1996, NASA announced that the 'Martian meteorite' appeared to contain fossilised lifeforms, thereby suggesting that there was once life on Mars. Less well known is the 'lunar meteorite' designated ALH81005, which, among others, is believed to have reached Antarctica from the **Moon**.

Alpha A former designation of the **International Space Station (ISS)**.

ALSEP An acronym for Apollo Lunar Surface Experiments Package. The ALSEP, carried on all **Apollo** landing missions except Apollo 11, was stored in the **descent stage** of the lunar module and powered by a plutonium-238 **radioisotope thermoelectric generator (RTG)**, designated SNAP-27 (an acronym for systems of nuclear auxiliary power). The package contained **seismometers**, a **magnetometer**, and **solar wind** and lunar heat flow experiments.

altimeter An instrument designed to measure **altitude** (sense (i)).

[See also **radar altimeter**]

altitude
  (i) The vertical height of a body above the surface of a planet (typically above sea level for Earth).
  (ii) In astronomy, navigation, etc., a measure of the angle above the horizon. See **elevation (angle)**.

altitude-azimuth mount A structure for the support and guidance of an astronomical telescope or a satellite **earth station** which uses the 'horizon system' of celestial coordinates – see **azimuth**. In satellite applications it is referred to as an **elevation**-over-azimuth mount if its lower axis is perpendicular to the ground, and 'X–Y' if its lower axis is parallel to the ground. The term 'Az-El mount' is also sometimes heard. The major alternative to the 'Alt-Az' mount for telescopes is the **equatorial mount**, known as a **polar mount** for earth stations.
  [See also **kinetheodolite**]

aluminium (Al) A low-density metal, widely used (when alloyed with other metals) in the aerospace industry. Historically, aluminium was alloyed with only a few elements close to it in the periodic table: magnesium, zinc, copper, silicon, manganese and lithium. However, later techniques, including rapid solidification technology, have trebled this number. Typical applications: spacecraft body-panel **face-skins**, mounting-brackets and fittings (machined), launch vehicle **adapter ring**s (forged).
  [See also **honeycomb panel, materials**]

AM – see **amplitude modulation**

Ames Research Center – see **NASA**

AMF – see **apogee motor firing**

ammonium perchlorate (NH$_4$ClO$_4$) A solid **oxidiser** used in **rocket motors**. See **solid propellant**.

amplifier An electrical device which increases the strength of an input signal and presents a magnified replica of the signal at the output.
  [See also **amplifier chain, HPA, LNA, SSPA, TWTA, linearity**]

amplifier chain A general term for a number of amplifiers, and associated hardware, linked together in series. In a practical amplification device (for instance a spacecraft **communications payload**) a number of discrete, specialised amplifiers (e.g. pre-amplifiers, **low noise amplifier**s and IF **amplifier**s) are commonly linked together to form a chain. Within the communications payload one finds equipment divided, by function, into a **receive chain** and a **transmit chain**.

amplitude modulation (AM) A transmission method using a modulated **carrier** wave, whereby the **amplitude** of the carrier is varied in accordance with the amplitude of the input signal; the **frequency** of the carrier remains unchanged.

[See also **modulation, frequency modulation (FM), phase modulation (PM), pulse code modulation (PCM), delta modulation (DM)**]

Andoya The location the Andoya Rocket Range (at approximately 69° N, 16° E), a Norwegian **launch site** used mainly for **sounding rockets**.

anechoic chamber A ground-based test-facility for the evaluation of **radio frequency (RF)** equipment on a spacecraft, which simulates the RF propagation characteristics of **free space**. The walls and all service equipment and mounts are covered with RF absorbing material to reduce reflections (or 'echoes') to a minimum [see figure A2]. The larger chambers admit the whole spacecraft, but it is quite common to test only the **communications payload** (**antenna**s and **transponders**) at an earlier stage in the design process.

[See also **thermal-vacuum chamber, acoustic test chamber, vibration facility**]

Angara A Russian **launch vehicle** developed in the late 1990s, in a number of variants, to replace the **Zenit** and **Proton**. Its smallest version was designed to lift about 2200 kg to **low Earth orbit (LEO)**, while medium and heavy-lift versions had approximate payload capabilities of 14 and 24.5 tonnes, respectively, to LEO (2.5 and 6.8 tonnes to **geostationary transfer orbit**).

angle modulation – see **modulation**

announcement of opportunity – see AO

annular nozzle – see **plug nozzle**

anoxia A lack of oxygen. See **hypoxia**.

antenna The part of a radio system that enables a radio signal to be transmitted and/or received; the 'interface' between the radio equipment and the environment, between a '**free space**' RF wave and a guided wave. A radio **transmitter** 'excites' electric currents in the conductive surface layers of an antenna leading to the propagation of an electromagnetic wave; conversely, an incident radio wave 'excites' similar currents which are conducted to the **receiver**.

There are many different types of antenna, but using one method of categorisation four main types can be identified: wire, horn, reflector and array antennas [see figures A3, A4]. For spacecraft applications, wire antennas operate chiefly at VHF and UHF frequencies, often taking the form of a helix, conical spiral or simple dipole. The other types operate mainly at **microwave**

A2: Inmarsat-2 **communications satellite** in an **anechoic chamber**. Note the cupped dipole array **antenna**s. [British Aerospace]

frequencies. Horn antennas are used by themselves on spacecraft to provide wide coverage of the Earth [see **global beam**], and as **feedhorn**s to illuminate reflector antennas. Both horns and reflectors are known as 'aperture antennas'. Another type of aperture antenna is the microwave lens, or 'dielectric lens', which, like an optical lens, can be designed to convert a

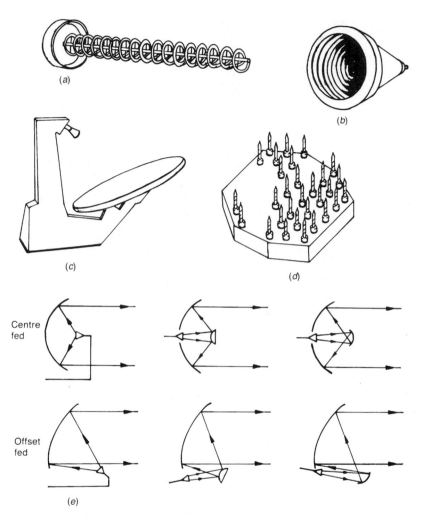

A3: **Antenna** types and reflector-antenna configurations: (a) wire antenna (helix); (b) horn antenna (conical corrugated); (c) reflector antenna (offset fed); (d) array antenna (phased array helices); (e) reflector antennas (from left to right): single reflector, **Cassegrain reflector**, **Gregorian reflector**.

spherical wave to a plane wave, thereby improving **directivity** [see **lens antenna**].

The limited **gain** and relatively wide **beamwidth** of horn antennas has led to the widespread use of reflector antennas, particularly on **communications satellites** where high gain and narrow **spot beams** have become increasingly desirable. Array antennas consist of a number of radiating elements designed

A4: TDRS-F tracking and data relay satellite [see **TDRSS**] showing **unfurlable antenna** (in furled position), two **deployable antenna**s and helix array **antenna**. [TRW]

to act together to form a particular **beam**. The array may comprise a number of slots in the wall of a **waveguide** (a 'slot-array antenna'), a number of **dipole**s, helices, horns or reflectors, depending on the frequency, required beamwidth, etc. See **phased array antenna**.

The word 'aerial' is still used in the space industry (mainly by veteran

British engineers), but is gradually being replaced by 'antenna'. Recommended plurals are 'antennas' for radio equipment, 'antennae' for insects.

[See also **frequency bands, communications payload, boresight, footprint, 'pointing', peak, axial ratio, f/D ratio, Cassegrain reflector, subreflector, antenna pointing mechanism, antenna radiation pattern, antenna platform, deployable antenna, unfurlable antenna, steerable antenna, elliptical antenna, parabolic antenna, shaped antenna, beam-forming network, isotropic antenna, omnidirectional antenna, TT&C antenna**]

antenna array An assembly of **antenna**s, not necessarily electrically coupled as they are in an **array antenna**.

[See also **antenna farm, antenna platform, antenna module**]

antenna beam The geometric distribution of **radio frequency** radiation formed by a **spacecraft** communications **antenna**.

antenna efficiency – see **aperture efficiency**

antenna farm A collective term for a number of **antenna**s. When applied to **space segment** hardware, it usually refers to an orbiting **platform** carrying an assembly of communications antennas, although it is sometimes extended to the **antenna platform** of a much smaller **satellite**. When applied to **earth segment** hardware, it refers to a collection of antennas at an **earth station** (sense (ii)).

[See also **antenna module, antenna array**]

antenna feed – see **feedhorn**

antenna gain The ratio of the signal power at the **output** of an **antenna** to that received at the **input**, usually measured in decibels (dB). The majority of antennas used in **satellite communications** are parabolic dishes which when used on the input side increase the gain of the signal passed to the **receiver**, and when on the output side increase the gain of the signal transmitted through space. This definition views the antenna as a gain-producing component in a **transmit chain** or **receive chain**; considering the antenna in a more theoretical sense, as an independent entity, the word 'gain' refers to an increase in signal-power over and above that which would be available from an **isotropic antenna**.

Two 'rule-of-thumb' expressions commonly used to calculate antenna gain are given below:

Peak gain: $10 \log_{10} \eta (\pi D/\lambda)^2$

where η is antenna efficiency (%); D is diameter (m); λ is wavelength (m); and

Half-power gain: $10 \log_{10} \eta(27,800/\theta \cdot \phi)$

where η is antenna efficiency (%); θ and ϕ are the orthogonal **half-power beamwidth**s in degrees. Alternative values for the constant may be found in other texts, where a distinction is made between antennas of different types. [See also **equivalent isotropic radiated power (EIRP)**, **power flux density (PFD)**, **directivity**, **decibel (dB)**]

antenna module A self-contained section of a modular spacecraft containing the communications **antenna** subsystem.
[See also **payload module**, **service module**]

antenna platform
- (i) The panel of a satellite's body, usually the Earth-pointing face, which supports the fixed **antennas**. Note, however, that antennas can also be mounted on other faces of the spacecraft [see **deployable antenna**].
- (ii) Another name for **antenna farm**.

[See also **space platform**, **antenna module**, **antenna array**]

antenna pointing mechanism (APM) A device which gives a spacecraft **antenna** a finer degree of **pointing** control than that offered by the body of the spacecraft itself, particularly important for **spot beam** antennas.
[See also **attitude control**, **phased array antenna**]

antenna radiation pattern A measure of the directional sensitivity of an **antenna**; graphically, a plot of the radiated field-strength against the angle from **boresight** [see figure A5]. The pattern is important since it determines the **beamwidth** and **directivity** of the antenna. According to the reciprocity theorem, the transmitting and receiving patterns of an antenna are identical at a given **wavelength**.

Sensitivity in a direction outside the main **beam** or main 'lobe' of the pattern is known as a 'side lobe'. This represents a detrimental attribute of an antenna, because it indicates that some of the power radiated from the antenna will not be contained within the main beam which, apart from being wasteful, could lead to **interference** with other systems. Equally, a signal received from the side-lobe direction could interfere with the system in question.

antenna taper The variation in electric field produced by a **feed** across an **antenna** surface. For a theoretical uniform illumination, the electric field is constant and the aperture taper efficiency is 1. However, most practical

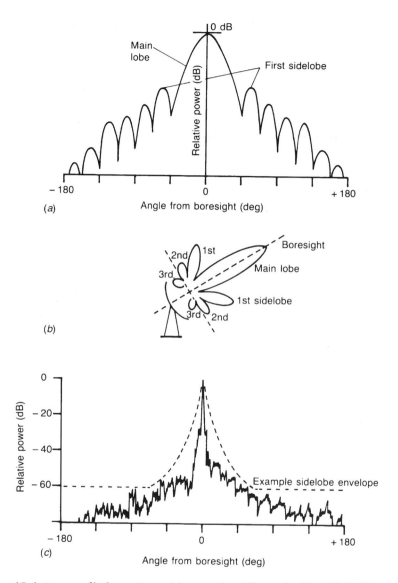

A5: **Antenna radiation patterns**: (a) conventional linear plot (simplified); (b) pattern in polar coordinates; (c) example pattern for a 3 m **earth station**.

antenna feeds produce a highly tapered distribution which decreases the electric field with distance from the centre: this increases the power in the main lobe and increases the beam efficiency of the antenna.

[See also **antenna radiation pattern**]

anti-Earth face The face of a **three-axis stabilised** spacecraft (defined as the minus-z face) which faces directly away from the Earth. See **spacecraft axes**.

anti-jam – see **jamming**

anti-slosh baffle A structure in a **liquid propellant** tank which damps out the motion of the liquid known as '**sloshing**', which can disturb a vehicle's flight dynamics [see figure A6]. Various different structures are used (e.g. vanes, rings, truncated cones, etc.). Some tanks also have anti-vortex baffles, which minimise the propellant's tendency to swirl as it flows out of the tank (like water down a plug-hole). Vortices can produce bubbles of gas which would otherwise pass to the **engine**(s) producing uneven combustion.

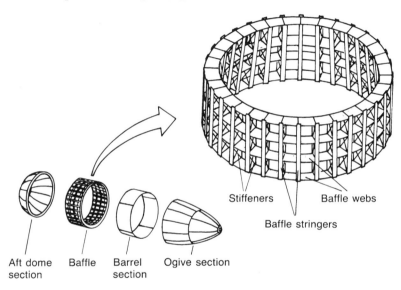

A6: **Anti-slosh baffle** in a **propellant tank**.

AO An abbreviation for announcement of opportunity (e.g. as in 'AO payload'). In space technology, the term refers mainly to payloads of science or **Earth observation** spacecraft and indicates the existence of an opportunity to propose, manufacture and fly a payload on that spacecraft.

AOCS – see **attitude and orbital control system**

AOS An abbreviation for acquisition of signal. Typically used to denote the moment that a tracking **earth station** acquires a **spacecraft**'s **telemetry** signal during **launch** and **transfer orbit** phases. Also used when a **communications** link with a spacecraft has been re-established after a period of 'radio silence' (e.g. following blockage by a **planetary body**, a period of radio **interference** or during **re-entry** [see **S-band blackout**]). The opposite of 'loss of signal' (LOS).

[See also **signal, carrier**]

AP An abbreviation for **ammonium perchlorate (NH_4ClO_4)**.

aperture antenna A 'horn antenna' or a 'reflector antenna'. See **antenna**.

[See also **synthetic aperture radar (SAR)**]

aperture blockage The reduction in the effective area of an **antenna** reflector by obstructions in the aperture, such as a **feed** or **subreflector** and its supports. Blockage decreases the **aperture efficiency** and degrades the **side lobe** performance owing to diffraction.

aperture efficiency The efficiency of an '**aperture antenna**'; a term used in the calculation of **antenna gain**, etc., which quantifies how effectively the antenna uses the **RF power** transmitted or received.

This can include 'illumination efficiency' or 'aperture taper efficiency' (the efficiency with which the power is distributed over the surface of an antenna reflector by the **feed**); the degree of **spillover**; the degree of **aperture blockage**; diffraction effects; phase errors; and **polarisation** and 'mismatch' losses.

aperture taper – see **antenna taper**

aphelion The furthest point from the Sun in an elliptical solar orbit (from the Greek for 'away from the sun'); the opposite of **perihelion**.

[See also **apoapsis**]

APM – see **antenna pointing mechanism**

apoapsis The point in an **orbit** furthest from the centre of gravitational attraction; the opposite of **periapsis**. From this general term, specific terms relating to a given **planetary body** can be derived: e.g. **apogee** for Earth, **aphelion** for Sun, apolune for Moon, apojove for Jupiter, etc.

apogee The point at which a body orbiting the Earth, in an **elliptical orbit**, is at its greatest distance from the Earth [see figure G1]; the opposite of **perigee**. The word is derived from the Greek 'apogaios': the prefix 'ap' meaning 'away from'; the suffix 'gee' referring to 'Earth'.

[See also **apoapsis, orbit, apogee kick motor**]

apogee boost motor (ABM) – see **apogee kick motor (AKM)**

apogee engine – see **liquid apogee engine**

apogee engine firing The **ignition** of a **liquid apogee engine (LAE)**, designed to transfer a satellite from a **geostationary transfer orbit (GTO)** to geostationary orbit (GEO) in several stages [see parasitic station acquisition]; the equivalent, for a solid propellant system, of an apogee motor firing.

[See also propellant, rocket motor]

apogee kick motor (AKM) A rocket motor used to transfer a satellite from geostationary transfer orbit (GTO) to geostationary orbit (GEO) [see figure R7]. Alternatively called an apogee boost motor (ABM) and sometimes called an apogee stage.

An AKM is fired when a spacecraft reaches the point in the GTO ellipse furthest from the Earth, the apogee. This type of solid propellant motor was an integral part of most early geostationary satellites and was usually installed inside a central thrust cylinder or other thrust structure. The majority of geostationary satellites now use a liquid bipropellant system which combines the functions of the AKM and the reaction control system – see combined (bipropellant) propulsion system. In a liquid propellant system, the equivalent device is known as a liquid apogee engine (LAE).

[See also rocket engine, perigee kick motor (PKM), inertial upper stage (IUS)]

apogee motor firing (AMF) The ignition of an apogee kick motor (AKM) – a solid propellant rocket motor – which occurs at the apogee of a geostationary transfer orbit (GTO); the equivalent, for a liquid propellant system, of an apogee engine firing.

[See also propellant, rocket engine]

apogee stage A launch vehicle stage which injects a spacecraft into geostationary orbit by firing its engine(s) at the apogee of the transfer orbit. Sometimes used as an alternative term for apogee kick motor.

[See also stage]

Apollo A series of American manned spacecraft designed to take three astronauts to lunar orbit and land two of them on the Moon. Apollo comprised two spacecraft: a command and service module (CSM) and a lunar module (LM), formerly called the LEM (for lunar excursion module) [see figures A7, A8]. After the third (S-IVB) stage of the Saturn V launch vehicle had injected the Apollo combination onto a lunar trajectory [trans-lunar injection (TLI)], the CSM separated and withdrew the LM from its launch fairing on top of the S-IVB stage. The service module's main engine was used to place the spacecraft in lunar orbit, to return it to Earth and for major mid-course corrections. It used the propellants UDMH and nitrogen tetroxide (N_2O_4) which were also employed by its four sets of RCS [reaction control

A7: **Apollo** 15 command and service module in lunar orbit, 30 July 1971. Note the **docking probe** on top of the command module. [NASA]

A8: **Apollo** 15 lunar module and **lunar roving vehicle** at the Hadley Rille landing site, 31 July 1971. **Astronaut** James Irwin salutes the Stars and Stripes. [NASA]

A9: **Astronaut** Edwin 'Buzz' Aldrin photographed by Neil Armstrong, the first man to set foot on the Moon, during the **Apollo** 11 mission, 16–24 July 1969. Note Armstrong and a leg of the Lunar Module 'Eagle' reflected in Aldrin's visor. [NASA]

system] thrusters, used for **rendezvous** and **dock**ing, minor **manoeuvre**s and as **ullage rocket**s.

The command module, the only part of the whole vehicle to return to Earth, was the crew **cabin** for the outward and return journeys as well as the **re-entry** vehicle: it had an ablative **heat shield** made from an epoxy resin bonded to a stainless steel honeycomb. A tunnel in the nose gave access to the lunar module, which comprised a **descent stage** and an **ascent stage**.

Following a number of unmanned launches of Apollo hardware, the first manned mission, Apollo 7, was launched by a **Saturn 1B** to test the CSM in

Earth orbit; Apollo 8, which was launched by a Saturn V without a LM, made ten orbits of the Moon; Apollo 9, which conducted **dock**ing tests in Earth orbit, was the first manned flight with a complete Apollo spacecraft; the Apollo 10 LM descended to within 15,240 m of the lunar surface in a 'dress rehearsal' for the landing; Apollo 11 made the first lunar landing in the Sea of Tranquillity (0° 41' N, 23° 26' E) at 8:17 pm GMT 20 July 1969, the first foot being placed on the surface at 2:56 am GMT 21 July (21.8 kg rock returned) [figure A9]. Apollos 12, 14, 15, 16 and 17 also made lunar landings, the latter three carrying a **lunar roving vehicle (LRV)** to extend their coverage of the surface [see figure A8]. Apollo 13 suffered an explosion of an oxygen tank in its service module en route to the Moon and failed to make a landing.

In summary of the Apollo missions, 12 men spent 160 man-hours on the Moon, collected over 2000 rock samples totalling 380 kg, and took over 30,000 photographs; 60 major experiments were placed on the lunar surface [see **ALSEP**] and 30 were carried out from lunar orbit.

[See also **extra-vehicular activity (EVA)**, **space insurance**]

Apollo–Soyuz Test Project (ASTP) A docking mission between an American **Apollo** and a Soviet **Soyuz** spacecraft which developed from a political agreement between President Nixon and Premier Kosygin. The spacecraft docked on 17 July 1975, Apollo 18 carrying three **astronaut**s, and Soyuz 19 carrying two **cosmonaut**s (launched 2 days earlier).

[See also **docking module**]

application The field of research or commerce for which a **satellite** or **spacecraft** is designed. The term is widely used in the space industry to differentiate between spacecraft designed, among others, for **communications**, meteorology, **remote sensing**, **space science**, **microgravity** research and **manned spacecraft** applications.

[See also **Earth observation**, **Earth resources satellite**, **navigation satellite**, **technology demonstrator**]

applications satellite An archaic term for a **satellite** with a particular application (e.g. **communications**, meteorology, **remote sensing**), as opposed to one carrying a scientific payload.

approach and landing test (ALT) A series of test flights conducted by the Space Shuttle **Enterprise** in 1977 in order to determine the **orbiter**'s aerodynamic characteristics.

APU – see **auxiliary power unit**

arcjet thruster A propulsive device that uses an electric arc to superheat
hydrazine propellant, approximately doubling its efficiency [see **specific
impulse**]; a type of **reaction control thruster**.
[See also **hydrazine thruster, hiphet thruster**]

area expansion ratio – see **expansion ratio**

arean Martian; of or pertaining to Mars (from Ares, the Greek counterpart to the
Roman god of war, Mars).

areocentric Centred on Mars (e.g. an **areostationary orbit**).

areostationary orbit A **stationary orbit** around Mars.
[See also **geostationary orbit (GEO)**]

Ariane A series of European **launch vehicles** developed under the auspices of the
European Space Agency (ESA). The Ariane programme was proposed by the
French space agency after the cancellation of the **Europa** launch vehicle and
was given the 'go-ahead' in 1973. The first flight, from the launch site at the
Guiana Space Centre, was in December 1979. The majority of its payloads are
commercial **communications satellites** placed in **geostationary orbit**,
although it can also launch spacecraft into **low Earth orbit** and **sun-
synchronous orbit**.

   The Ariane family has evolved through several versions, leading to the
Ariane 4, which itself is available in six **variants** with different **payload**
capabilities. Ariane 1, 2, 3 and 4 deliver(ed) up to 1850 kg, 2175 kg, 2700 kg
and 4460 kg, respectively, to **geostationary transfer orbit (GTO)**, the most
common delivery orbit for commercial satellites. Ariane 1 used the
propellants **nitrogen tetroxide** and **UDMH** in its first and second stage Viking
engines (UDMH being replaced by **UH25** from Ariane 2 onwards), and all
variants used **liquid oxygen/liquid hydrogen** in the third stage HM7/7B
engines. Ariane 1, 2 and 3 used the same first and second stages; 2 and 3 had a
'stretched' third stage; and 3 used two **strap-on** solid rocket boosters to
augment the first stage **thrust**. Ariane 4 [see figure G6] has a stretched first
stage, but uses the same second and third stages, although strengthened to
withstand increased launch **loads**. A number of solid and liquid **propellant**
strap-ons can be added to form the six different configurations: no boosters
(Ariane 40); 2 solids (42P); 4 solids (44P); 2 liquids (42L); 2 solids and 2 liquids
(44LP); and 4 liquids (44L).

   In the 1990s, the entirely new Ariane 5 launch vehicle was developed and,
following three demonstration launches, entered commercial operation in

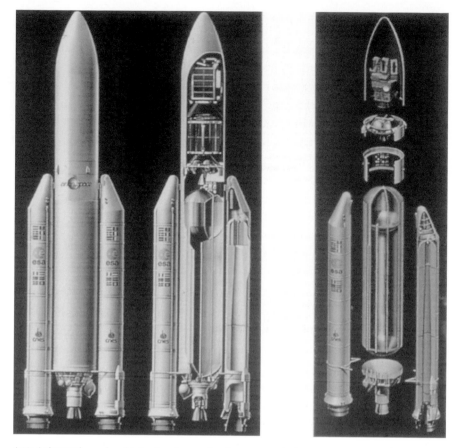

A10: **Ariane** 5 launch vehicle showing **payload** accommodation, **upper stage** and cutaways of main **propellant tank** and strap-on **solid rocket boosters**. [ESA-CNES-Arianespace/D. Ducros]

1999. It was initially capable of launching 6800 kg to GTO, but enhancement programmes were expected to increase this initially to about 9000 kg and subsequently to 11,000 kg with the development of a cryogenic **upper stage**. See figures A10, A11.

[See also **SYLDA, SPELDA, SPELTRA, countdown**]

ARPA An acronym for Advanced Research Projects Agency, a US-based organisation involved mainly in civil space research. [See also **DARPA**]

array – see **solar array, antenna array**

array antenna – see **antenna**

array blanket – see **solar array**

A11: **Ariane 503 launch**, 21 October 1998 (the four masts carry lightning rods). [Aerospatiale]

array degradation – see **degradation (of spacecraft materials)**

array shunt regulator A device which switches sections of a spacecraft **solar array** on and off in accordance with the demand for **power**.
[See also **shunt dump regulator, heat rejection**]

ARS An abbreviation for the American Rocket Society, which was founded in 1931 to promote interest in **astronautics**, particularly **rocket** propulsion, in part by publishing the *Journal of the American Rocket Society* (later known as *Jet Propulsion*) and *Astronautics* magazine.
[See also **BIS, VfR**]

artificial intelligence (AI) The ability of a machine, usually a **computer**, to imitate intelligent human behaviour; sometimes called machine intelligence. Systems incorporating artificial intelligence are also referred to as autonomous control systems, expert systems or remote agents (mainly to avoid the futuristic/fictional connotations of AI). AI is particularly useful for spacecraft designed to operate far from **Earth**, where autonomy is important because of the long travel-time of radio signals.
[See also **signal delay, light (velocity of)**]

artificial satellite An archaic term for a **spacecraft** in **orbit** around a **celestial body**, such as the Earth, the Sun or any other of its planets and moons. The term arose before the beginning of the **Space Age** to distinguish between natural and man-made satellites.

[See also **satellite**]

ascending node – see **nodes**

ascent engine A spacecraft **rocket engine** used for an ascent from the surface of a **planetary body**; the main engine in a spacecraft's **ascent stage** (e.g. the ascent engine of the **Apollo** lunar module).

[See also **descent engine**]

ascent stage The upper part of the **Apollo** lunar module, designed to transport the Apollo astronauts from the lunar surface to the command and service module in lunar orbit.

[See also **descent stage**]

ASE – see **airborne support equipment**

ASIC An acronym for application specific integrated circuit.

[See also **MMIC**]

ASLV An abbreviation for Augmented Satellite Launch Vehicle, an Indian **launch vehicle** based on the earlier SLV-3. First launched in 1987, the ASLV's **payload capability** was about 150 kg to **low Earth orbit (LEO)**.

[See also **PSLV, GSLV, Sriharikota**]

assembly

(i)   A blanket term for any group of components assembled to perform a particular function (e.g. **bearing and power transfer assembly (BAPTA)**, **thruster** assembly, **reaction wheel** assembly, **focal plane assembly**).

(ii) Construction (as in **assembly, integration & test (AIT), vehicle assembly building** (VAB)).

assembly, integration & test (AIT) A recognised **spacecraft** manufacturing function comprising these three main activities: assembly refers to the construction of individual units or **subsystems**; integration refers to the 'higher level' construction of **modules** (see **spacecraft integration, payload integration**); and test refers to the various checks carried out on a completed spacecraft (see **vibration facility, vibration table, acoustic test chamber, thermal-vacuum chamber, anechoic chamber**).

asteroid A small **planetary body** in orbit around the **Sun**, mainly as part of the asteroid belt between the orbits of Mars and Jupiter, but also elsewhere in the solar system. Also known as planetoids or minor planets.

[See also **NEO**]

ASTP – see **Apollo–Soyuz Test Project**

astrogeology The 'geology' of planetary bodies other than the **Earth**. Strictly, geology is the scientific study of the origin, structure and composition of the Earth, but this term extends this to the geological features of other bodies. [See also **planetary body, asteroid, meteorite, ALH84001**]

astromast A type of extendable truss structure used to **deploy** items, such as solar arrays and **deployable antenna**s, from spacecraft [figures A12, A13]. See **solar array**.

A12: An **astromast** deploys a **solar array** from the **payload bay** of the **Space Shuttle**. [Rockwell International]

astronaut A person trained for space travel. Before the introduction of the **Space Shuttle**, the term was applied to any person who travelled in space [but see also **cosmonaut**], all of whom were pilots specifically trained for the duties implied by spaceflight. The only exception, in the history of American spaceflight (pre-Shuttle), was 'scientist-astronaut' and **Apollo** 17 lunar module pilot, Harrison ('Jack') Schmitt, a geologist who became an astronaut. Now that any 'reasonably fit' person is physically capable of flight aboard the Shuttle, the term 'astronaut' tends to be confined to a person who has trained

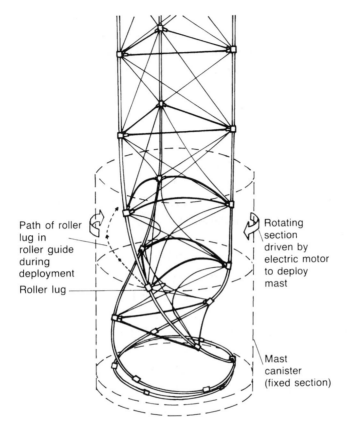

Path of roller
lug in
roller guide
during
deployment

Roller lug

Rotating
section
driven by
electric motor
to deploy
mast

Mast
canister
(fixed section)

A13: **Astromast** deployment method.

as a 'career astronaut'. Other participants in a Shuttle flight are called
mission specialists or **payload specialists**.

astronaut manoeuvring unit (AMU) – see **manned manoeuvring unit (MMU)**

astronautics The study of the fundamental physical principles governing
spaceflight (i.e. **orbits**, trajectories, **gravitational perturbations**, etc.),
although the word has come to be used as a blanket term for the science and
technology of spaceflight, **space science** and the exploration of space. In
either case, 'astronautics' is not confined to the activities of astronauts.
[See also **cosmonautics**]

astronomical satellite An orbiting spacecraft concerned with astronomical
observations, usually based in **low Earth orbit**. The main advantage of

astronomy from orbit is the satellite's position above the **atmosphere**, the limiting factor for many branches of terrestrial astronomy.

[See also **space astronomy, Hubble Space Telescope, VLBI**]

astronomical unit (AU) The mean distance between the **Earth** and the **Sun** (i.e. the radius of the Earth's orbit, 1.495×10^{11} m); a unit of distance used in astronomy.

[See also **parsec**]

asymmetric feed – see **offset-fed (antenna)**

asymmetric link A **satellite communications** link with different capacities on the **uplink** and **downlink**. For example, a typical **Internet** link requires a low **bit rate** on the uplink (to transmit the request for **data**) and a much higher rate on the downlink (to **download** the data).

[See also **capacity**]

Athena An American solid propellant **launch vehicle** developed in the 1990s as a small satellite launcher; originally known as the Lockheed Launch Vehicle (LLV) and later the Lockheed Martin Launch Vehicle (LMLV). The Athena 1 **variant**, which was first launched in 1995, uses the Castor-120 solid **rocket motor** (SRM) for its first **stage** and the Orbus 21D SRM for its second; the Athena 2 uses two Castors and an Orbus to form a three-stage rocket. Their respective payload capabilities are 800 kg and 2000 kg to **low Earth orbit (LEO)**. A larger variant, Athena 3, is designed to deliver about 5500 kg to LEO.

[See also **Vandenberg Air Force Base (VAFB), solid propellant, Atlas, Delta, Titan, Scout**]

Atlantis The name given to the fourth '**flight model**' of the Space Shuttle **orbiter** (Orbiter Vehicle OV-104), which was first launched on 3 October 1985 (STS 51-J). Atlantis was named after a sailing ship used for ocean research from 1930 to 1966.

[See also **Space Shuttle, space transportation system (STS), Columbia, Challenger, Discovery, Endeavour, Enterprise**]

Atlas An American **launch vehicle** developed from an intercontinental ballistic missile (**ICBM**), which had its first flight in June 1957 and became an operational weapon system in 1959. The original Atlas was considered a 'one-and-a-half' **stage** vehicle. Its three main engines were ignited simultaneously at launch, but only one, known as the 'sustainer', remained attached to the propellant tanks for the full duration of the flight. The other two, called 'boosters', were jettisoned along with their **fairing**s, etc., part way through the flight. This configuration became the first stage of successive versions of the Atlas launch vehicle, which used the **propellant**s **liquid oxygen** and **kerosene**.

The Atlas-D variant, for instance, was used to orbit the **Mercury** capsule; the Atlas-Able and Atlas-Agena two-stage vehicles launched the Ranger probes to the Moon, the early Mariners to Mars and Venus, etc.; and the Atlas-Agena D carried the Agena target docking vehicle for the **Gemini** programme. The **Centaur** stage used on the Atlas-Centaur was America's first liquid oxygen/**liquid hydrogen** stage: it first flew in May 1962, and became **operational** in May 1966 with the launch of the Moon **probe** Surveyor.

The Atlas became a commercially operated launch vehicle in the late 1980s [see figure A14]. The **payload capability** of the Atlas I, II, IIA and IIAS, respectively, was about 2255 kg, 2810 kg, 3065 kg and 3720 kg to **geostationary transfer orbit (GTO)**, the most common delivery orbit for commercial satellites. The Atlas IIIA and IIIB, formerly designated IIAR and IIARS, have respective capabilities of 4060 kg and 4500 kg to GTO. The Atlas V, a version also known as the **Evolved Expendable Launch Vehicle (EELV)**, is designed with a payload capability of up to 8200 kg.

[See also **balloon tank, Delta, Titan, Scout**]

Atlas-Centaur An Atlas **launch vehicle** with a Centaur **upper stage**. See **Atlas**.

atmosphere The gaseous envelope surrounding a **celestial body** (or **planetary body**). For example, the **Earth**'s atmosphere is composed of molecular nitrogen, N_2 ($>$78.0%); molecular oxygen, O_2 ($>$20.9%); argon, Ar (0.93%); carbon dioxide, CO_2 (0.03%); and neon, Ne (0.002%). It has an average atmospheric pressure of 101,325 pascal (14.7 psi) at sea level. It is divided into several layers or regions called 'spheres', the boundaries between which are known as 'pauses' (a nomenclature introduced by British mathematician and geophysicist Sydney Chapman):

- troposphere: from the surface to the tropopause (at an altitude of 10–12 km)
- stratosphere: from the tropopause to the stratopause (at about 50 km)
- mesosphere: from the stratopause to the mesopause (at about 85 km)
- thermosphere: from the mesopause to about 600 km
- exosphere: above about 600 km, where individual atoms can reach **escape velocity** without collisions.

Those regions above the troposphere as generally known as the upper atmosphere.

[See also **cabin pressure, magnetosphere, ionosphere, heliosphere**]

A14: Launch of an **Atlas** IIA from Cape Canaveral, Florida, carrying **Eutelsat**'s Hot Bird 2 **communications satellite** (21 November 1996). Note the **launch vehicle's Centaur** upper stage below the **payload fairing**. [Lockheed Martin]

atmospheric attenuation The weakening of a radio **signal** due to its passage through a planet's **atmosphere**.

The attenuation of a **radio frequency** signal in the Earth's atmosphere is of everyday importance for **satellite communications** and is used in the calculation of a communications **link budget**. It is partly due to oxygen and

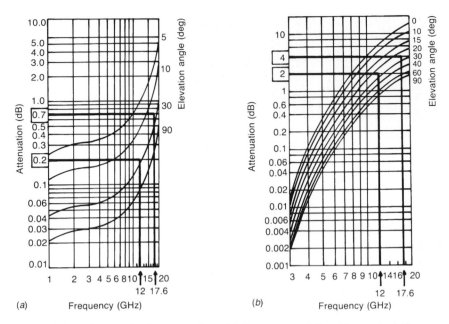

A15: Variation of **atmospheric attenuation** with **frequency** and **elevation angle**: (a) for oxygen and water vapour; (b) for rainfall (with example values for direct broadcasting **uplink** and **downlink** frequencies).

water vapour, whose molecules are excited into higher energy states by absorption of the radiation, and partly due to precipitation in its various forms, which also acts as a radiation absorber [see **rain attenuation**]. The degree of attenuation is heavily dependent on the **frequency** of the radiation: attenuation increases with increasing frequency [see figure A15]. In a satellite communications system, the **elevation angle** of the **satellite** as measured from the **earth station** also has an effect: attenuation decreases with increasing elevation angle (since the atmospheric path is shortest at the **zenith**).

atmospheric drag The resistance to the motion of a body due to a planet's **atmosphere**. The orbital velocity of **spacecraft** in very **low Earth orbit**s may be decreased to such an extent that they re-enter the atmosphere and 'burn up'. An increase in solar activity [see **solar flare**] tends to increase the average height of the atmosphere, thereby increasing the drag at a particular height. This can decrease the **lifetime** of spacecraft in LEO, particularly those with large surface areas (e.g. the USA's **Skylab** space station, which re-entered earlier than expected for this reason).
[See also **drag compensation, aerobraking, aerocapture**]

ATOX An acronym for atomic oxygen; single atoms of oxygen (O) as opposed to molecular oxygen (O_2). Atomic oxygen is generated in the **upper atmosphere** by solar ultraviolet radiation; its abundance is therefore related to the 11-year solar cycle. It has an erosive effect on spacecraft surfaces, such as **solar cells** and **thermal insulation**, and can affect their thermal properties (e.g. Kapton material on the **Space Shuttle** has exhibited a grass-like structure which has changed its thermal properties).

[See also **solar flare, coverslip, materials**]

attenuation

   (i) The function of an **attenuator**.

   (ii) The natural weakening of a **signal** as it passes through an attenuating medium such as an **atmosphere**. See **atmospheric attenuation**.

attenuator A device which provides a measured degree of **attenuation**, or weakening, of a **signal**. This may be necessary as a fine adjustment where standard equipment provides too great a degree of amplification, or to terminate the open end of a **transmission line** which can occur at a switch when power is routed along the alternative course. Attenuators can be of the fixed or variable type, and are colloquially known as 'pads'. The variable type can be adjusted from the controlling **ground station**.

attitude The orientation of a spacecraft's axes with reference to a number of known points, either on Earth or in space.

[See also **spacecraft axes, attitude control**]

attitude and orbital control system (AOCS) A spacecraft **subsystem** which combines the functions of **attitude control** and **orbital control**.

[See also **attitude determination and control system (ADCS)**]

attitude control The ability to maintain or change a spacecraft's **attitude**; the subsystem or process by which this control is effected.

Attitude control is required to ensure that **sensors**, communications **antennas** and other instruments remain pointing in the right direction. Most spacecraft have an attitude control subsystem, since they need to counter the perturbing forces experienced in orbit [see **perturbations**] and maintain a specified orientation. The system usually comprises a set of **reaction control thrusters** and may include other **actuators** such as **reaction wheels**.

Attitude control is usually combined with **orbital control** in a common subsystem, the 'attitude and orbital control system' (AOCS), but see **attitude determination and control system (ADCS)**.

[See also **pointing, combined (bipropellant) propulsion system**]

attitude determination and control system (ADCS) A spacecraft subsystem which determines a spacecraft's **attitude** by means of **sensors**, and then controls its attitude using **reaction control thrusters** and/or **reaction wheels**. An ADCS is generally specified for **interplanetary** spacecraft, since they travel on a **trajectory** as opposed to in an **orbit**; the equivalent subsystem for orbiting **spacecraft** (**satellites**) is the **attitude and orbital control system (AOCS)**.

[See also **attitude control, orbital control, pointing**]

ATV – see **automated transfer vehicle**

AU An abbreviation for:
 (i) **astronomical unit**, the mean distance between the **Earth** and the **Sun**;
 (ii) **accounting unit**, a monetary measure used by the European Space Agency (**ESA**) to simplify financial dealings with member states.

automated transfer vehicle (ATV) An ESA spacecraft designed as a resupply **vehicle** for the **International Space Station (ISS)**. The ATV, launched by **Ariane** 5, will also be used to boost the ISS into a higher orbit, a periodic necessity since friction with the Earth's **upper atmosphere** will tend to lower its orbital **altitude**.

[See also **CTV, CRV**]

autonomous control system – see **artificial intelligence (AI)**

autopilot An automatic guidance and control system on a **launch vehicle**, which calculates the vehicle's **attitude** and position using an **inertial platform** and flies the rocket according to preprogrammed **data**. Usually part of the vehicle's **instrument unit**.

auxiliary power unit (APU) A device which provides a source of power additional to and separate from that provided by a primary power source.

An example would be one of three devices in the Space Shuttle **orbiter** that provides hydraulic power for main engine steering ('**thrust vector** control'), landing gear **actuators**, brakes, nose-wheel steering, **elevon** and rudder/**speedbrake** control, etc. The three APUs (one to control each **Space Shuttle main engine**) are used during launch and for the approach and landing. The APU is fuelled by **hydrazine** which is decomposed in a **gas generator**; the gas is used to drive a turbine which drives a hydraulic pump.

[See also **power supply, fuel cell**]

avionics Electronics technology applied to aeronautics (from *aviation electronics*), and by extension to **astronautics**; electronic equipment on board a **spacecraft** – usually a manned spacecraft by analogy with aviation. Avionics equipment can include guidance and navigation systems, **communications** equipment, and items such as **gyroscopes**, **radar altimeter**s, etc.

axes (of a spacecraft) – see **spacecraft axes**

axial ratio A measure of **polarisation** circularity in a **cross-polarised** antenna. [See also **cross-polar discrimination**]

axial thruster A **reaction control thruster** on a **spin-stabilised** spacecraft mounted in alignment with the spacecraft's spin axis (colloquially and collectively termed 'axials'). It produces **thrust** in a direction parallel to the spin axis and is typically used for north–south station-keeping and **precession** control [see **orbital control**].
[See also **radial thruster**]

axisymmetric feed – see **centre-fed (antenna)**

azimuth A measure of the angle between the observer's meridian and the meridian of the subject under observation, in the horizon system of celestial coordinates. See also **elevation (angle)**: knowledge of a **satellite**'s azimuth and elevation angle allow an **earth station** to be aligned with it. An alternative 'equatorial' or 'geographical' system uses **declination** and **hour-angle**. Azimuth is measured clockwise from due-south for satellite applications, as in astronomy; it is measured clockwise from due-north in navigation circles.

B

back off (from saturation) In **telecommunications**, the degree to which the output **power** of an **amplifier** is reduced from the point where it is described as 'saturated' to its operational level (it is said to be 'backed off from saturation'); the process of reducing the input power level to a point where the amplifier is operating in its 'linear' region (i.e. where the output power bears a linear relationship to the input power).

A satellite **transponder** is a non-linear device, in that its **output** power is not simply proportional to its **input** power (it is represented graphically by a curve approximating a third order polynomial). For multiple **carriers**, this non-linearity generates harmonics that produce intermodulation interference among neighbouring **channels**. To mitigate the effect the transponder is 'backed off'. Back off is not necessary when only one carrier occupies the transponder (e.g. an **FM** video channel or a **TDMA** carrier) [see **single channel per carrier (SCPC)**].

[See also **saturated output power**, **linearity**]

back-contamination The contamination of the terrestrial environment, which may occur as a result of returning samples from another **planetary body**.

[See also **sample return**, **quarantine facility**, **forward contamination**]

background noise A colloquial term for '**noise**'; a blanket term for any undesired electrical disturbance in a circuit or communications system independent of its origin. See **thermal noise**, **broadband noise**, **shot noise**.

backhaul link A terrestrial **telecommunications** link from a satellite **ground station** to a local terrestrial switching centre, which channels the signal into the public network; also called a 'terrestrial tail'.

back-up A spare, substitute or **redundant** component, **system**, **spacecraft**, etc.

[See also **prime system**, **prime spacecraft**, **ground spare**, **in-orbit spare**]

baffle

(i) Any device which restrains the motion of a fluid to reduce or eliminate vortices and/or **sloshing**, e.g. an **anti-slosh baffle** which restrains the motion of a **liquid propellant** in a **propellant tank**.

(ii) Any device which restricts the passage of light or other radiation in the **electromagnetic spectrum**, e.g. a 'light baffle' which reduces the radiation from undesired sources entering a **telescope** or other **imaging device**.

[See also **sun-shade**]

Baikonur Cosmodrome A former-Soviet **launch site**, established in 1955 near the town of Tyuratam, in Kazakhstan, to the east of the Aral Sea (at approximately 45.6° N, 63.4° E). Following the dissolution of the Soviet Union, Russia was obliged to lease the facilities from Kazakhstan. Although the Soviets referred to it as the 'Baikonur Cosmodrome', it is some 300 km from the town of Baykonyr (alternative spelling) and only about 20 km from Tyuratam. The site was developed to provide **ICBM** test and launch facilities and the first space-related launch from the cosmodrome was that of **Sputnik** 1 on 4 October 1957. It was also the launch site for **Vostok** 1 (on 12 April 1961) and has been used since then for both manned and unmanned launches.

[See also **Northern Cosmodrome, Plesetsk, Kapustin Yar, Svobodny, Gagarin Centre**]

balance mass The additional mass added to a completed spacecraft before **launch** to ensure its static and dynamic balance during all phases of its mission. For example, static balance is important for **attitude control** in ensuring that **thruster** firings, etc., have the desired effect; dynamic balance is important in reducing **nutation**, when a spacecraft is **spin-stabilised**.

[See also **mass budget**]

ballistic fly-by – see **ballistic trajectory**

ballistic trajectory The path described by a body moving in three dimensions under the influence of gravitational forces and any resistive forces due to the medium of travel (although ideally thrust, drag and lift would be zero); the **trajectory** followed by a body moving in accordance with the laws of ballistics, the study of projectile flight dynamics.

For example, a **launch vehicle** without **lifting surfaces** travels along a ballistic trajectory once **propulsion** ceases (like a 'ballistic missile') – during **guidance** or propulsive phases the vehicle is not considered a ballistic body. A spacecraft using the gravitational field of a planet as a means of propulsion is undertaking a 'ballistic fly-by' (e.g. **Pioneer, Voyager**, etc.).

[See also **free-return trajectory, trans-lunar trajectory, sub-orbital trajectory**]

balloon tank A colloquial term for the **propellant tank**s designed by Karel J. Bossart and used for the **Atlas** ICBM (and **launch vehicle**) and **Centaur** upper stage. The pressure- stabilised, thin steel, **monocoque** structure formed a 'steel balloon', which served as part of the **primary structure** of the **stage** as well as the **skin** of the **propellant tank**. The **pressurant** kept the tank rigid.

ballute A combination of balloon and parachute sometimes used to assist in the deceleration of a **spacecraft** prior to **touchdown**.
[See also **retro-rocket**]

band (frequency) – see **frequency band**

band-pass filter A **filter** with both high- and low-frequency cut-offs which allows a specified band of signal-frequencies to pass. This is the generic term for a group of filters designed to confine a signal to a specified **bandwidth** (e.g. **input filter**, **output filter** and **channel filter**).

bandwidth The width of a **frequency band**, typically measured in MHz. The bandwidth is one of the factors which gives a measure of the amount of information (telephone calls, data-streams, etc.) that can be transmitted over a communications link, for instance through a satellite **transponder**.
[See also **capacity**]

BAP TA (bearing and power transfer assembly) A rotating interface between a spacecraft body and a **solar array**.

On a **three-axis stabilised** spacecraft the BAPTA keeps the array pointing towards the Sun and transfers the power generated to the spacecraft equipment. Each array has its own BAPTA, which accepts pointing commands from sun sensors mounted on the arrays. On a **spin-stabilised** spacecraft the cylindrical array and the majority of the vehicle rotate while the **communications payload** is de-spun to allow the **antenna**s to point towards Earth – the single BAPTA transfers power to the payload.

Aside from the electronics, a typical BAPTA consists of a bearing, a drive-motor and a slip-ring unit. The mechanical part of the device is alternatively known as a solar array drive mechanism (SADM), which in combination with the solar array drive electronics (SADE) is called a solar array drive assembly (SADA).

barbecue roll A slow **roll** given to a spacecraft to equalise the heating effect of the Sun over its surface; a **passive** method of thermal control which reduces the thermal gradient across the spacecraft. Used particularly for spacecraft on a **trajectory** between two astronomical bodies, where the Sun's direction is approximately constant (e.g. used for **Apollo** on a lunar trajectory: rotation rate about $0.1°\ s^{-1}$, or one full turn per hour).
[See also **thermal control subsystem**]

baseband The **frequency band** that a **signal** occupies when initially generated; the signal **bandwidth** prior to **modulation** and subsequent to **demodulation**. The signal may be the output of an item of equipment such as a telephone, TV camera, computer, etc.

bathtub curve A graphical plot of failure-rate, or probability of failure, against time. Figure B1 depicts two generalised curves, one for electronic components, another for mechanical devices. The probability of failure decreases during the 'burn-in' or 'infant mortality' phase, is at a minimum (by design) during the operational **lifetime**, and increases again in the 'wear-out' phase. The figure shows that, although both types of device have a similar burn-in phase, mechanical devices begin to wear out much sooner, thus decreasing the optimum **operational** lifetime.

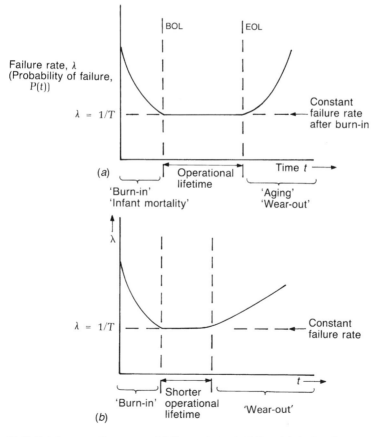

B1: **Bathtub curve** of component failure-rate against time: (a) electronic components; (b) mechanical devices.

In the case of a **spacecraft**, most of the mechanical functions (e.g. **antenna** and **solar array** deployments) are carried out early in the mission and ageing is usually irrelevant. The period between a spacecraft's **beginning of life (BOL)** and **end of life (EOL)** exhibits a constant failure-rate, due simply to random, unpredictable failures.

Using a burn-in process for component selection can enhance **reliability** by a factor of a hundred at little extra cost. Failure-rates can also be improved by derating a component, for example by reducing maximum **power** consumption, voltage, current, operating temperature, or a combination of these, to retard the ageing process. Conversely, the performance of components over a projected lifetime can be simulated by increasing such factors.

[See also **probability distribution**, **MTBF**]

battery A source of **DC power** carried aboard most **spacecraft**. For spacecraft which derive their power from **solar array**s, the battery subsystem is referred to as the **secondary power supply** since it is subsidiary to the arrays. Batteries are also used, as a **primary power supply**, for short duration missions or activities, e.g. on board **launch vehicles** and planetary **entry-probes**.

The batteries themselves are defined as either primary (non-rechargeable) or secondary (rechargeable). The most common type of primary for space-related use is the silver–zinc battery (which can also be used as a secondary, albeit with very short cycle life). There are also a variety of lithium-based primaries, such as lithium sulphur dioxide, lithium carbon monofluoride and lithium thionyl chloride; although lithium has a higher energy density, silver–zinc is easier to handle and can be discharged at a higher rate. Secondary batteries are required to store power for use during **eclipse**, when power cannot be generated from the arrays, and during the **transfer orbit** for satellites that have no array capability during this phase. They can also be used to supply peaks of power greater than that provided on a continuous basis by the solar array.

The fundamental battery component is the voltaic cell, a number of which are connected in series to provide the required operating voltage [see **bus voltage**]. The principal types of battery used in Earth-orbiting **satellite**s are based on the nickel–cadmium (NiCd) cell and the nickel–hydrogen (NiH$_2$) cell; the latter has gradually replaced the former for spacecraft applications [see figure B2]. Lithium-ion (Li), silver–zinc (AgZn) and nickel-metal-hydride (NMH) batteries are specified for more limited applications.

A battery is rated in terms of its **capacity**, which is measured in ampere

B2: Nickel–hydrogen **battery** module for a Globalstar spacecraft. [Space Systems/Loral]

hours (Ah). Its energy, measured in watt hours (Wh), is the product of the capacity and the bus voltage. The **specific energy** (or energy density) is measured in Wh kg^{-1}. A spacecraft battery subsystem can account for up to 10% of the total **mass budget**.

[See also **depth of discharge (DOD)**, **duty cycle**, **radioisotope thermoelectric generator (RTG)**, **fuel cell**, **regulated power supply**]

battery reconditioning The controlled discharging and subsequent recharging of a **battery** conducted on a regular basis to improve battery performance. For example, the batteries on satellites in **geostationary orbit** are reconditioned during the **solstice** period, when battery **power** is not required.

baud The transmission or signalling rate on a **data** channel, where a baud is a signalling element per second. The data transfer rate (measured in bits per second) is usually a simple multiple of this. Named after the French inventor J.M.E. Baudot.

[See also **bit**, **bit rate**]

BDR An abbreviation for baseline design review. It is followed, at a later stage in the programme, by a critical design review (**CDR**) at which the design is finalised.

beacon
> (i) A low-power **carrier** which may be unmodulated for **propagation** tests or modulated with **telemetry** or **tracking** data. See **modulation**.
>
> (ii) A device on a **spacecraft** which transmits a signal that allows it to be tracked by a **ground station** or another spacecraft. Also used to conduct propagation tests which ascertain the effect of the atmosphere on the beacon signal.
>
> [See also **tracking, atmospheric attenuation**]

beam
> (i) The term used to describe the geometric distribution of **radio frequency** radiation formed by a **spacecraft** communications **antenna** (e.g. if the antenna aperture is circular, the beam may be conical).
>
> [See also **beam area, beamwidth, global beam, hemispherical beam, zone beam, spot beam, multiple beam**]
>
> (ii) See **electron beam**.
>
> (iii) See **beam-builder**.

beam area An area on the surface of the Earth, or another **planetary body**, which is within the **half-power beamwidth** of a spacecraft **antenna**.

In **satellite communications**, for which the term was originally defined, the periphery of the beam area corresponds to the -3 dB points on a satellite antenna **radiation pattern** and is therefore bounded by the -3 dB contour. In many cases the beam area practically coincides with the **coverage area**, but the two are defined from different premises and should not be confused.

beam centre The point on the surface of the Earth, or another **planetary body**, where the signal power in the **beam** of a spacecraft **antenna** is at a maximum – its **peak**.

[See also **beamwidth, footprint, boresight**]

beam edge Often used in an inexact manner but, most usefully, the beam edge is the locus of points on the surface of the Earth, or another **planetary body**, where the signal power in the **beam** of a spacecraft **antenna** is half of its **peak** value.

[See also **beamwidth, beam centre, footprint**]

beam waveguide A type of **waveguide** used in **earth stations** between the **high power amplifier (HPA)** and the **antenna** to reduce RF losses [see figure B3]. Instead of propagating the normal 'waveguide modes', which are represented by current distributions in the accurately fashioned walls of the guide itself, the beam waveguide simply provides a 'weather-proof tunnel' for a shaped

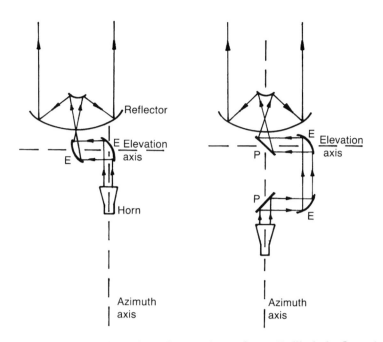

B3: **Beam waveguide** configurations: P, plane reflector; E, elliptical reflector (or offset paraboloid).

beam of RF radiation. A narrow beam is formed by a **feedhorn** and reflected 'optically' through any bends in the guide until it reaches the antenna, where a **subreflector** intercepts it and reflects it onto the main reflector. The main reflector forms the **uplink** beam in the usual way.

beam-builder An automated machine intended to construct framework structures in space, possibly for future **space platform**s or **space station**s. Although such a device has yet to be used operationally in space, a development model has been demonstrated. The machine supplies three continuously extruded longitudinal beams formed from flat spools of aluminium alloy, which are spot-welded into a triangular-section structure using pre-formed cross-braces. The length of the beam is theoretically limited only by the length of the spools and the number of cross-braces the machine can hold.

beam-forming network An arrangement of **feedhorn**s designed to illuminate an **antenna** reflector to form a given shaped beam.
[See also **shaped reflector**, **multiple beam**]

beam-hopping The practice of rapidly switching a satellite **downlink** transmission between different **beam**s in a **multiple beam** antenna. See **satellite switching**.

beamwidth The width of the **beam** of radiation shaped by a communications antenna [units: degrees]. Unless an antenna is designed to be omnidirectional, it concentrates the radiated signal into a beam with a certain **directivity**. Naturally, the beam widens as it leaves the proximity of the antenna, so that by the time the signal reaches its destination it has spread out considerably. The measure of spreading is represented by the beamwidth of the antenna, important in that it indicates whether two ends of a link will be able to communicate: they must lie within the beamwidths of each other's antennas to do so. Moreover, the converse tends to reduce the possibility of **interference** between two systems which do not wish to communicate.

The most useful way to define the concept of beamwidth involves a numerical statement of how the radiated power decreases away from the beam centre or **boresight**: the 'half-power beamwidth' (HPBW) is the width of the beam between the points where the radiated power is half of its **peak** value. The concept is applicable to all communications antennas, not just the commonly seen parabolic dishes. See **antenna**.

The power decreases from the centre of the beam (the peak) towards the edge, as shown in figure B4: at the half-power point (-3 dB) the beamwidth can be determined. For circular parabolic antennas the beamwidth is related to antenna diameter and **wavelength** by the following expression:

$$\text{HPBW} \simeq 72\lambda/D$$

where λ is the wavelength and D is the **earth station** antenna diameter (both in m). The constant 72 in the above expression assumes a circular aperture with parabolic illumination; if the illumination is uniform across the aperture, the constant is 58.4. The same expression can be used for linear apertures, but the constants are 66 for parabolic illumination and 51 for uniform illumination.

The projection on the Earth of a satellite antenna's HPBW delineates the **beam area**. The HPBW is a benchmark in the radiation pattern of an antenna since, to a certain extent, it allows the spacing of satellites and earth stations to be arranged so that they do not interfere with other services. The true

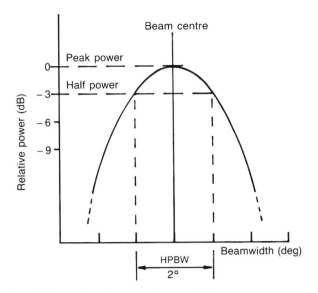

B4: An illustration of antenna **beamwidth**.

criterion by which antennas are judged is, however, the sidelobe
performance.

[See also **antenna radiation pattern, footprint, omnidirectional antenna**]

beanie cap A conical attachment on a **swing-arm** (known as the gaseous oxygen
arm) on the fixed **service structure** at the Space Shuttle's launch pad which
extracts vented oxygen vapour from the nose of the **external tank (ET)** prior
to launch. The release of the **cryogenic** vapour from the **liquid oxygen** tank
would cause air to freeze around the nose of the ET, and subsequent launch
vibrations could cause the resultant ice to fall and damage the tiles of the
orbiter's thermal protection system. The beanie cap, which seals the ET's
nose with an inflatable collar, removes the vented gas before it can have this
effect. The swing-arm is retracted about two minutes before launch.

[See also **vent and relief valve**]

bearing and power transfer assembly – see BAPTA

beginning of life (BOL) The beginning of a **spacecraft**'s operational **lifetime**. The
term is used to specify the magnitude of a parameter at the beginning of a
spacecraft's lifetime, as opposed to at **end of life (EOL)** when it may be
different (e.g. the output power from a **solar array** [see **degradation**], the

mass of **station-keeping** propellant, etc.). Also refers to components – see **bathtub curve**.

Bell X-1 – see **X-1**

bends – see **decompression sickness**

bent-pipe repeater – see **transparent repeater**

BER – see **bit error rate**

beryllium (Be) A low-density metal increasingly used (when alloyed with other metals) in the aerospace industry. It can tolerate high temperatures and has high specific stiffness, high specific heat capacity and good thermal conductivity. It is, however, susceptible to surface damage from machining which can lead to brittle fracture. This, coupled with its toxicity, has made it more difficult to use than, say, **aluminium** or **titanium**. Typical applications: components, mechanisms, casings and mirrors.

[See also **materials**]

binary phase shift keying – see **BPSK**

binder A substance which binds the components of a composite solid propellant together. See **solid propellant**.

bipropellant One of two chemical **propellants** (**fuel** or **oxidiser**) which are stored separately and combined for combustion to take place (e.g. **liquid hydrogen** and **liquid oxygen**; UDMH and **nitrogen tetroxide**).

Bipropellants are predominantly **liquid propellants**, but a liquid oxidiser and a solid fuel are used together in a 'hybrid' propulsion system [see **hybrid rocket**].

[See also **bipropellant thruster, combined (bipropellant) propulsion system, liquid apogee engine**]

bipropellant propulsion system – see **combined (bipropellant) propulsion system**

bipropellant thruster A propulsive device utilising two chemical **propellants**, an **oxidiser** and a **fuel**; a type of **reaction control thruster** [see figures B5, S13]. Common bipropellants are monomethyl hydrazine (MMH: $CH_3HN_2H_2$) as the fuel and nitrogen tetroxide (N_2O_4) as the oxidiser. Since they are hypergolic (i.e. ignite spontaneously when mixed), the thruster is simplified by the lack of an **ignition system**.

[See also **hydrazine thruster, hiphet thruster, orbital manoeuvring system (OMS)**]

BIS An abbreviation for the British Interplanetary Society, which was founded in 1933 to promote interest in **astronautics**, in part by publishing the *Journal of*

B5: Bipropellant thruster showing two propellant valves (at top), **plume shield** and **nozzle**. [Mark Williamson]

the British Interplanetary Society (JBIS) and *Spaceflight* magazine. One of the society's founding members was Arthur C. Clarke, who publicised the potential application of **geostationary orbit (GEO)**.

[See also **ARS, VfR**]

bit The smallest unit of information in a computer or communications system which can have one of two values (1 and 0) in binary notation. A contraction of *bi*nary dig*it*. Usually seen with a prefix (e.g. kbit, Mbit, Gbit).

[See also **bit rate, baud, byte**]

bit error rate (BER) A concept used to express the accuracy of digital **demodulation** or decoding; a measure of how truthfully the received signal represents the signal originally transmitted.

bit rate A measure of the rate at which information passes through a **computer** or **communications system**: a **data** transfer rate [units: bits s^{-1}, kbits s^{-1}, Mbits s^{-1}, etc. (also written bps, kbps, Mbps, etc.)]. For example, the most common rate for a 1990s domestic **modem** was 56 kbps, while an **integrated services digital network (ISDN)** line typically provided 128 kbps, and a digital subscriber line (DSL) 1.544 Mbps (known as a T1 circuit) or 2.048 Mbps (E1).

Examples of the bit rates required for various applications show that, logically enough, bit rate is related to the complexity of the information carried. For example, speech via mobile radio and reasonable quality **telephony** can be provided at rates as low as 2.4 and 9.6 kbps respectively; **slow-scan TV** and fast **facsimile** require 64 kbps; hi-fi sound 250 kbps; a **teletext** picture about 1 Mbps; a 625-line colour TV signal 68 Mbps; and a high definition TV (HDTV) system about 144 Mbps. However, with **digital compression**, the required bit rate can be reduced (e.g. a common rate for digital satellite broadcasting services of the early 2000s was 8.448 Mbps). [See also **bit, baud, bit error rate**]

blast-off – see **lift-off**

block A modifier indicating a specific level of development of **hardware** (e.g. block-I and -II **spacecraft**); a method of distinguishing one spacecraft **generation** from another, usually applied to military satellites [see **Global Positioning System (GPS)**].

blockage – see **aperture blockage**

blockhouse – see **launch control centre (LCC)**

blowdown system A closed **liquid propellant** delivery system which uses a fixed mass of **pressurant** throughout its period of operation; a system in which the pressure derived from the pressurant decreases as the tank is emptied (i.e. the **ullage pressure** decreases as the **ullage** volume increases).

Some **combined (bipropellant) propulsion system**s used in geostationary satellites operate with a renewable supply of pressurant for the apogee engine firing and then switch to 'blowdown mode' for the **reaction control thruster** firings made throughout the satellite's **lifetime**. Other delivery systems use one or other method all the time.

Blue Streak A British intermediate range ballistic missile (IRBM) developed in the 1950s (**propellant: liquid oxygen/kerosene**). When the United Kingdom's effort to develop an independent nuclear deterrent was disbanded in 1960, Blue Streak was proposed as the first **stage** of a European three-stage launcher called **Europa**, which was then developed under the auspices of the **European Launcher Development Organisation (ELDO)**. Following multiple failures of the upper two stages supplied by France and West Germany, Britain withdrew from ELDO but continued to supply Blue Streak boosters. By the early 1970s Europa had still not succeeded in orbiting a payload, so the programme was cancelled and ELDO was disbanded. In 1975, ELDO and ESRO (European Space Research Organisation) were amalgamated to form the European Space Agency (**ESA**), which subsequently initiated the **Ariane** programme and delegated its management to CNES (the French national space agency). [See also **Woomera, ICBM**]

BNSC An abbreviation for British National Space Centre, which was formed in November 1985 to coordinate civil UK space activities and develop future policy. The BNSC brought together the civil activities of a number of established bodies: the Department of Trade and Industry (DTI), Ministry of Defence (MOD), Science and Engineering Research Council (SERC) and Natural Environment Research Council (NERC).

body stabilised – see **three-axis stabilised (spacecraft)**

boilerplate A term applied to a version of a **spacecraft** used only for test purposes; an engineering test vehicle (e.g. in the early American manned spaceflight programmes, **launch vehicles** were fitted with (unmanned) 'boilerplate **capsules**' which were the same size, shape and weight as the real capsules but contained few internal fittings or equipment).

boil-off The evaporation of **cryogenic propellant**, usually from a **launch vehicle**, into the atmosphere. The evaporant is released through a 'boil-off' valve prior to final **pressurisation** and **launch** [see **vent and relief valve**].

BOL – an abbreviation for **beginning of life**

bolometer A type of resistance thermometer, used particularly for the detection of radiant energy at infrared **wavelengths** – on Earth in infrared **telescopes** and in space as **earth sensors** (as part of an **attitude control** system).

bolt-cutter – see **pyrotechnic cable-cutter**

Boltzmann's constant A constant named after Ludwig Edward Boltzmann [1844–1906], an Austrian physicist, which relates kinetic energy and temperature.

It is, in effect, a 'conversion factor'. In the same way that there is an absolute reference for temperature – 'absolute zero' ($-273.15°$ C or 0 K) – there is an absolute reference for 'noise' in a communications system. The very lowest level of **background noise** is **thermal noise**, due to the motion of atoms and molecules which constitute any physical object. As the absolute temperature on the **kelvin** scale approaches zero, molecular motion ceases and the thermal noise approaches its zero, absolute reference. This is represented by Boltzmann's constant ($k = -228.6$ dBW $Hz^{-1} K^{-1}$), a constant of proportionality between the molecular kinetic energy and temperature, such that $KE = 3/2(kT)$. The units of the constant describe, in terms of power (dBW), a level of noise contained in every 1 Hz of **bandwidth** for every 1 K of temperature.

Boltzmann's constant (in SI units, 1.381×10^{-23} J K^{-1}) is the ratio of the universal gas constant (8.315 J K^{-1} mol^{-1}) to Avogadro's number (6.022×10^{23} mol^{-1}), and should not be confused with the Stefan–Boltzmann constant ($\sigma = 5.67 \times 10^{-8}$ W $m^{-2} K^{-4}$), which is used to define the energy radiated by a black body.

boom A spar or pole deployed from a spacecraft to position an instrument or detector at a set distance from the main body of the vehicle. For example, mounting a **magnetometer** on a boom ensures that the spacecraft's contribution to the measured magnetic field is minimised.

boost phase

(i) The portion of a **launch vehicle**'s flight when **thrust** is being exerted by the **propulsion system**; a period of powered flight, as opposed to **coasting**.

(ii) The powered portion of a ballistic vehicle's **flight**.

[See also **ballistic trajectory**]

booster

(i) A propulsive component of a **launch vehicle**, usually a **strap-on** booster added to the **core vehicle** to augment its **thrust**. Sometimes called a 'booster stage'.

(ii) A colloquial term for 'launch vehicle'.

[See also **solid rocket booster, rocket engine, rocket motor, retro-rocket**]

boresight The main axis of an **antenna**: a distinction is made between the **radio frequency (RF)** boresight [see **peak**] and the geometric boresight (physical axis) of the antenna, the angle between the two being known as the '**squint**' angle.

[See also **beamwidth**]

boresighting A colloquial term for the pointing of an **antenna**. See **boresight**.

box Jargon: an area in the sky, as viewed from an **earth station**, which bounds the excursions of a **satellite** in **geostationary orbit** from its mean position. The satellite is not precisely stationary with respect to the earth station, mainly owing to the gravitational forces acting upon it: the satellite drifts both above and below the equatorial plane within the limits of its **orbital inclination**. It executes a figure-of-eight pattern in the sky with the cross-over in the eight at its mean position as viewed from the ground [see figure B6]. This pattern is repeated daily. A common specification requires the satellite to remain within a box subtending an angle of $\pm 0.1°$ from the Earth's centre, which equates to about 125 km (77.5 miles) at geostationary height.

[See also **orbital control**]

BPSK (binary phase shift keying) A method for modulating the radio frequency **carrier** in digital **satellite communications** links; a type of **phase modulation**. In BPSK or 'two-phase **modulation**', where sample carrier waveforms are 180 degrees out of phase, one **bit** of data is encoded on the carrier for each phase change, allowing the representation of a '1' or '0'.

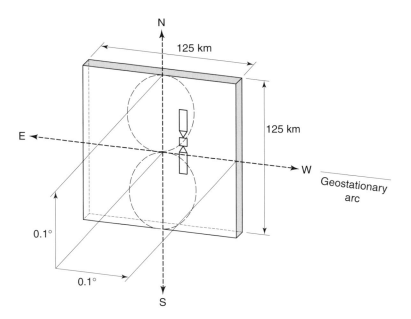

B6: The station-keeping 'box' of a satellite in **geostationary orbit**, showing the tolerances on its **orbital position**.

[See also **radio frequency**, **modulation**, **QPSK (quadrature phase shift keying)**, **digital compression**]

breadboard A lab-bench assembly of components or circuits of a preliminary design intended to prove the feasibility of the design, without regard to the final packaging or **configuration**. The term is derived from the practice of early electronics engineers, who mounted their experimental circuits on a wooden board usually used for cutting bread.

broadband A modifier indicating the broad **bandwidth** of the quantity under consideration (e.g. broadband **signal**, **broadband noise**, broadband **multimedia**). Essentially the same as **wideband**.

broadband noise Noise which is spread across the **band** of a communications system; noise which occupies the full **bandwidth**. Also called **wideband noise**. In general, the wider the band the greater the noise, so in most cases the minimum bandwidth capable of accommodating the **signal** is used.

broadcasting The transmission of a **radio frequency** signal carrying radio or television programmes (for example) over a wide area of reception. The transmission medium is a specified part of the **electromagnetic spectrum**, there being no physical connection between transmitting and receiving **antennas**, in contrast to **cable** or **optical fibre** networks.

[See also **narrowcasting, multicasting, unicasting, link, cable TV, direct broadcasting by satellite, digital compression, digital video broadcast (DVB)**]

broadcasting-satellite service (BSS) ITU terminology for a **satellite communications** system in which signals transmitted by the satellite are intended for direct reception by the general public. In the formal International Telecommunication Union (**ITU**) definition the term 'direct reception' encompasses both individual and community reception.

[See also **direct broadcasting by satellite (DBS), fixed-satellite service (FSS), mobile-satellite service (MSS)**]

BSS – see **broadcasting-satellite service**

built-in hold A pre-planned halt in the **countdown** to a **launch**, included to allow time for any unforeseen procedural delays which may occur and for any final checks (e.g. on the weather) which may be required.

[See also **hold, recycle countdown**]

bumper shield – see **Whipple shield**

Buran The name given to the first **flight model** of the Soviet space shuttle fleet [see figure B7]. Buran, which means 'Snowstorm', was launched on its first unmanned test flight on 15 November 1998 by an **Energia** launch vehicle.

B7: Soviet space shuttle **Buran** on its Antonov An-225 carrier aircraft. [Mark Williamson]

burn Jargon: the operation of a **rocket engine** to alter the **attitude**, **trajectory** or velocity of a **spacecraft** (e.g. 'engine burn', 'de-orbit burn', etc.). Tends to be used only for propulsion in vacuo (i.e. not for **launch vehicle**s).
[See also **apogee motor firing, mid-course correction, plane change, de-orbit**]

burn-in – see **bathtub curve**

burn-out The moment at which a propulsive device ceases to provide **thrust**, usually because its **propellant** supply has been exhausted; the cessation of propellant **combustion** (also called 'engine cut-off'). Used particularly for **launch vehicle** stages (e.g. 'third **stage** burn-out').
[See also **throttling, engine re-start (capability)**]

burst A very-short-duration transmission in a communications system. For example, in **time division multiple access (TDMA)** the basic repetitive unit is the 'frame' which consists of a number of bursts, one from each **earth station** connected to the system. A typical system, using **QPSK** at a burst-rate of 60 Mbits s^{-1}, may have a frame duration of 15 ms, which is equivalent to 900,000 bits per frame. The first burst in each frame is transmitted by a controlling earth station and followed by bursts from the other stations in sequence. Although individual bursts are separated by 'guard times' to avoid overlapping at the receiving station, it is essential that all earth stations are time-synchronised.

burst test A method of testing the structural integrity of a **propellant tank** – and, by extension, its design – which involves introducing a gas at high pressure. If the tank survives a pressure sufficiently above its operating pressure before bursting, the design may be considered proven.

bus

    (i) Jargon for the part of a spacecraft that 'carries' or supports the **payload**. For example, on a **communications satellite**, the **service module** is known as the 'bus'; it may also be referred to as 'the **platform**'. Spacecraft manufacturers tend to produce standardised spacecraft buses as part of a product range: e.g. the American manufacturer Hughes produced the HS376, HS601 and HS702 buses in a number of variants (where HS is an abbreviation for Hughes satellite).

    (ii) Jargon for the power supply and distribution system of a spacecraft. See **power-bus**.

    (iii) Jargon for the part of a computer that transports **data** and interconnects components or **subsystem**s (e.g. data bus, serial bus). Its capacity is measured in **bits** (e.g. a 32-bit bus can transport data more quickly than a 16-bit bus).

bus voltage The voltage rating of a spacecraft's main electrical power distribution circuit, or **power-bus**. The range of bus voltages acceptable to spacecraft equipment is dependent on whether the power subsystem is 'regulated' or 'unregulated' (see **regulated power supply**). The great majority of spacecraft power supplies are of the direct current type (see **DC power**).

DC bus voltages have increased gradually over the years, more or less in line with the general rise in available DC power. A 16 V standard was developed in the 1960s and used until the early 1970s, when a 28 V system became the predominant system for satellites manufactured in the USA. When the European Space Agency (**ESA**) made plans for satellites with up to 10 kW of onboard power, it developed a 50 V standard to reduce the power-losses and resultant thermal dissipation problems that would be suffered with the 28 V system; European manufacturers also used a 42.5 V system, which was later adopted by some US companies. In the 1990s, as satellite powers increased further, bus voltages increased to around 100 V.

An 18 V alternating current (AC) supply was proposed in the early 1970s, but the equipment was heavy and inefficient, and the concept was shelved. By the 1980s, technology developments had decreased the mass and the power losses, but the application of AC has been limited to a few scientific satellites and spacecraft payloads (e.g. the **Hubble Space Telescope** uses a 20 kHz AC standard for secondary power distribution).

byte A unit of information in a computer or communications system typically six or eight bits in length; a contraction of *binary term*. The smallest unit of information that can be addressed or given a memory location. Normally two or four bytes are assembled to form a 'word', one of the basic units of data. Computer memory is measured in multiples of bytes (e.g. kilobytes (1024 bytes), megabytes (1,048,576 bytes) and gigabytes (1,073,741,824 bytes)). [See also **bit**, **bit rate**, **baud**]

C

C/A-code Coarse acquisition-code, a mode of operation of the **Global Positioning System (GPS)** providing degraded position accuracy.

cabin A general term for the habitable volume of a **manned spacecraft**, used particularly for small vehicles (e.g. 'spacecraft cabin'); for larger vehicles the alternative term '**flight deck**' is more common.
[See also **capsule**]

cabin pressure The pressure of the atmosphere maintained within a **manned spacecraft**. Early American spacecraft (i.e. **Mercury**, **Gemini** and **Apollo**) were supplied with a pure oxygen atmosphere at a pressure of about 38 kPa (5.5 psi). The Space Shuttle **orbiter** uses an oxygen/nitrogen atmosphere at 101.4 kPa (14.7 psi – 'sea-level pressure'). Russian spacecraft (i.e. **Vostok**, **Voskhod**, **Soyuz**, **Salyut** and **Mir**) have always used an oxygen/nitrogen atmosphere at sea-level pressure.
[See also **pre-breathe, decompression sickness, environmental control and life support system (ECLSS), atmosphere**]

cable A medium for the transmission of television, data, etc. based on lengths of **coaxial cable** that provide a physical link between transmitting and receiving locations. The 'cable' may, increasingly in future, be a fibre-optic cable.
[See also **cable TV, optical fibre**]

cable TV A television service transmitted to the subscriber by **coaxial cable**; sometimes abbreviated to **CATV**.
[See also **SMATV, broadcasting**]

cable-cutter – see **pyrotechnic cable-cutter**

capacity
    (i) In telecommunications, a measure of the ability of a circuit, **transponder**, etc. to carry information or communications **traffic** (telephone calls, TV channels, data, etc.). The capacity of a **communications satellite**, for example, is a function of many factors, such as the available **bandwidth** of its transponders, the **modulation** scheme, the satellite transmission power and the size of the **earth station** antenna.
        [See also **bit rate**]

(ii) The total stored charge in a **battery**: since charge is the product of current and time, capacity is measured in ampere hours (Ah).

(iii) The **payload** mass that can be lifted by a **launch vehicle**: launch capacity (an alternative for **launch capability**).

(iv) In **space insurance**, the maximum amount of cover (typically measured in US dollars) that underwriters are willing to attach to an individual launch at a given time.

capcom An abbreviation for 'capsule communicator'; a person assigned the task of relaying messages between '**mission control**' and a **manned spacecraft** (coined when spacecraft were called **capsules**).

Cape Canaveral – see **Kennedy Space Center**

Cape Canaveral Air Station (CCAS) The US Air Force **launch** facility (located at approximately 28.5° N, 80.3° W), formerly known as Cape Canaveral Air Force Station.

[See also **Kennedy Space Center**]

Cape Kennedy – see **Kennedy Space Center**

capsule

(i) A name given to early **manned spacecraft** (or those carrying animals), due largely to their small size and their 'enclosing and protecting' nature (e.g. '**Mercury** capsule', '**Apollo** capsule', etc.).

(ii) The section of an **unmanned spacecraft** designed to return part of a **payload** to Earth (e.g. 'film capsule', 're-entry capsule').

[See also **cabin, flight-deck**]

capsule communicator – see **capcom**

carbon composite A material composed of threads or fibres of the element carbon set in a 'matrix' of thermo-setting polymers such as the polyesters, phenols and epoxies.

Carbon fibre, one of the many **materials** used to reinforce polymer matrices, was developed in 1963 at the Royal Aircraft Establishment (RAE), Farnborough (UK). In the modern factory process carbon fibres are produced, for example, from specially treated fibres of polyacrylonitrile (PAN), a carbon-based polymer, known as acrylic precursors. The controlled oxidation of the PAN precursor converts the acrylic fibres to a heat-resistant polymer which retains its textile character, making it possible to wind, weave or knit the final product [see **filament winding**]. The level of oxidation can be adjusted to determine the precise properties of the resultant carbon fibre, dependent upon its eventual application.

Many similar terms will be found in the literature appertaining to composite materials used in spacecraft: carbon fibre, carbon composite, graphite composite, graphite epoxy and carbon fibre reinforced plastic (CFRP). In the main these descriptions refer to the same class of materials, since they are all based on the chemical element 'carbon'. However, in a more accurate sense, carbon products tend to be based on the relatively untreated 'amorphous carbon', whereas graphite-based products have been subjected to high temperature and/or high pressure processing to produce a more ordered structure with the properties of higher electrical and thermal conductivity. The differences between particular products may also include the degree to which the precursor has been oxidised, the proportions of carbon and graphite, the constitution of the matrix, etc. Typical applications include spacecraft **face-skins**, **antenna** reflector surfaces, structural components and propellant tanks [see figure C1].
[See also **materials**]

C1: A selection of **carbon composite** propellant tanks. Fibres of material are visible as a result of the **filament winding** process.

carbon dioxide absorber A device in a manned spacecraft's **life support system** which filters carbon dioxide from the air and recirculates it for further use. Colloquially called a 'CO_2 scrubber'. See **environmental control and life support system (ECLSS)**.
[See also **carbon dioxide narcosis**]

carbon dioxide narcosis Unconsciousness induced by an excess of carbon dioxide in the atmosphere (i.e. more than about 2 kPa (0.3 psi) partial pressure, the maximum recommended concentration being about 0.7 kPa/0.1 psi). The normal sea level partial pressure of carbon dioxide is about 0.03 kPa (0.004 psi), compared with the total air pressure of 101.4 kPa (14.7 psi). [See also **carbon dioxide absorber, environmental control and life support system (ECLSS)**]

carbon fibre reinforced plastic (CFRP) A carbon composite material used widely in spacecraft structures, mainly because of its characteristics of low density and high strength. See **carbon composite**.
[See also **materials**]

carbon–carbon composite A material composed of a carbon matrix reinforced by threads or fibres of carbon. The carbon matrix can be produced by 'hydrocarbon cracking' or by pyrolysing a resin to leave a carbon residue into which the fibres are embedded. This class of composites has been developed for their high temperature resistance: in excess of 1400 °C in an inert (non-oxidising) environment. Typical applications include rocket motor **nozzles** and **thermal protection systems**.
[See also **composites, carbon composite, ceramic–metal composite, metal–matrix composite, materials**]

cargo bay – see **payload bay**

carrier In **telecommunications**, an electromagnetic wave of fixed **amplitude, phase** and **frequency** which, when modulated by a **signal**, is used to convey information through a radio transmission system. The frequency of the carrier is the frequency at which the **transmitter** operates, so the carrier conveys all the information held by the **baseband** signals and the **subcarriers** which are modulated onto it.
[See also **carrier-to-noise ratio, signal, modulation**]

carrier generator – see **local oscillator**

carrier wave – see **carrier**

carrier-to-interference ratio – see **protection ratio**

carrier-to-noise power density ratio (C/N₀) The ratio of the power-level of a signal **carrier** wave to the **noise power density (NPD)**, represented by the expressions:

$$\frac{C}{N_o} = \frac{P_C}{(P_N/B)} = \frac{P_C}{kT} = \frac{C}{kT}$$

where P_C is the carrier power, P_N is the **noise power**, B is the **bandwidth**, k is **Boltzmann's constant** and T is the **absolute temperature**. See **noise power density**.

The difference between C/N (**carrier-to-noise ratio**) and C/N_0 is that C/N_0 considers only the **noise** in 1 Hz of bandwidth (noise/Hz), whereas C/N takes account of the total noise in the full bandwidth.

carrier-to-noise ratio The ratio of the power-level of a signal **carrier** wave to the power-level of the **noise** at the same time and place in a radio transmission system. Abbreviated to C/N or, less frequently, CNR, this quantity refers to a **signal** at RF or IF as opposed to **baseband**.

[See also **noise power**, **signal-to-noise ratio**, **carrier-to-noise power density ratio**, **radio frequency**, **intermediate frequency**]

Cassegrain reflector A style of dual reflector **antenna** using a convex hyperboloidal **subreflector** and paraboloidal main reflector. The **feedhorn**, subreflector and main reflector are coaxial in the standard design [see figure C2], but the feed can be offset [see **offset-fed (antenna)** and figure A3]. This geometry was first described for optical astronomical telescopes in 1672 by Prof. N. Cassegrain of the Collège de Chartres to the Paris Académie des Sciences, although the first practical example was not constructed until the mid-eighteenth century, by James Short [see figure N1].

[See also **Gregorian reflector**, **Newtonian telescope**, **Schmidt telescope**, **Coudé focus**, **space telescope**]

C2: A 6-m-diameter **earth station** antenna with a centre-fed **Cassegrain reflector**. [Matra Marconi Space]

CATV Originally an abbreviation for Community Antenna Television, but now also used as an abbreviation for cable television.
[See also **cable TV, SMATV**]

cavity A component of a **klystron** power amplifier.
[See also **coupled cavity**]

C-band: 4–8 GHz – see **frequency bands**

CCAS An abbreviation for **Cape Canaveral Air Station**.

CCD – see **charge-coupled device**

CCIR A French abbreviation for the International Radio Consultative Committee, a body established within the International Telecommunication Union (ITU) in 1927 to consider and issue recommendations on technical, operational and tariff matters relating to **radio frequency** use. On 1 March 1993, the ITU's telecommunication standardisation sector took responsibility for the standard-setting activities of both the CCIR and **CCITT** [see **ITU**].

CCITT A French abbreviation for the International Telegraph and Telephone Consultative Committee, a body established in 1956, by amalgamating existing bodies within the International Telecommunication Union (ITU), to consider and issue recommendations on technical, operational and tariff matters relating to international telegraph and telephone services. On 1 March 1993, the ITU's telecommunication standardisation sector took responsibility for the standard-setting activities of both the CCITT and **CCIR** [see **ITU**].

CDMA – see **code division multiple access**

CDR An abbreviation for critical design review, a major programme review at which a design is finalised. It is preceded, at an earlier stage in the programme, by a preliminary design review (PDR) or baseline design review (BDR).

celestial body Any astronomical body (e.g. stars, planets, moons). Historically, the Earth has not been included, since it has not been thought of as a 'heavenly body' – it is, after all, the centre of the 'celestial sphere'. However, from a point in outer space, the Earth is just another celestial body.
[See also **planetary body**]

Centaur A rocket **upper stage** used with the **Atlas** and, later, **Titan** launch vehicles, powered by the **cryogenic propellants** **liquid oxygen** and **liquid hydrogen** [see figure U2]. The first Centaur flight occurred in May 1962; the improved Centaur-1A (for Atlas) and Centaur-1T (for Titan) first flew in April 1973 and February 1974, respectively. Two variants, designated Centaur G and Centaur G-prime, were derived for use with **NASA**'s **Space Shuttle**, but as a result of safety measures instigated following the destruction of the Space Shuttle **Challenger** their use was confined to **expendable launch vehicles** (ELVs), such as the Titan IV.

[See also **payload assist module (PAM)**, **inertial upper stage (IUS)**]

central processing unit (CPU) The part of a computer that performs logical and arithmetical operations on the **data**. [See also **microprocessor**]

centre of gravity The point through which the resultant of the gravitational forces on a body acts. Commonly abbreviated to 'c-of-g'. Where gravitational forces are very low (i.e. in **weightlessness**), the term **centre of mass** is substituted.

[See also **gravity**, **low gravity**, **microgravity**]

centre of mass The point through which the resultant of any applied forces on a body acts. Commonly abbreviated to 'c-of-m'. For example, for best effect the thrust vector of a spacecraft's **thruster** should act through the c-of-m; when subject to an **attitude control** manoeuvre, a spacecraft rotates about an axis through its c-of-m.

[See also **spacecraft axes**]

Centre Spatiale Guyanais (CSG) – see **Guiana Space Centre**

centre-fed (antenna) An **antenna** system in which the **feedhorn** is placed on the principal axis of the antenna reflector. Although a centre-fed design is simpler than the alternative offset-fed design, a horn which intercepts an antenna **beam** causes **aperture blockage** and diffraction [see figures A3, C2]. Both **spacecraft** and **earth station** antennas can be either offset-fed (asymmetric) or centre-fed (axisymmetric).

centrifugal force A force on a body moving in a circle or along a curved path which acts directly away from the centre of curvature (literally 'centre-fleeing'). The opposing force is called 'centripetal', from the Latin for 'seeking the centre'; since the acceleration of a body in circular motion is defined mathematically as 'acting towards the centre of rotation', it is termed 'centripetal acceleration'.

In physics, centrifugal force is known as a 'fictitious force', because its effects occur only as a reaction to the opposing centripetal force (a 'true force'). Take, as an example, a spacecraft travelling in accordance with Newton's First Law: it will continue to move in a straight line until a force opposes that motion. If a planet were to 'suddenly appear' to one side of the spacecraft, its gravitational attraction would pull the spacecraft into an orbit around it. The true force here is obviously that due to gravity, which acts towards the centre of the planet. Of course the spacecraft does not move in the direction of this force, because it also has an orbital velocity. It is this which creates the apparent outward, 'balancing' force which stops it moving towards the planet – this is the fictitious centrifugal force.

If the planet's gravity were to be instantaneously removed, the spacecraft would continue at a tangent to its previous motion, in a straight line, thereby proving that the radial, centrifugal force existed only as a result of the centripetal (gravitational) force.

In a **centrifuge**, or a spacecraft spun to induce 'artificial gravity', the centrifugal force is again a reaction against the centripetal acceleration produced by circular motion. The bodies involved would fly off at a tangent were it not for a seat or floor pushing against them – it is this force which gives the impression of 'weight'.

[See also **weightlessness**]

centrifuge

(i) A device which simulates the varying accelerations experienced in **spaceflight**, particularly during **launch**. Although modern manned **launch vehicles** such as the Space Shuttle are designed not to exceed about '3g' during launch, early vehicles used for the launch of **manned spacecraft** greatly exceeded this value (e.g. '9g'): the centrifuge not only allowed **astronauts** to experience the accelerations in a controlled manner, but also allowed them to develop a tolerance which would avoid a 'black-out' at a critical phase of the mission.

(ii) Any rotating device for separating liquids of differing density or liquids from solids by the action of **centrifugal force**.

centripetal force/acceleration – see **centrifugal force**

CEPT A French abbreviation for European Conference of Postal and Telecommunications Administrations, an international commission designed to strengthen relations between the European post and **telecommunications** administrations (PTTs).

ceramic matrix composite (CMC) A material composed of a ceramic matrix reinforced by threads or fibres of a ceramic material (e.g. silicon carbide/silicon carbide). Fibre-reinforced CMCs offer high temperature resistance without the inherent fragility of **ceramics** [see figure C3].

[See also **composites, carbon composite, metal matrix composite, materials**]

C3: **Ceramic matrix composite** rocket **nozzle** components. [SEP]

ceramics A class of materials made from clay and similar substances; inorganic
compounds or mixtures requiring heat treatment to fuse them into an
homogeneous mass. The majority of ceramics can withstand a much wider
temperature range than most metals and are generally resistant to
temperatures up to and above 2000 °C (in an oxidising environment): e.g.
alumina, used in ablative shields, melts at 2054 °C.
[See also **ceramic–metal composite**, **composites**, **materials**, **ablation**]

CEU – see **control electronics unit**

CFRP – see **carbon fibre reinforced plastic**

Challenger

(i) The name given to the second '**flight model**' of the Space Shuttle **orbiter**
(Orbiter Vehicle OV-099), named after a sailing ship which explored the
Atlantic and Pacific Oceans in the 1870s. Challenger was originally built
as a 'Structural Test Article' (STA-099), but after the completion of the
design verification it was converted to an orbiter capable of **spaceflight**. It
was first launched on 4 April 1983 (STS 31-B). On 28 January 1986 (mission
51-L, the 25th Space Shuttle mission) the failure of an 'O-ring' seal on one

of the **solid rocket booster**s caused the vehicle to explode, with the consequent destruction of the spacecraft and the loss of its crew.

(ii) The name given to the **Apollo** 17 Lunar Module.

[See also **Space Shuttle, space transportation system (STS), Atlantis, Columbia, Discovery, Endeavour, Enterprise**]

channel A **band** of radio frequencies with a pre-determined upper and lower limit designed to carry a specific quantity or quality of information. Thus every channel has a **bandwidth** and, depending how it is to be used, can carry a certain **bit rate**.

[See also **channel filter, single channel per carrier (SCPC), multiplexing**]

channel filter A specific type of **band-pass filter** used to define the **bandwidth** of a communications **channel**. This **filter** is placed immediately before the transmit section of a **transponder** chain, which is largely responsible for amplifying the signal.

channel isolation A measure of separation between **radio frequency (RF)** channels. When used qualitatively, it is expressed in terms of **frequency separation**: the channel centre-frequencies are separated from one another by a number of hertz (Hz). The band of frequencies separating the two nominal channel **bandwidth**s is known as a guard-band.

In a more quantitative sense, the isolation of RF channels is quoted in terms of the relative **power**s of their **carriers**: one channel is a number of dB above the other. This is otherwise known as **protection ratio**.

[See also **channel, isolation, guardband**]

channel separation The practice of providing different **radio frequency** channels to avoid **interference** between communications services which operate in the same geographical area or which use **satellite**s in the same **nominal orbital position**. This is especially important if the services operate on the same **polarisation**. Essentially the same as **frequency separation**.

[See also **channel, channel isolation, interference, co-channel interference**]

channelisation The process of dividing up the **radio frequency** spectrum into specifically defined portions of **bandwidth**, or **channel**s, for the efficient and non-interfering transmission of information.

characteristic velocity

(i) A parameter which characterises the **combustion** performance of a **rocket engine** or **rocket motor**; the figure of merit for the **propellant**.

Characteristic velocity (denoted by C^* or 'C-star' and measured in m s^{-1}) is given by the expression:

$$C^* = P_c A_t / m'$$

where P_c is the combustion chamber pressure (N m^{-2}), A_t is the nozzle throat area (m^2) and m' is the propellant mass flow-rate (kg s^{-1}). The higher the value of C^*, the higher the combustion efficiency.

 If a rocket **nozzle** was to be cut off at the **throat**, the **exhaust velocity** would be equal to the characteristic velocity, thus allowing the effectiveness of the **propellant** to be demonstrated independent of the design of the exit cone.

(ii) The velocity a vehicle would acquire due to the **thrust** of its engines under ideal conditions (i.e. no **gravitational attraction**, no **drag**, etc.).

[See also **thrust coefficient**]

charge

 (i) An attribute of matter responsible for electrical phenomena which exists in two forms (positive and negative charge); a quantity of electricity determined by the product of an electric current and the time for which it flows, measured in coulombs (C). See **depth of discharge (DOD)**, **battery**, **duty cycle**.

(ii) An alternative term for **propellant grain**.

charge-coupled device (CCD) A light-sensitive semiconductor device, or 'chip', used to produce images in **telescope**s, **remote sensing** payloads, etc; an alternative to or replacement for a **vidicon** tube or photographic film. In response to incident light, a linear array of **silicon** detector elements releases **electron**s, which are 'read out' by a **microprocessor** that assigns brightness values to individual picture elements, or 'pixels', in accordance with the electron count. Since the read-out mode involves addressing individual CCD detector elements in series, and coupling the charge between adjacent elements, the device was termed 'charge-coupled'.

[See also **pushbroom scanner**]

charge–discharge cycle The cyclic variation in the **charge** of a **battery**; the period from the onset of charging, through discharging to recharging.

[See also **duty cycle**, **depth of discharge (DOD)**]

chase-plane An aircraft which follows and observes a spacecraft, particularly the Space Shuttle **orbiter** during its descent and landing phase.

chemical propulsion A form of **rocket** propulsion in which chemical propellants either undergo **combustion** or are decomposed to produce **thrust**.
[See also **propellant, rocket engine, rocket motor, hydrazine thruster, electric propulsion, nuclear propulsion**]

chilldown Jargon: the process of preparing or 'conditioning' **launch vehicle** and ground-based propellant systems for the loading of **cryogenic propellant**. When the equipment has reached a sufficiently low temperature, the propellant can be loaded without damaging the equipment. A 'slow-fill' typically precedes a nominal 'fast-fill' to further acclimatise the tanks.

chugging A form of **combustion** instability in a **rocket engine** which manifests itself as a low frequency pulsing sound.

cigarette combustion – see **propellant grain**

circular orbit An **orbit** with a nominally constant radius, e.g. **geostationary orbit** and many **low Earth orbits**.
[See also **elliptical orbit, parking orbit, drift orbit, halo orbit**]

circular polarisation – see **polarisation**

cislunar Of or relating to the **space** between the **Earth** and the **Moon**'s orbit (cis, 'this side of').

clamp band A restraining device which holds two **hardware** components together and can be released (for example, pyrotechnically) on command. A clamp band may be used, for example, to secure a **spacecraft** to the uppermost **stage** of a **launch vehicle**.

clamp release A term used in **space insurance** to define the moment of **launch**; the moment at which a device holding a **launch vehicle** on its **launch pad** is released to allow **lift-off**.
[See also **hold-down arm**]

Clarke orbit – see **geostationary orbit**

cleanroom A room in which the standard of cleanliness, measured by the allowed number of particles of a certain size, is higher than the general indoor environment; an area typically used for the **integration** of **spacecraft**, **subsystems** and other instruments whose performance could be detrimentally affected by 'dirt' [see figures G4, H4, S2].

Various systems are used to quantify cleanliness, typically by specifying limits on the number of particles of a given size in a given volume. For example, a system used in the United States is based on the number of 0.5 μm particles per cubic foot (0.028 m^3), such that a 'class 1000' environment has fewer than a thousand 0.5 μm particles per cubic foot. **Satellite** and **launch**

vehicle integration and testing is typically performed in class 100,000 facilities, though science spacecraft often demand cleaner environments. Class 10 and class 100 facilities are used for assembly of sensitive items such as **gyroscope**s and optical equipment.

[See also **white room**]

closed-ecology life support system (CELSS) – see **life support system**

closure panel A section of a spacecraft's exterior which can be removed to allow access to components during the later stages of assembly. In practice, the fitting of these panels is one of the final activities in the spacecraft **integration** process since they are used to 'close up' the spacecraft. For this reason they tend to be called closure panels rather than 'access panels', but use of the terms is open to personal interpretation.

CMC – see **ceramic matrix composite**

CMG – see **control moment gyro**

CNES An acronym for Centre National d'Etudes Spatiales, the French national space centre; a government organisation, founded in March 1962, responsible for the implementation of French space policy.

CO_2 scrubber A colloquial term for a **carbon dioxide absorber**.

coast earth station An **earth station** in the maritime **mobile-satellite service (MSS)**, usually located near the coast since the **satellite**s of the MSS **space segment** are located (in **geostationary orbit**) centrally over the world's oceans.

[See also **ship earth station, land earth station**]

coasting The motion of a **spacecraft** or **launch vehicle** due to its momentum or a force of **gravitational attraction** (i.e. when its **propulsion system** is inactive). The term is typically applied to launch vehicles during **stage separation** and to spacecraft travelling between one **planetary body** and another (NB if a spacecraft is in **orbit**, it is said to be 'orbiting', not coasting).

[See also **boost phase, ballistic trajectory**]

coax Colloquial abbreviation for **coaxial cable**.

coaxial cable An electrical **transmission line** with two coaxial conductors (on a common axis) – known as the 'inner' and the 'outer' – separated by an insulator (or 'dielectric' material) [see figure T5]. The inner is solid or stranded, often of copper-coated aluminium, and the outer is a flexible wire braid which allows bending to a certain degree. A domestic example is the cable used to carry the TV signal from an aerial to a television receiver.

    Coaxial cable, often abbreviated to 'coax', is capable of carrying relatively low powers compared with **waveguide**, and is commonly used, therefore, on the RF input side of a **travelling wave tube (TWT)**.

coaxial switch A device designed to isolate part of a circuit and divert a signal carried in **coaxial cable** (one of two types of commonly used electrical **transmission line**). Devices are either switched magnetically using solenoids, causing a mechanical deviation of the line, or electronically using diodes. [See also **waveguide, waveguide switch**]

co-channel interference Interference from a transmitter operating on or near the same channel **frequency**. Under normal conditions this type of interference is rare, since telecommunications and broadcasting services are allocated widely spaced channel frequencies if they are in close physical proximity. However, when meteorological conditions create an inversion in the lower atmosphere, signals can be 'guided' over larger than normal distances by a phenomenon known as 'ducting', where atmospheric layers act as a 'natural **waveguide**'.

code division multiple access (CDMA) A coding method for information transmission between a potentially large number of users, whereby all the users occupy the total transponder **bandwidth** all the time [compare **time division multiple access (TDMA)** and **frequency division multiple access (FDMA)**]; also referred to as **spread spectrum** since the signal is spread across a relatively broad section of the **electromagnetic spectrum**. In CDMA the signals are encoded (modulated by a pseudorandom noise code) and can only be received by authorised receivers. Originally used primarily for military communications, but increasingly used for commercial systems. [See also **encryption**]

codec An electronic device used in **telecommunications** for **coding** and **decoding** digital transmissions (e.g. of **television**). A contraction of *co*der–*deco*der.

coding – see **encryption, scrambling, scrambler**.

COF An acronym for Columbus Orbital Facility, a **module** of the **International Space Station (ISS)** provided by the European Space Agency (**ESA**).

cold-gas thruster A type of **reaction control thruster** which uses a gas (e.g. nitrogen, argon, freon, propane) stored at high pressure and fed, through a pressure-reducer, to an expansion **nozzle** to produce **thrust**. It is termed 'cold' because there is no **combustion** or heating of the **propellant**. [See also **hydrazine thruster, hiphet thruster, bipropellant thruster, ion thruster**]

cold-soak Exposure to an extended period of low temperatures (e.g. to reduce the temperature of a spacecraft component or **subsystem** to the local ambient, usually in an **environmental chamber** which simulates the conditions experienced in space).

[See also **heat-soak**]

cold-welding A process whereby two metal surfaces 'diffuse into each other', owing to a lack of lubricant or even an air gap between them. Although cold-welding can occur on Earth when two surfaces are held in close contact under pressure, the **space environment** is especially conducive to the effect.

collector That component of a **travelling wave tube (TWT)** which collects, and absorbs the energy of, the **electron beam** after its interaction with the **RF-wave** in the **slow-wave structure**. The **klystron** high power amplifier incorporates a collector which operates in a similar way. A collector typically consists of several 'stages' which collect electrons with different kinetic energies. To decelerate the electrons, collector stages are operated at negative potentials (i.e. voltages below 'ground'). This is known as a 'depressed collector'. The efficiency of the device can be enhanced by increasing the number of stages (although rarely beyond five) and tube efficiencies of over 60% are possible.

Decelerating the electrons causes their kinetic energy to be converted to heat, which the collector must then remove from the TWT either by conduction to a baseplate or by direct radiation. In the case of a high-power space 'tube' the collector enclosure projects through the spacecraft panel to radiate directly into space [see figure T7].

collocate To group together, or 'co-locate'; a term used to express a similarity in position of two or more **satellites** in **geostationary orbit**, such that satellites in the same **nominal orbital position** are said to be 'collocated'. Collocated satellites are sometimes referred to as a **constellation**, though the term has a wider meaning.

Columbia

(i) The name given to the first '**flight model**' of the Space Shuttle **orbiter** (Orbiter Vehicle OV-102), named after a sailing frigate launched in 1836, one of the first Navy ships to circumnavigate the globe. Columbia was completed and transported to **Kennedy Space Center** in March 1979 but, due to technical problems, was not launched until 12 April 1981. This was the first of four development flight tests (STS-1 to STS-4).

(ii) The name given to the **Apollo** 11 command module.

[See also **Space Shuttle, space transportation system (STS), Atlantis, Challenger, Discovery, Endeavour, Enterprise**]

Columbus Orbital Facility (COF) A **module** of the **International Space Station (ISS)**, provided by the European Space Agency (**ESA**).

combined (bipropellant) propulsion system A spacecraft **propulsion system**, using liquid **bipropellants**, which combines the functions of otherwise separate **orbital injection** and **attitude control** systems; also known as a unified propulsion system [see figure S13]. By comparison, most early **communications satellites**, and some other **spacecraft**, featured a separate **apogee kick motor (AKM)** and **reaction control system (RCS)** [see figure M3].

   The **bipropellant** system's **liquid apogee engine (LAE)** is typically capable of several separate **burns**, allowing the accurate transfer of spacecraft to **geostationary orbit (GEO)**, for example. By varying the timing and magnitude of the burns to adjust the periods of intermediate **transfer orbit**s, the final burn can be performed at the desired on-station longitude, injecting the satellite directly into its correct **orbital position**. This technique, known as **parasitic station acquisition**, can save **station-keeping** fuel and increase the satellite's **lifetime**. NB By contrast, a solid propellant AKM is capable of only a single burn [see **rocket motor**]. Another advantage of the LAE is that, once in GEO, it can be isolated from the rest of the system so that all propellant not used for the apogee burns can be made available to the **attitude and orbital control system**, which also increases the satellite's potential lifetime.

   For planetary spacecraft which are required to make a given number of orbital injection burns, and then point towards targets of scientific interest, a combined system also provides greater flexibility in mission planning [see figure C4].

   [See also **blowdown system**]

combined cycle engine A **rocket engine** that uses oxygen from the atmosphere as an **oxidiser** for part of its **flight** and relies on oxidiser carried in its **propellant tank**s for the remainder. Also known as an air-breathing rocket. [See also **scramjet**]

combustion Any process of burning (e.g. the combustion of a rocket **propellant**; the oxidation of a rocket **fuel**). The substances which result from the combustion process are usually called 'combustion products'. [See also **combustion chamber, propellant grain, solid propellant, liquid propellant, oxidiser**]

C4: **Combined (bipropellant) propulsion system** of the **Galileo** Jupiter orbiter. Note the main engine at centre and thrusters at far left and right. [MBB]

combustion chamber The part of a **rocket engine** in which **propellants** are burned to produce 'combustion gases' which subsequently pass through an exhaust **nozzle** to provide the **thrust** for a **spacecraft** or **launch vehicle**. [See also **exhaust velocity**, **regenerative cooling**]

command An instruction transmitted to a **spacecraft** or **launch vehicle** from a controlling **earth station** or **mission control centre** to maintain or adjust its **operational** status; an abbreviation for telecommand. See **telemetry**, **tracking and command**.

command and service module – see Apollo

command module – see Apollo

commercial off-the-shelf – see COTS

common carrier A national organisation or private company granted the authority to provide public telecommunications services in its parent country, in Europe generally the **PTT**.

commonality The practice of common design. Equipment or systems with a high degree of commonality are largely interchangeable and can operate together.

communication The imparting or exchange of information. Note: the corresponding noun in technical circles is **communications**, which may also communications

be used as an adjective, e.g. **communications satellite**, communications link, etc. Careful technical users do not refer to 'communication satellites'.

communications The collective term for the information transmitted via a telecommunications system (e.g. a **communications satellite**). In the widest sense of the definition, the type of information is immaterial: it may include, among others, **telephony, television, facsimile** (photographs or photocopies) or computer **data**. Note: the word often functions as singular, e.g. 'communications is an important application of satellites'.

[See also **communication, link**]

communications carrier assembly – see **snoopy cap**

communications link – see **link**

communications module – see **payload module**

communications package – see **communications payload**

communications payload On a spacecraft, an assembly of electronic and **microwave** equipment designed to provide communications between the spacecraft and a **planetary body**, or another spacecraft (see **inter-satellite link**) [see figures C5, C6]; the revenue-earning part of a commercial **communications satellite**, comprising both **transponders** and **antennas**; also called a 'communications subsystem' [see **subsystem**]. Sometimes called a communications package, especially on a spacecraft other than a communications satellite where the 'package' exists to handle spacecraft data rather than provide a revenue-based communications service (e.g. manned spacecraft and planetary probes).

[See also **repeater, input section, output section, receive chain, transmit chain, payload module, antenna module**]

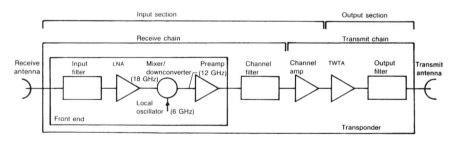

C5: A satellite **communications payload**; a **single conversion transponder** (LNA, low noise amplifier; TWTA, travelling wave tube amplifier).

70

C6: Inmarsat 3 **communications payload** showing **travelling wave tubes** across centre of panel and **electronic power conditioners** below. [Matra Marconi Space]

communications repeater – see **repeater**

communications satellite A spacecraft designed to provide a communications service which may include, for example: telephony, television, **TV distribution**, **facsimile**, **data communications** and **videoconference** services.

Historically, most communications satellites have occupied a position in **geostationary orbit**, which offers the advantages that small **earth terminals** do not need to track the satellites and that three spacecraft can provide complete coverage of the Earth, except for the high latitude (polar) regions. For many years, however, the USSR used the high inclination, elliptical **Molniya orbit** for communications satellites, since its territory was predominantly at high latitudes. From the late 1990s, **constellation**s of satellites designed primarily for mobile communications were launched to low and medium altitude orbits.

Note: communications satellite is often abbreviated to 'comsat', not to be confused with the organisation Comsat.

[See also **communications payload**, **payload module**, **service module**, **hybrid satellite**]

communications subsystem – see **communications payload**

communications system Any assembly of hardware and associated infrastructure designed to enable communications. A satellite communications system comprises a **space segment** and an **earth segment** (satellites and earth stations).
[See also **communications payload, subsystem, personal communications system, GMPCS**]

companding A technique used in communications systems to improve the **signal-to-noise ratio** by compressing the **signal** at the **transmitter** and expanding it at the **receiver** – the word compand is a contraction of *com*press and ex*pand*.
[See also **digital compression**]

composite propellant – see **solid propellant**

composites A class of **materials** in which fibres of a reinforcing material are set in a 'matrix' of another material. See **carbon composite, Kevlar composite, ceramic–metal composite, metal–matrix composite**.

compression – see **digital compression**

computer An electronic device (**hardware**) that processes **data** according to a set of instructions (**software**).
[See also **microprocessor, central processing unit (CPU), data bus, artificial intelligence (AI)**]

computer enhancement – see **image intensification**

comsat
(i) A colloquial abbreviation for **communications satellite**.
(ii) When capitalised, an abbreviation for the Communications Satellite Corporation, the US signatory to **Intelsat**. Comsat was established to provide operational and technical services to Intelsat, and to own and operate its own domestic satellite system.

configuration The arrangement of a number of parts. In space technology, commonly used with reference to a **vehicle** of some description: e.g. a '**launch vehicle** configuration', the arrangement of its **stages**, **strap-on** boosters, etc.; or a **manned spacecraft**, particularly the more complex, such as a **space station**. Also used as a verb: e.g. 'the way the vehicle is configured'.
[See also **variant**]

conic sections A family of curves produced by cutting a cone in various ways. If the cut is made parallel to the base of the cone, the shape of the resulting cross section is that of a circle; if the cut is inclined slightly (at a shallower angle to the cone's base than to its side), it produces an ellipse; if the cut is made parallel to the side of the cone, it produces a **parabola**; if the cut is inclined at a steeper angle to the cone's base than to its side, it produces a **hyperbola**. In the limits, a cut through the tip of the cone produces a point and a cut along its side produces a straight line, extreme cases of a very small circle and a very narrow parabola, respectively. The importance of the conic sections in **astronomy** and **spaceflight** is that any body that moves in an unperturbed **orbit** (or **trajectory**) will follow a path described by one of the conic sections. Although no orbit or trajectory is mathematically perfect, the equations describing these curves provide the basis for orbital calculations. [See also **perturbations, triaxiality, luni-solar gravity, solar wind, atmospheric drag, gravitational boost**]

coning An instability experienced by a **launch vehicle** which combines motion in the **pitch** and **yaw** axes such that the vehicle's **roll axis** precesses about the direction of flight (like the **precession** of a **gyroscope**'s spin axis). The term is derived from the fact that the vehicle's longitudinal (roll) axis sweeps out the surface of a cone as it precesses.

constant conductance heat pipe (CCHP) – see **heat pipe**

constant-wear garment A one-piece zippered undergarment made of 'porous-knit' cotton, worn by **Apollo** astronauts beneath either a Teflon in-flight coverall or a **spacesuit**.

constellation In **satellite communications**, a number of **satellite**s with approximately equal spacing around an **orbit**, or orbits, designed to provide maximum coverage of the Earth. The simplest constellation comprises three satellites in **geostationary orbit** with **orbital position**s about 120 degrees apart. More complex constellations have a greater number of satellites in lower altitude orbits: e.g. the Navstar **Global Positioning System (GPS)** is designed to comprise 18 **operational** satellites, equally divided between six 12-hour circular orbits (altitude 20,200 km, **inclination** 55°).

In the 1990s, the term was predominantly applied to constellations of satellites designed primarily for mobile communications and placed in low and medium altitude orbits. These systems, such as Iridium, ICO Global and Globalstar [see figure C7], were categorised as global mobile personal communications systems [see **GMPCS**].

[See also **spatial diversity, dispenser**]

constructive total loss A term used in **space insurance** to indicate that a **spacecraft** has suffered a **total loss** of capability as a result of a failure of

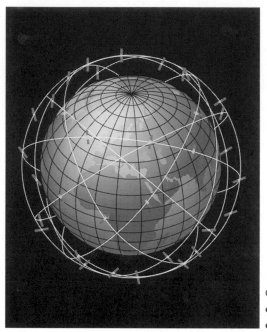

C7: Globalstar 48-satellite constellation in low Earth orbit (LEO). [Globalstar]

another (external) **system**. For example, a mission may be declared a constructive total loss if a spacecraft is unable to complete a significant part of its intended **mission** because its **launch vehicle** has placed it in a lower than intended **orbit**. This would require a boost from the spacecraft's **propulsion system**, thereby reducing its in-orbit **lifetime**. The degree of reduction that would qualify would be defined in the launch insurance contract, a typical figure being 'more than 50% of a satellite's intended lifetime'.

[See also **partial loss, TLO, deductible**]

consumable Any item or supply which is consumed in an irreversible process, particularly in connection with manned spaceflight when used in a **life support system** (e.g. oxygen, water and food). It also includes **propellant**, some **coolants**, etc.

control electronics unit (CEU) A device concerned with the overall control of a spacecraft – its central computer. It accepts **inputs** from **sensors** and directs **outputs** to **actuators**, as well as accepting **commands** from, and returning **telemetry** to, the Earth. A typical spacecraft on-board control system continually checks both input and output **data** and its own operation, so that it can correct errors and switch to **redundant** units in cases of failure. Some CEUs possess a degree of **artificial intelligence**, which allows them to make autonomous decisions; this is particularly useful for **deep space** missions, for example.

control law A set of mathematical instructions which governs the operation of a device, mechanism or **system**.

control moment gyro (CMG) A gimballed **momentum wheel** used to provide **attitude control** on a large **spacecraft** such as a **space station** (e.g. **Skylab**). The wheel is mounted with its spin axis perpendicular to the **gimbal**, so that a torque applied at the gimbal produces a change in angular momentum perpendicular to that of the wheel, and thus a reaction torque on the spacecraft. CMGs are heavier than **reaction wheel**s but can improve attitude control by a factor of a hundred. Those used on the Russian **Salyut** and **Mir** stations are known as **gyrodyne**s.

control system Any mechanism or computer program which regulates the operation of a device or assembly of devices (e.g. a **system** which controls the **attitude** of a **spacecraft** or the **flight-path** of a **launch vehicle**).
[See also **attitude control, orbital control, control electronics unit (CEU), autopilot**]

conus An acronym for *Con*tiguous *United States*; the 48 states of the US mainland (i.e. excluding Alaska and Hawaii). The geographical arrangement of the USA has an impact on the design of many American **communications satellite**s in that **spot beam**s have to be provided for coverage of the outlying states.

coolant Any fluid used as a heat exchange medium to transfer heat from one part of a system to another (e.g. water in the cooling system of a **spacesuit**, or **liquid hydrogen** in a **rocket engine** [see **regenerative cooling**]).
[See also **cooling system, environmental control and life support system (ECLSS)**]

cooling system Any device or structure which conducts or radiates heat in an effort to reduce the temperature of another device or structure.
 (i) In spacecraft thermal design the term is usually applied to a device which refrigerates the component it protects. The devices most commonly in need of active cooling are those designed to detect heat itself (i.e. infrared **sensor**s) especially when they are used to detect very low levels of heat. The IR detectors used in infrared telescopes, for instance, must be cooled to liquid helium temperatures – a few degrees above **absolute zero** – so that the received signal is not completely drowned by the 'thermal noise'

of the molecules in the detector itself. The term also includes non-refrigerating 'pumped fluid loops' used on **manned spacecraft**, **space platforms**, etc. Open-cycle systems, which vent evaporated coolant to space, are not widely used because of their tendency to pollute the spacecraft's local environment and degrade optical surfaces and **solar cells**. Closed-cycle systems resemble the domestic refrigerator in that they feature a coolant circulated through a closed loop, used as a heat exchange medium. Spacecraft systems, however, are more compact and usually cool to much lower temperatures.

(ii) The cooling system of a **spacesuit** is typically an undergarment worn next to the skin which is laced with capillaries through which water is pumped; otherwise known as a liquid cooling and ventilation garment (LCVG).

[See also **heat rejection**, **heat pipe**, **thermal control subsystem**, **cryogenics**, **bolometer**, **degradation**, **environmental control and life support system (ECLSS)**]

co-polar(ised) Having the same **polarisation**. For example, two **RF-waves** with right-hand circular polarisation are said to be co-polarised. If one of the waves suffers **polarisation conversion**, it is said to have a co-polar component and a **cross-polar** component.

cordon sanitaire A volume of space around a **spacecraft** (typically a **space station**) which acts as a buffer zone, or zone of protection.

core vehicle The part of a **launch vehicle** which is the same for all **variants**. Variants, which allow different payload sizes and masses to be launched, may be formed by the addition of **upper stages**, **strap-on** boosters and/or different **payload fairings**, etc.

corner-cube reflector – see **laser ranging retroreflector**

corona

(i) In physics, an electrical discharge appearing around the surface of a charged conductor. See **electrostatic discharge (ESD)**.

(ii) In astronomy, the outermost region of the Sun's atmosphere.

cosmic radiation **Radiation** from outside the solar system comprising high energy electrons and nucleons (chiefly protons); also called 'cosmic rays'.

[See also **space radiation**, **single event effect**]

cosmodrome A Russian term for **launch site**. See **Baikonur Cosmodrome**, **Northern Cosmodrome**.

cosmonaut The Russian term for a person who has been trained for **spaceflight**;

the title used by any person flying in a **spacecraft** operated by the former USSR. The term is also used independently by the French.

[See also **astronaut, astronautics, cosmonautics**]

cosmonautics The Russian term for '**astronautics**'. Considered 'in translation', the Russian term 'flight through space' is more general than the alternative 'flight to the stars' and more applicable to Man's present endeavours, but there is no sign of etymology taking precedence over common usage.

Cosmos The name given to a series of Soviet/Russian spacecraft and launch vehicles; alternative spelling **Kosmos**. The two-stage Cosmos-3M commercial **launch vehicle** is converted from the former-Soviet R-14 intercontinental ballistic missile (**ICBM**) and operated under a Russian joint venture. It has a **payload capability** of about 1500 kg to **low Earth orbit (LEO)**, and made its first commercial launch, from **Kapustin Yar**, in April 1999.

[See also **Rockot, Dnepr, Start**]

COSPAR An acronym for Committee on Space Research, part of the International Council of Scientific Unions (ICSU).

COSPAS-SARSAT An acronym for Kosmicheskaya Systyema Poiska Avariynych Sudov – Search and Rescue Satellite-Aided Tracking. The COSPAS-SARSAT system, declared **operational** in 1984, comprises a number of satellites in low, near-polar orbits which receive transmissions from emergency location beacons (known on aircraft as emergency location transmitters (ELTs) and on ships as emergency position indicating radio beacons (EPIRBs).

[See also **navigation satellite, Global Positioning System (GPS)**]

COSTAR An acronym for corrective optics space telescope axial replacement, an optical instrument designed to correct the spherical aberration of the primary mirror of the **Hubble Space Telescope (HST)**.

COTS An acronym for commercial off-the-shelf (**hardware** or **software**). Using COTS hardware or software is an alternative to designing and producing it specifically for a given **application**. COTS hardware, for example, is selected from available products and tested to ensure **reliability**, rather than designing 'hi-rel' (high-reliability) parts from scratch. It may also offer a cheaper alternative to 'mil-spec' (military **specification**) or 'rad-hard' (**radiation**-hardened) parts.

Coudé focus (telescope) A type of astronomical reflecting telescope, based on the design of the **Cassegrain reflector**, in which light is reflected from a paraboloidal primary mirror onto a hyperboloidal secondary. The two designs differ in that the Coudé has an additional (plane) mirror below the secondary which reflects the light through an aperture in the side of the telescope to the eyepiece [see figure N1]. The second mirror moves to compensate for the motion of the telescope, ensuring that the exit-beam remains fixed. Large terrestrial telescopes can have more than one usable focus, such as the 200″ (5 m) Palomar telescope which has Cassegrain, Coudé and prime foci.

[See also **space telescope**, **Newtonian telescope**, **Schmidt telescope**, **Gregorian reflector**]

countdown The period, timed by a clock running in reverse or 'counting backwards', leading up to a **launch**; the act of 'counting down' to the moment of **lift-off**, particularly the verbal '5 . . . 4 . . . 3 . . . 2 . . . 1 . . . etc.' immediately prior to launch. The first use of regressive counting appeared in the 1929 Fritz Lang film '*Die Frau im Mond*' (The Woman in the Moon).

The exact timing of events around 'zero' varies from one **launch vehicle** to another: e.g. the first **stage** engines of the **Ariane** 5 launcher are ignited at 'zero' and the vehicle is released at $T + 6.7$ seconds when the **solid rocket boosters** are ignited, whereas the **Space Shuttle main engines** are ignited at $T - 6.6$ s and the vehicle is released at 'zero', when its SRBs are ignited. The time delay in both cases is that required for the **thrust** of the respective **liquid propellant** engines to build up to the level required to achieve lift-off.

[See also **countdown sequencer**, **hold-down arm**]

countdown sequencer A computerised system which controls the sequence of events preceding a **launch**. A typical system initiates specific actions at pre-programmed times, or prompts a human operator to do so, and indicates the status of all **launch vehicle** systems and **ground support equipment** to operators in a **launch control centre**. Owing to the complexity of most launch vehicles, much of the process is automatic: only if the sequencer detects that a parameter is outside its pre-determined limits will it initiate a 'hold' in the **countdown**.

coupled cavity A type of **slow-wave structure**; a component of a **travelling wave tube**.

coverage area An area on the surface of the Earth, defined for the purposes of **satellite communications**, within which the **power flux density (PFD)** of the radiated satellite signal is sufficient to provide the desired quality of reception for a telecommunications service in the absence of **interference** [see figures C8, F3].

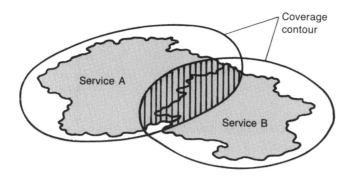

Key:
Service area: protection against interference can be demanded by the administration originating the service

Coverage area: PFD may be sufficient for service, but protection is not guaranteed

Given the above, an area where the 'overspill signal' from the adjacent administration would be likely to suffer interference

C8: The relationship between the **coverage area** and **service area** of a **satellite communications** service as defined by the **ITU** (International Telecommunication Union).

The PFD may be sufficient for service but, in contrast to the **service area**, protection against interference is not guaranteed. One of the provisions of the **ITU** Radio Regulations states that the coverage area must be the smallest area which encompasses the service area, but since no satellite beam (**footprint**) can adhere exactly to a country's political boundary, this necessarily leads to **overspill**.
[See also **beam area, protection ratio**]

coverglass – see **coverslip**

coverslip A protective cover for a **solar cell**, bonded to its top surface; otherwise known as a coverglass. Solar cells are susceptible to damage from the Sun's high energy radiation and particle flux. The coverslip filters out short wavelength UV to prevent darkening of the coverglass adhesive and IR wavelengths, which give rise to undesired heating of the cell. More generally, coverslips offer protection against **micrometeoroid**s and, in **low Earth orbit**, the impact of residual atmospheric oxygen [see **ATOX**].
[See also **degradation**]

cowling – see **skirt**

CPU – see **central processing unit**

cradle A U-shaped framework structure used to support a **spacecraft**, typically a **communications satellite** with a **perigee kick motor** attached, in the **payload bay** of the Space Shuttle. Also known as a 'cradle assembly' and sometimes referred to as **airborne support equipment (ASE)**. The cradle also supports a retractable **sunshield** which protects the satellite before its release from the payload bay.

[See also **pallet**]

Crawler A vehicle used to transport the Space Shuttle, mounted on its **mobile launch platform (MLP)**, from the **vehicle assembly building (VAB)** to the **launch pad** [see figure C9]; also known as the 'Crawler transporter'. It is 40 m long, 35 m wide and 6.1 m high, and has an unladen weight of 2700 tonnes. The Crawler and its payload, an MLP and an unfuelled Space Shuttle, weigh

C9: A **Crawler** transporter carries a **Space Shuttle** and **mobile launch platform** up the ramp to **launch pad** 39A at **Kennedy Space Center**. The Crawler compensates for the incline to keep the launch platform level. [NASA]

almost 5000 tonnes. It has a dual-tracked drive unit at each corner and is powered by two 2750 horsepower diesel engines driving four 1000 kW generators which provide power to 16 traction motors: its maximum speeds are 3.2 km h^{-1} (unloaded) and 1.6 km h^{-1} (loaded). A levelling system keeps the Shuttle vertical as it negotiates the 5% gradient up to the launch pad, where it deposits the MLP and withdraws. Two Crawlers, based on the design of a stripping shovel used for surface coal-mining in Kentucky, USA, in the 1960s, were built for the **Apollo** programme, to move assembled **Saturn V** launch vehicles from the VAB to the pad.

crawlerway A road at **Kennedy Space Center** which links the **vehicle assembly building (VAB)** with **launch pad**s 39A and 39B, so called because it is designed to be used by the '**Crawler**'. It comprises two 12-m-wide lanes 15 m apart and measures about 5.6 km from the VAB to launch pad 39A and about 6.8 km to 39B.

crew The men and women aboard a **manned spacecraft**. See **astronaut**, **cosmonaut, mission specialist**.

crosslink A **communications** link between **spacecraft** in **orbit**. The term is imprecise in that it does not differentiate between an **inter-satellite link (ISL)**, a **link** between spacecraft in the same orbit, and an **inter-orbit link (IOL)**, a link between spacecraft in different orbits.

cross-polar discrimination (XPD) – see **discrimination**

cross-polar(ised) Having the opposite **polarisation**. For example, an **RF-wave** with right-hand circular polarisation and another with left-hand circular polarisation are said to be cross-polarised. If one of the waves suffers **polarisation conversion**, it is said to have a **co-polar** component and a cross-polar component. Since the waves were of opposite **polarity** to begin with, it is the co-polar component which may lead to **interference** between the two **signal**s.

crossrange capability The ability of a **Space Shuttle**, or other winged space vehicle, to **manoeuvre** to the left or right of the planned flightpath on returning from orbit. The Space Shuttle **orbiter**, for example, has a crossrange capability of about 2000 km left or right of its nominal **glide path**. It was designed with delta wings to provide this, since it has no engines to allow a second approach attempt.
[See also **range, downrange**]

cross-strapping The practice of interconnecting two or more **redundant** chains of equipment to allow switching between the chains in the event of a failure.

Cross-strapping is common in **propulsion subsystem**s and in chains of electronic and **microwave** equipment (e.g. in **communications payload**s, **attitude and orbital control system**s, **telemetry, tracking and command** systems, etc.).

[See also **single point failure (SPF)**]

CRV An abbreviation for crew return vehicle, or crew rescue vehicle, a **spacecraft** designed to return the **crew** of the **International Space Station (ISS)** to **Earth** in an emergency.

cryogen A substance at very low temperature or used to produce very low temperatures (e.g. liquid helium used to cool infrared detectors, etc.]

[See also **cryogenic propellant**]

cryogenic propellant A **liquid propellant** (**fuel** or **oxidiser**) formed by the liquefaction of a gas when cooled to very low ('cryogenic') temperatures.

The most widely used cryogenic fuel and oxidiser, respectively, are **liquid hydrogen** (LH_2) which boils at about -253 °C, and **liquid oxygen** ('LO_2' or 'LOX') which boils at about -183 °C. The LH_2/LOX combination is used for the **Space Shuttle main engine**s, the Centaur **upper stage** and **Ariane** 4's third **stage**, for example.

Cryogenic propellants, in general, are difficult to handle, since they evaporate at environmental temperatures, and their use requires propellant tanks to be insulated, which adds to the structural weight of the vehicle. Hydrogen, in particular, has a very low density, requiring large tanks, which also add to the weight. However, their relatively high **specific impulse** offsets the disadvantages.

A possible alternative to LOX is the potentially more potent oxidiser fluorine (F_2), but it is extremely corrosive and toxic, and therefore difficult to work with. The potential oxidiser ozone (O_3) produces the maximum energy release with any fuel, but is highly combustible and decomposes explosively in the presence of trace impurities.

[See also **rocket engine, regenerative cooling**]

cryogenics The branch of physics concerned with very low temperatures and the phenomena occurring at those temperatures.

[See also **cryogenic propellant**]

CSG A French abbreviation for Centre Spatiale Guyanais. See **Guiana Space Centre**.

CTPB An abbreviation for the **solid propellant** carboxyl-terminated polybutadiene. See **polybutadiene**.

CTV An abbreviation for crew transfer (or transport) vehicle, a **spacecraft** designed to carry a **crew** to the **International Space Station (ISS)**.

cupola An observation **module** on the **International Space Station (ISS)** provided by the European Space Agency (**ESA**). It has seven windows to allow observations of the **Earth** and the stars, and monitoring of work on, and **spacecraft** movements around, the station [see figure J1].

cupped dipole – see **dipole**

Cyclone – see **Tsyklon**

D

Dacron A low conductance/insulating material (a trademark) used, for example, as an interlayer between successive layers of **Mylar** film in spacecraft **thermal insulation**.

[See also **multi-layer insulation (MLI)**]

DAMA – see **demand assignment multiple access**

DANDE An acronym for despin active nutation damping electronics; an **attitude control** subsystem on a **spin-stabilised** spacecraft which senses **nutation** using an **accelerometer** and uses the spacecraft's despin motor to provide a cancellation force to reduce the nutation.

DARPA An acronym for Defense Advanced Research Projects Agency, a US-based organisation founded in 1958 to conduct defence-related and civil space research.

data Information (generally used to mean computerised information, a coded **bit**-stream), or the result of observations or measurements (e.g. **remote sensing** data). Widely used as a singular noun (e.g. the data has been received), but more properly as the plural of 'datum' (e.g. the data have been received). In satellite **telecommunications**, the term 'data' refers to a category of **traffic** commonly handled by communications satellites, differentiating it from others such as **telephony** and **television**. This type of traffic is often called **data communications**.

[See also **link**]

data bus The part of a computer that transports **data** and interconnects components or **subsystem**s. See **bus** (iii).

data communications A type of **traffic** handled by a communications link, consisting generally of computerised or digital information as opposed to other traffic commonly handled by satellites, such as **telephony** and **television**. It is often abbreviated to 'datacoms'.

data link – see **link**

data rate The rate of information transmission through a communications system measured in bits per second (bits s^{-1}) or any multiple (e.g. kbits s^{-1}, Mbits s^{-1}, Gbits s^{-1}).

[See also **bit, byte, data, data communications**]

datacoms – colloquial abbreviation for **data communications**

datum
    (i) A single piece or **bit** of information (the singular of **data**).
    (ii) A known point or line from which other measurements or estimates can be made.

dBc ('dB relative to **carrier**') – see **decibel**

dBHz ('dB-hertz') – see **decibel**

dBi ('dB-isotropic') – see **decibel**

dBK ('dB-kelvin') – see **decibel**

DBS – see **direct broadcasting by satellite**

dBW ('dB-watts') – see **decibel**

DC power A source of electrical power supplied with direct current. See **power**. The great majority of spacecraft power systems use DC, alternating current (AC) being used only for special applications.

dead-band Jargon: the extent of any measured quantity between specified upper and lower bounds. For example, the extent of a geostationary satellite's apparent motion as viewed from the ground, otherwise known as its station-keeping '**box**'.
[See also **orbital control, guardband, bandwidth**]

debris – see **space debris, orbital debris**.

debris mitigation The reduction or limitation of **space debris**, an increasingly important aspect of space development and exploration. See, for example, **de-orbit, graveyard orbit, passivation**.
[See also **orbital debris, spacecraft debris**]

decay – see **orbital decay**

decibel (dB) In communications technology, a logarithmic ratio between two signal power levels used, among other things, to denote the **gain** of an **antenna, amplifier**, etc. (a gain of 3 dB is equivalent to a doubling in signal strength). A numerical value (in dB) can be found using the following expression:

No. of dB $= 10 \log_{10} P_1/P_2$

where P_1 and P_2 are the output and input powers respectively. A positive dB-value represents a gain; a negative value is a loss (e.g. 0.5 is equivalent to -3 dB, colloquially referred to as '3 dB down'). If two voltages are being compared instead of two powers, V_1 and V_2 replace P_1 and P_2 and '20' replaces '10' in the above expression.

    As an aid to usage, the table below provides rounded values of P_1/P_2 against their equivalents in dB. It shows why engineers prefer to use the decibel: it is easier, for example, to write 60 dB than 1,000,000, which is a typical amplification factor for a **travelling wave tube amplifier (TWTA)**. Intermediate values are obtained as follows: e.g. for 21 dB, simply take the ratio value for 20 dB and multiply it by that for 1 dB; $100 \times 1.3 = 130$.

dB	power ratio (P_1/P_2)	dB	power ratio (P_1/P_2)
0	1.0	10	10
1	1.3	20	100
2	1.6	30	1000
3	2.0	40	10,000
4	2.5	50	100,000
5	3.2	60	1,000,000
6	4.0	70	10,000,000
7	5.0	80	100,000,000
8	6.3	90	1,000,000,000
9	7.9	100	10,000,000,000

dBi is a measure of gain relative to an **isotropic** source; dBc denotes a measurement 'relative to the **carrier**'; dBW is the ratio of **power** output to a reference signal at 1 watt $(P_2 = 1)$, expressed in decibels. In the same way that power is referenced to the watt in 'dBW' and the milliwatt in 'dBm', frequency is referenced to the hertz in 'dBHz' and temperature to the kelvin in 'dBK'. Conversion from the numerical quantity in SI units to its equivalent in dB[x] is performed simply by taking $10 \log_{10}$ (e.g. as in the **bandwidth** of a **transponder**, 36 MHz = 75.56 dBHz; or the **noise temperature** of a **receiver**, 4 K = 6.02 dBK).

[See also **antenna gain** and **power flux density (PFD)** for examples]

declination A measure of the angle north or south of the celestial equator, in the equatorial or geographical system of celestial coordinates (north is positive, south is negative). Knowledge of a **satellite**'s declination and **hour-angle** allows an **earth station** to be aligned with it. An alternative 'horizon system' uses **altitude** (or **elevation**) and **azimuth**.

[See also **polar mount, altitude-azimuth mount**]

decoding – see **decryption, descrambling, descrambler**.

decompression The loss of atmospheric pressure from a **manned spacecraft** or

spacesuit which may lead to **decompression sickness, hypoxia, anoxia** or **ebullism**.

decompression sickness A medical disorder caused by a sudden and substantial change in atmospheric pressure: characterised by severe pain, cramp and breathing difficulties. Otherwise known as 'dysbarism' or 'the bends'. The disorder, usually associated with deep-sea divers, develops when nitrogen absorbed in the blood and body tissues is suddenly released in bubbles as the ambient pressure on the body is reduced.

The problem was faced for the **Apollo–Soyuz Test Project** when spacecraft with different atmospheres were joined for the first time. The **Apollo** had an atmosphere of pure oxygen at about 37 kPa (5.4 psi), substantially lower than the average sea level pressure of 101.4 kPa (14.7 psi). The **Soyuz**, however, was pressurised with an oxygen/nitrogen mixture at about 70 kPa (10 psi). A direct transfer of **cosmonauts** from the Soyuz to the Apollo would have given them 'the bends', so an intermediate **docking module** was designed to seal off the Apollo and equalise the pressure to that of the Soyuz before a meeting could take place. In addition, to simplify the operation, the Soviets reduced the pressure in the Soyuz to 65 kPa (9.5 psi).

In June 1997, the **Mir** space station suffered a collision with an unmanned Progress supply vehicle and the subsequent depressurisation of one of its modules. In general, the observed increase in **orbital debris** has increased the potential for decompression of a **spacecraft** or **spacesuit**.

[See also **hypoxia, ebullism, cabin pressure**]

decryption Decoding. Most military and some commercial radio transmissions are 'encrypted', to make them unintelligible to unauthorised receivers. They must be 'decrypted' by a receiver with the appropriate 'key' in order to be understood.

deductible A term used in **space insurance** to indicate a specified, otherwise allowable, type of loss for which the insurer makes no payment. For example, if a series of **spacecraft** is insured together, 'one loss deductible' means that the first loss will not be paid for, while any subsequent losses will. On a smaller scale, a policy could specify 'one **transponder** deductible' (i.e. two transponders must fail before any insurance payment is made). This is analogous to the domestic policy which carries an 'excess'.

de-emphasis In **telecommunications**, the restoration of a flat **baseband** response after **demodulation**; the opposite of **pre-emphasis**.

deep space The term is inexact, but typically refers to space outside the Earth–**Moon** system: that volume of **space** which is not 'in the vicinity of Earth'; not **near-Earth space**. **Spacecraft** on interplanetary trajectories are usually tracked by a **deep space network (DSN)**, but a DSN can, of course, also track spacecraft within the Earth–Moon system.

Some authorities have defined deep space as 'more than 2 million kilometres from Earth', but this is not yet widely accepted. In a less 'geocentric' definition, deep space could be defined as the volume of space outside the solar system (e.g. 'the **Voyager** spacecraft left the solar system and headed into deep space'); however, the question of where the solar system begins and ends would then arise.

[See also **NEO**]

deep space network (DSN) A network of **ground stations** used for **tracking** spacecraft in **deep space**, generally those on interplanetary trajectories. Although other nations operate ground stations capable of handling spacecraft in deep space, the term Deep Space Network usually refers to the DSN operated for NASA by the Jet Propulsion Laboratory (JPL) in Pasadena California. It has stations at Goldstone, California, Madrid, Spain and Canberra, Australia, sites which are also part of NASA's **space tracking and data network (STDN)**.

degradation (of spacecraft materials) The deterioration of spacecraft materials under the influence of the space environment and through the normal operation of the spacecraft. For example, degradation of the **solar cell**, due to solar protons and electrons, energetic radiation and micrometeoroid impact, is observed to decrease the power output of **solar array**s by 2 or 3% per year [see figure D1]. For this reason it is common to quote two different values for spacecraft solar array output: one for beginning of life (BOL) and another for end of life (EOL). What this means in practice is that arrays have to be oversized by a considerable amount, based on BOL power, to ensure that the satellite will be fully operational at EOL.

Spacecraft materials may also be degraded by deposition following **outgassing** or thruster firings [see **plume impingement**]. For example, degradation of thermal control surfaces [see **surface coatings**] can lead to an increase in overall spacecraft temperature as the surfaces become less reflective, while the degradation of optical surfaces has an obvious effect on image quality.

[See also **coverslip, plume shield**]

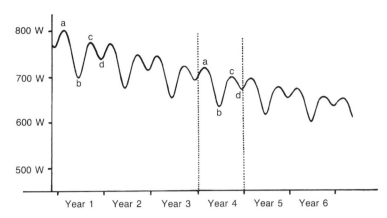

D1: Variation of example **solar array** capacity for a satellite in **geostationary orbit**. Overlaid on a general decrease, due to array **degradation**, is an annual variation related to the Sun–Earth distance and the angle of solar illumination at the array surface (a, spring **equinox**; b, summer **solstice**; c, autumn equinox; d, winter solstice).

degree of freedom An attribute of a body capable of linear or angular motion; an ability to move along an axis or rotate about it. A **spacecraft**, for example, has six degrees of freedom: three linear, along each of three orthogonal axes (roll, pitch and yaw), and three rotational about those axes. A **launch vehicle** possesses similar degrees of freedom.
  [See also **spacecraft axes**, **frame of reference**]

delay – see **signal delay**

delayed repeater satellite – see **store-and-forward**

delivery in orbit An option in commercial satellite contracts whereby the client accepts delivery of the satellite once it is **operational**, in **orbit**, as opposed to direct from the manufacturer, on the ground. The option was introduced in the mid-1980s by the former Hughes Aircraft Company for British Satellite Broadcasting (BSB), whose Marcopolo satellites were launched in 1989 and 1990.

Delta An American **launch vehicle** developed from the **Thor** rocket and first launched in May 1960. It used **liquid propellant** in its first and second stages (**liquid oxygen/kerosene** and **nitrogen teroxide/aerozine-50** respectively), and **solid propellant** in its third stage, which was typically a **payload assist module (PAM)**. First stage thrust was augmented by up to nine **strap-on** solid rocket boosters: six ignited at lift-off, burnt out after about a minute and were jettisoned; the remaining three were then ignited.

D2: **Delta** II launch vehicle: note strap-on **solid rocket booster**s. [McDonnell Douglas]

In a similar configuration, the Delta became a commercially operated launch vehicle in the late 1980s. The **payload capability** of the Delta II variant [see figure D2] is about 1880 kg to **geostationary transfer orbit (GTO)**, the most common delivery orbit for commercial satellites, while that for the Delta III is about 3820 kg. The Delta IV, a version also known as the **Evolved Expendable Launch Vehicle (EELV)**, is designed with a payload capability of up to 6670 kg.

[See also **Atlas**, **Titan**, **Scout**]

delta modulation A transmission method using a modulated radio-frequency carrier wave, whereby the **amplitude** of the original analogue input signal is sampled at discrete time-intervals to create a representative digital translation of the signal. To this extent delta modulation is similar to **pulse code modulation**, except that the sampling rate is typically 24,000–40,000 times a second (for good quality speech) compared with 8000 Hz for PCM. However, the major difference is that, having obtained the samples, a one-**bit** code is then used to transmit the change in input levels between samples. The 'delta' (Δ), used in the mathematical sense, refers to the change.

delta V (ΔV) Mathematical terminology for a change in velocity. Typically used with reference to **launch vehicles** and **spacecraft** (e.g. when quoting the velocity-change required to transfer a spacecraft from one **orbit** to another). [See also **rocket equation**]

demand assignment multiple access (DAMA) A coding method for information transmission between a number of users, whereby satellite **capacity** is assigned according to demand and not on a rigid, 'pre-defined' basis. A spectrum-efficient method of dynamically allocating telephony **channels** in a **transponder**.

demodulation The reverse process to modulation whereby a signal is recovered from a higher frequency modulated **carrier** wave. See **modulation**.

demodulator A device whose function is to demodulate a signal from a **carrier** wave. See **modulation**.

de-orbit To cause a **spacecraft** to leave its **operational** orbit, typically by firing a **thruster** to reduce its orbital altitude (usage e.g. 'to de-orbit a spacecraft', 'a de-orbit **burn**'); a so-called '**debris mitigation** measure'. De-orbiting usually entails removing a spacecraft or other space object completely from orbit by causing it to re-enter a planet's atmosphere, as opposed to simply changing its **orbital parameters**. Some **constellation** satellites are designed to be de-orbited at the end of their operational lives. [See also **orbital debris, re-entry, graveyard orbit**]

deploy
(i) To move or extend a component, device or **assembly** away from the main body of a **spacecraft** (e.g. the 'deployment' of a **solar array, antenna, boom**, etc.).
(ii) To place a **satellite** system 'on station' (chiefly military).
(iii) To release a **payload** from the **payload bay** of the Space Shuttle **orbiter**, or other similar **vehicle**.
[See also **deployable antenna**]

D3: GE-1's **deployable antennas** under test in **anechoic chamber**. [Lockheed Martin]

deployable antenna A spacecraft **antenna** which 'unfolds' from a position of
storage once the vehicle reaches its planned **orbit** or **trajectory** [see figures
D3, D4, T4].

Many spacecraft (e.g. **communications satellites**) require large diameter
antennas, which produce narrow **spot beam**s. Since they are generally too
large to be mounted on the satellite's **antenna platform**, and also have to fit
within the **payload fairing** of the **launch vehicle**, the only solution is to
clamp them to the side of the satellite during launch and **deploy** them once
in orbit. The **solar arrays** of the **three-axis-stabilised** satellite are stored and
deployed in a similar fashion.

[See also **unfurlable antenna**]

deployable battery A spacecraft **battery** which 'unfolds' from a position of
storage once the vehicle reaches its planned **orbit** or **trajectory**. The
advantage of deploying a battery and its dedicated **radiator** is that it reduces
the thermal radiation demands of the spacecraft's main body, which has a

D4: TDRS-E in **cleanroom**. Note **deployable antennas** supported by **ground support equipment** and **unfurlable antennas** in furled position. [TRW]

limited area. It is also obviates the need to transport heat across the hinge line of the alternative **deployable radiator**.

deployable radiator A spacecraft **radiator** which 'unfolds' from a position of storage once the vehicle reaches its planned **orbit** or **trajectory**, to further increase the surface area available for **radiation**. See figure I2.

depth of discharge (DOD) The amount by which a **battery** can be discharged without detrimentally affecting its future performance, measured as a percentage of its rated **capacity**. For example, a DOD capability of 60% means that 60% of the maximum charge can be used. A battery's DOD capability has an effect on a spacecraft's **mass budget**, since if one battery has a capability twice that of another, a spacecraft will have to carry twice as many of the second battery to provide the same usable **power**.

    The life-limiting factor for a battery is the number of charge–discharge cycles performed at a given DOD. For example, while a battery used in a geostationary satellite may have a lifetime of 12 years, that same battery might be rated for only five years in a low Earth orbiting spacecraft, because it is required to perform some 25,000 cycles at a 50% DOD as opposed to some 1080 cycles at 80% DOD.

[See also **duty cycle**, **eclipse**]

descending node – see **nodes**

descent engine A spacecraft **rocket engine** used for a descent to the surface of a **planetary body**; the main engine in the spacecraft's 'descent stage' (e.g. the descent engine of the **Apollo** lunar module).

[See also **retro-rocket**, **ascent engine**]

descent stage The lower part of the **Apollo** lunar module, designed to execute a landing of the combined ascent and descent stages on the lunar surface. The astronauts were housed in a cabin forming the larger part of the **ascent stage**.

descrambler An electronic device used to restore radio transmissions, made unintelligible to unauthorised receivers by an electronic device known as a **scrambler**, to their original form.

descrambling A process by which radio transmissions, made unintelligible to unauthorised receivers by an electronic device known as a **scrambler**, are restored to their original form. Also used in telephone systems, etc.

design driver Jargon: any concept, practicality or physical law which fundamentally governs the design of a component or **system** (e.g. for most **spacecraft** the concept of 'minimum mass' is a design driver, due to the high cost-per-kilogram of launching any mass into space).

design lifetime – see **lifetime**

despin active nutation damping electronics – see **DANDE**

de-spun platform An **antenna platform** on a **spin-stabilised** spacecraft which remains stationary with respect to the Earth to maintain the **pointing** of the antenna **beam**s. The shelf containing the communications **repeater** may also be de-spun, depending on the design [see figure S13].

[See also **BAPTA**]

detector Any device that receives a **signal** or stimulus and responds to it with a proportional signal which can be used in another device or display; an alternative term for **sensor**, although in general usage, 'detectors' tend to be used for scientific **application**s and 'sensors' for engineering applications.

[See also **bolometer, charge-coupled device (CCD)**]

dielectric A substance of very low electrical conductivity; an insulator. The best dielectric medium is a vacuum.

dielectric lens – see **lens antenna**

differential navigation A navigation technique in which satellite navigation signals are used to define a receiver's position with respect to a known, fixed base station. It is considerably more accurate than standard, absolute navigation techniques, which determine position with respect to map coordinates. For example, using the Global Positioning System (GPS), differential techniques can produce average accuracies of 2.5 m (with even greater accuracy – down to 0.5 cm – for receivers that remain stationary for short periods). See **Global Positioning System (GPS)**.

diffusion member A stiff structural component used to 'diffuse' concentrated **load**s over a wider area of structure (e.g. in the **thrust structure** of a launch vehicle **propulsion bay**).

digital compression The compression of video, audio or other electronic **signal**s performed in order to reduce the **bandwidth** required for their transmission. A technique used in **satellite communications** and **broadcasting** to allow a large number of **channel**s (e.g. a hundred or more) to be carried by a single **transponder**.

[See also **companding, bit rate**]

digital elevation model A collection of digital data which allows the derivation of topographical information including altitude/elevation, slope and surface roughness of a given terrain.

digital video broadcast (DVB) An **encryption** standard for **broadcasting** which allows a satellite **transponder** to carry a large number of **channel**s [see **digital compression**]. Its other advantages include the use of smaller receiving **antenna**s and secure transmission schemes.

diplexer A two-channel multiplexer. See **multiplexer**.

dipole A simple type of 'wire antenna' [see **antenna**], commonly called a 'half-wave dipole', which consists of a linear conductor approximately* half a **wavelength** long with the input connection in the middle (so-called 'centre-fed'). Sometimes mounted in an enclosure known as a cup (a 'cupped dipole') and grouped with others to form an array antenna [see figure A2].

direct broadcasting by satellite (DBS) A **satellite communications** system whereby **television**, and associated audio and data channels, are broadcast direct to the home, as opposed to via a landline or **microwave link**. It obviates the need for a large number of terrestrial transmitters and the inherent problems in providing full coverage in difficult terrain. The TV signal is transmitted from a single uplink **earth station** to the satellite, then retransmitted to domestic antennas throughout the **service area**.

The home is equipped with a small parabolic **antenna** (typically between about 0.3 m and 1 m in diameter), coupled to indoor and outdoor electronics units and permanently pointed towards the satellite in **geostationary orbit**. A **downconverter** attached to the antenna converts the high-frequency satellite signals to the lower frequencies accepted by the indoor electronics (**receiver**) linked to, or incorporated with, the TV set.

In the 1990s, direct **broadcasting** systems came to be known as **direct-to-home (DTH)** systems. In fact, there is a technical difference between DBS and DTH which relates to their respective usage of **frequency bands**, as allocated by the International Telecommunication Union (**ITU**). DBS refers to applications using the **broadcasting-satellite service (BSS)** frequencies allocated specifically for direct broadcast satellite applications, while DTH refers to DBS-like services using the **fixed-satellite service (FSS)** band.

Early DBS satellites differed from other **communications satellite**s mainly in that they utilised high-powered transmitters (high-power **travelling wave tubes**) to provide high **equivalent isotropic radiated power (EIRP)**. The desired simplicity of the receiving equipment led to the requirement for TWTAs transmitting up to 260 W of **RF power**, hence the term 'high-power DBS'. These high powers enabled the small antennas and relatively inexpensive receiving installations to receive a good TV picture. Later systems, using TWTAs of around 100 W, tended to be called 'medium-power DBS', and those around 50 W 'low-power DBS'. Satellite systems which used standard telecommunications TWTs to provide a DBS service to small antennas were termed 'Quasi-DBS' (now known as DTH).

[See also **head-end unit**, **link**, **link budget**]

---

* Resonance is typically attained at 0.49 wavelengths.

direct orbit – see **prograde orbit**

directivity The ratio of the radiation intensity from an **antenna** in a particular direction to that available from an **isotropic antenna**. A perfectly isotropic antenna can be said to have 'no **gain**' and 'no directivity'; any other antenna has definable values of both. In the theoretical case of a 'lossless antenna', the directivity and gain are the same.

direct-to-home (DTH) A **satellite communications** system whereby **television**, and associated audio and data channels, are broadcast direct to the home, as opposed to via a landline or **microwave link**. Also known as direct broadcasting by satellite (DBS), although there is a technical difference between the two terms: DBS refers to applications using the **broadcasting-satellite service (BSS)** frequencies allocated specifically for direct broadcast satellite applications, while DTH refers to DBS-like services using the **fixed-satellite service (FSS)** band.
[See also **direct broadcasting by satellite (DBS)**]

Discovery The name given to the third '**flight model**' of the Space Shuttle **orbiter** (Orbiter Vehicle OV-103), which was first launched on 30 August 1984 (STS 41-D). Discovery was named after two sailing ships: Henry Hudson's which attempted a search for a Northwest Passage between the Atlantic and Pacific Oceans in 1610–11, but instead discovered Hudson Bay; and Captain Cook's which discovered Hawaii and explored Alaska, etc.
[See also **Space Shuttle**, **space transportation system (STS)**, **Atlantis**, **Columbia**, **Challenger**, **Endeavour**, **Enterprise**]

discrimination A measure of the ability of a communications system or a component within that system to separate wanted from unwanted **signal**s. Commonly used in the form 'cross-polar discrimination' (abbreviated to XPD) when comparing two **carrier** waves on opposite **polarisation**s: the term relates to the degree to which the **feed** discriminates against the cross-polar component of the unwanted carrier.

dish A colloquial term for an antenna reflector, especially an **earth terminal**. See **antenna**.

dispenser A device, forming the interface between a **launch vehicle**'s uppermost **stage** and its **payload**s, which is designed to release the payloads into different **orbit**s (e.g. for the delivery of **constellation** satellites to differing **orbital plane**s) [see figure D5].
[See also **engine re-start (capability)**, **restartable upper stage**]

Dnepr A commercial **launch vehicle** converted from the former-Soviet SS-18 'Satan' heavy intercontinental ballistic missile (**ICBM**) and operated under a

D5: Four Iridium satellites mounted on a **dispenser** (and a fifth being prepared) in advance of integration with a **Delta** launch vehicle. Note **solar array** panels (mounted face to face) and **phased array antenna**s (facing out). [Lockheed Martin]

Russian–Ukrainian–Kazakh joint venture. It is available in two versions: the Dnepr-1, with a **payload capability** of about 4500 kg to a 200 km **low Earth orbit (LEO)**, and the Dnepr-M which delivers 4100 kg to 300 km.
[See also **Rockot**, **Cosmos**, **Start**]

dock To link two **spacecraft** together in space. The term is used irrespective of the size of the spacecraft, and therefore includes devices such as the **manned manoeuvring unit (MMU)**. In general there are two degrees of 'docking': the initial attachment, termed a 'soft dock'; and the establishment of a good mechanical contact which locks the spacecraft together, termed a 'hard dock'. For a **manned spacecraft** a hard dock is required before the crew can be safely transferred, since it forms an 'airtight seal'.

docking module A part of a **spacecraft** containing one or more **docking port**s; alternatively termed a 'docking adapter'. Examples include the **Apollo-Soyuz Test Project** Docking Module, used to join together the two spacecraft which had incompatible docking mechanisms, and the **Skylab** Multiple Docking Adapter which allowed two **Apollo** spacecraft to dock with the space station (although it was never done). The docking modules on the **International Space Station (ISS)** are known as 'nodes'.
[See also **dock**, **docking probe**]

docking port An access point on a **spacecraft** to which another spacecraft can be attached or 'docked' for the transfer of **crew** or materials.
[See also **dock**, **docking probe**]

docking probe The 'male half' of a **spacecraft** docking system, typically attached to the 'active vehicle' (the spacecraft which executes the docking manoeuvres) [see figure A7]. The 'passive vehicle' contains a **drogue** into which the probe is inserted. Typically, once the spacecraft are 'hard-docked' together, the probe and drogue can be removed to allow access through a docking tunnel.
[See also **dock**, **docking port**]

DOD
(i) An abbreviation for the United States Department of Defense (written DOD or DoD), a major US user of **space technology** from **launch vehicles** to **satellites**.
(ii) An abbreviation for **depth of discharge**.

Doppler shift A change in **frequency** due to the relative motion of a transmitter and receiver. The largest Doppler shifts experienced in communicating with **spacecraft** are typically those with **deep space** probes which have high velocities relative to the Earth. Satellites communicating between different orbits via an **inter-orbit link** also experience a Doppler shift.

double conversion transponder – see **dual conversion transponder**

double hop A signal route in a **communications satellite** system which includes

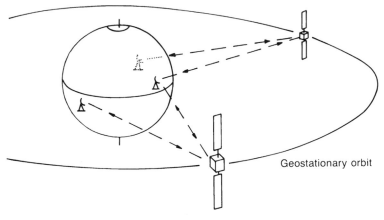

Geostationary orbit

D6: A **double hop** satellite communications **link**.

a pass through two **satellites**: i.e. the signal is **uplink**ed to one satellite, **downlink**ed to an **earth station**, uplinked to a second satellite and downlinked to a second earth station [see figure D6]. Since the signal travel-time to **geostationary orbit** is about 0.12 s for each of the four links, the whole process takes almost half a second. Since this makes normal conversation difficult, 'double hopping' is not recommended for two-way telephone calls and one of the satellite links is often replaced by a terrestrial link.

[See also **inter-satellite link**, **terrestrial tail**]

double-base propellant – see **solid propellant**

doubler A structural element which increases the thickness of a panel; local reinforcing for highly stressed areas.

[See also **thermal doubler**]

downconverter A device for reducing the **frequency** of a **signal**. It contains a circuit called a **mixer**, which 'mixes' the incoming signal with a **local oscillator** frequency. This process, known as the heterodyne process or 'heterodyning', produces frequencies corresponding to the sum and the difference of the two original frequencies. The output of the downconverter is the difference signal; in a **transponder** with a downconverter and an upconverter this signal is otherwise known as the IF or **intermediate frequency**.

[See also **upconverter**]

downlink The communications path or **link** between a **spacecraft** and an **earth**

station, 'from **space** to **Earth**'. Also used as a verb: 'to downlink'. Opposite: **uplink**.

[See also **download**]

download The process of copying **data** from the **Internet** to a **computer**, or 'downlinking' data or **telemetry** from a **spacecraft** to an **earth station** or earth terminal. Opposite (for **commands**): **upload**.

downrange The distance between a **launch vehicle** and the **launch pad** (or the ground station tracking the vehicle) measured on the surface of the Earth (e.g. 'the launch vehicle is 3 km downrange'); the distance measured along the projection of the vehicle's **flight path** on the ground, its **ground track**.

[See also **slant range, range**]

drag The retarding force acting on a body in motion through a fluid, particularly air. Drag acts in a direction parallel to the direction of motion in opposition to the **thrust**.

[See also **atmospheric drag, drag compensation**]

drag compensation The practice of firing a spacecraft **thruster** to compensate for the effects of **atmospheric drag** experienced in **low Earth orbit**. More colloquially called 'drag make-up'.

drag make-up A colloquial term for **drag compensation**.

DRAM An acronym for dynamic random access memory. See **RAM**.

drift

(i) A gradual change in a spacecraft's **orbital parameters**, particularly the change in the longitudinal position of a spacecraft in **geostationary orbit**, due mainly to gravitational **perturbations** and the **solar wind**. If this undesirable drift remained uncorrected, all geostationary satellites would tend to move away from their allocated **orbital positions** towards one or other of the stable equilibrium points. Drift is normally corrected by the periodic firing of the spacecraft's **reaction control thrusters**. See **equilibrium point, orbital control**.

(ii) The motion of a **satellite** in a **drift orbit**.

(iii) A lateral divergence of a **flight path** (due primarily to crosswinds).

(iv) A slow change in the **frequency** of a **transmitter** or other device containing a frequency-source.

(v) The angular deviation of the spin axis of a **gyroscope** from its fixed reference in **inertial space**.

drift orbit An orbital path into which a satellite is injected by its **apogee kick motor** or **liquid apogee engine** en route to its final position in **geostationary**

orbit. The drift orbit is an approximation to geostationary orbit in terms of height and circularity, but allows the satellite to **drift** towards its final **orbital position** without using its limited supply of **reaction control thruster** propellant.

[See also **equilibrium point**, **orbital control**]

drift tube A component of a **klystron** power amplifier.

drogue

(i) Drogue parachute: a small parachute designed to pull a larger parachute from its container (e.g. for use with recoverable spacecraft and rocket boosters, etc.).

(ii) Docking drogue: the 'female half' of a **spacecraft** docking system. See **docking probe**.

drop tower A terrestrial facility used to produce several seconds of **microgravity** for experimental payloads. The payloads are allowed to fall inside the tower, which is usually evacuated to simulate the **space environment**.

[See also **gravity**, **weightlessness**]

drum-stabilised – see **spin-stabilised**

dry mass The mass of a **spacecraft** or **launch vehicle** without **propellant** or **pressurant**. In the case of a launch vehicle, where the concept of weight has a meaning, it is sometimes called 'dry weight'. Although **solid propellant** is, in effect, 'dry', it is treated in the same way as **liquid propellant** in this case.

[See also **mass budget**, **launch mass**, **in-orbit mass**]

dry weight The weight of a **launch vehicle** without **propellant** or **pressurant**. See **dry mass**.

Dryden Flight Research Center – see **NASA**

DSN – see **deep space network**

DTH – see **direct-to-home**

dual conversion transponder A **transponder** which converts an **uplink** frequency to an **intermediate frequency (IF)** before converting that to a **downlink** frequency. Alternatively called a double conversion transponder. See **upconverter**.

[See also **downconverter**]

dual-gridded reflector An **antenna** reflector designed for polarisation **frequency re-use** which comprises two reflector surfaces mounted one in front of the other. The front grid, which on the **uplink** reflects one polarity of radiation into its **feedhorn**, is transparent to the opposite polarity which is reflected from the rear grid into another horn. Angling the grids slightly with respect to each other allows a small separation between the horns while maintaining a similar **footprint** for both polarities on the **downlink**. Dual-gridded antennas have been widely used on spin-stabilised communications satellites which generally have room for only one antenna on their top platform.

ducting The entrapment of an electromagnetic wave between layers of the Earth's atmosphere or between the atmosphere and the Earth (effectively a 'natural **waveguide**'). The effect allows radio signals to be received at greater distances than they would be by 'line-of-sight'.

dummy payload A **payload** mock-up which simulates the size and mass of a real payload, but contains few or no working parts. Sometimes used on new **launch vehicle**s when it is felt unwise to risk a real payload.

dummy stage A launch vehicle **stage** which, as part of a test-vehicle, contains no **propellant** and is therefore incapable of powered flight; a stage which simulates a launch vehicle **payload** for the purposes of a test-flight.
[See also **live stage**]

Dutch roll A simultaneous rotation about the roll and yaw axes. See **spacecraft axes**.

duty cycle The cyclic variation in the state of a device; the period from 'switch-on' through 'switch-off' to 'switch-on' again. For example the duty cycle of a spacecraft **battery** is dependent on the spacecraft's **orbit**: in **geostationary orbit** a battery undergoes about 90 charge–discharge cycles per year, with varying periods [see **eclipse season**], whereas one in **low Earth orbit** typically undergoes around 5000 cycles per year, each with a 35-minute discharge period and a 65-minute charge period. Duty cycle is sometimes expressed as a percentage, describing the proportion of time a device is 'on duty': e.g. a device with a 50% duty cycle is on for half the time.
[See also **depth of discharge (DOD)**]

DVB – see **digital video broadcast**

DVS An abbreviation for direct via satellite.

dynamic pressure The pressure on a vehicle due to its motion through the atmosphere. The point in the **flight** of a **launch vehicle** at which it experiences the most severe aerodynamic forces is the point of 'maximum dynamic pressure', sometimes abbreviated to 'Max-Q'. Since the **aerodynamic stress** is due to a combination of the ambient air pressure and the speed of the vehicle, most launch systems are designed to reduce **thrust**

slightly as Max-Q approaches and then increase it when the air pressure has dropped sufficiently (e.g. Max-Q for the Space Shuttle is reached at an altitude of about 10,200 m, about 60 seconds after **launch**).

[See also **thrust profile, throttling, propellant grain**]

dysbarism – see **decompression sickness**

E

Early Bird A colloquial name for Intelsat I, the first commercial geostationary **communications satellite**; launched in 1965 for the **Intelsat** organisation. [See also **geostationary orbit (GEO), space insurance**]

Earth The third planet from the **Sun**. Usage: when used as a proper name (when referring directly to the planet), Earth should have an initial capital (e.g. **Earth observation, Earth-lock**); when used as a common noun (to describe an object or system), it should have a lower case initial (e.g. **earth station, earth segment, earth sensor**). A related adjective is 'terrestrial' ('of the Earth', from terra, Latin for Earth) – see, for example, **terrestrial tail**.

  Physical data: equatorial diameter 12,756 km; polar diameter 12,712 km; mass 5.97×10^{24} kg; **escape velocity** 11.18 km s^{-1}; surface **gravity** 9.8 m s^{-2}; average solar distance 149,597,870 km (1 **astronomical unit**); average light travel time from Sun 8 minutes 19 seconds; axial inclination 23.45°; **orbital inclination** 0.0° [see **ecliptic**]; **orbital period** 365.2422 days; **sidereal period** 23h 56m 4s.

[See also **light (velocity of), NEO**]

Earth observation A satellite **application** which involves obtaining **images** and other **data** on any aspect of the **Earth** for scientific or commercial use. In general usage, the description 'Earth observation **satellite**' has tended to replace '**remote sensing** satellite', which itself had previously replaced 'Earth resources satellite' as the spacecraft began to be used for a broader range of applications.

Earth resources satellite A class of **remote sensing** satellite concerned primarily with land-use. The term 'remote sensing' has tended to replace 'Earth resources' and, in turn, has generally been superseded by the term '**Earth observation**'.

earth segment The terrestrial part of a **satellite**-based communications system or **deep space** communications network. Its major constituent is the **earth station**, but the earth segment may include several stations as well as the infrastructure linking them to the user network. Earth segment is generally taken as synonymous with **ground segment**.

[See also **space segment**]

earth sensor A device used to establish a spacecraft's **attitude** relative to the **Earth** by detecting the limits of the Earth's disk at infrared wavelengths; also called an **infrared sensor**. Sensing is either static or dynamic.

The static **sensor** uses a common direction-sensing technique involving an arrangement of sensors grouped around the pointing axis so that they each receive the same illumination when the sensor is exactly aligned with the target-source. An infrared image of the Earth is projected onto an array of thermopiles, which converts the incident IR radiation to an electric current. When the sensor is directly in line with the Earth the output from each of the thermopiles is the same; if the sensor is tilted with respect to the Earth the output changes accordingly. This type of differential output can be used to control a satellite's **reaction wheels**, **reaction control thrusters**, etc. This is described as a 'null-seeking' technique. Since the unit has fixed optics, it can only be used in an orbit in which the Earth's apparent size remains constant (it is most prevalent in **geostationary orbit**, from which the Earth subtends an angle of about 17.4°). This allows the Earth's disk to cover the sensing elements in identical proportions when the sensor axis is pointing towards the centre of the Earth, and to provide difference-signals when it moves off-axis. Since the temperature of the Earth's atmosphere varies with the time of day, a filter limits the incoming radiation to the carbon dioxide absorption band (around 15 μm) where the radiance is nearly constant.

The dynamic sensor operates by scanning across the Earth's disk and detecting the temperature difference between space and Earth. It is otherwise known as a 'horizon sensor' or 'horizon scanner'. When dynamic sensing is used on **three-axis stabilised** spacecraft, usually those in **low Earth orbit**, the scanning motion is provided by a rotating mirror. **Spin-stabilised** spacecraft use two types of earth sensor, both of which make use of the satellite's inherent spin: the telescope type has a limited field of view and relatively high accuracy; the fan-beam type senses the Earth for a longer period but is less accurate.

[See also **sun sensor**]

earth station

(i) An installation on the **Earth** containing the equipment necessary for **communications** with a **spacecraft**, i.e. chiefly an **antenna** system capable of transmission and reception, a **receive chain**, a **transmit chain** and the necessary interfaces with other terrestrial equipment [see figure E1]. An earth station is the major component of the **earth segment** (or

E1: **Earth station** at Fucino, Italy. [Inmarsat]

ground segment) and is synonymous with **ground station**. Smaller versions are usually called earth terminals, a term sometimes used to include receive-only installations (e.g. **TVRO**). To differentiate between types of earth station, an installation may be called a **land earth station** or **ship earth station**.

(ii) The term 'earth station' is also used to refer to a facility comprising several earth station antennas (e.g. Goonhilly earth station in Cornwall, England). 'Earth terminal' can be used in a similar way but usually refers to a smaller installation.

[See also **VSAT**]

earth terminal A small earth station – see **earth station**.

Earth-lock Jargon: 'locked onto the **Earth**'. A term used for **spacecraft** which require part of their structure (typically an **antenna platform**) to remain **pointing** towards the Earth. This is realised using infrared **earth sensors** and an **attitude control** system.

[See also **spin-up** (sense iii)]

Earth-pointing face The face of a **three-axis stabilised** spacecraft (defined as the plus-z face) which faces directly towards the **Earth**; the surface on which an Earth-pointing **antenna, sensor** or instrument is mounted. See **spacecraft axes**.

earthshine Thermal energy absorbed from the **Sun** and re-radiated by the **Earth**. See **albedo**.

Eastern Test Range – see **Kennedy Space Center**

east–west station keeping – see **orbital control**

EBU An abbreviation for European Broadcasting Union, an association of **broadcasting** organisations whose members operate national broadcasting systems in Europe, North Africa and North America.

ebullism The vaporisation of fluids from the human body under conditions of very low pressure (e.g. as experienced during a 'catastrophic' or explosive **decompression** of a **spacecraft** or **spacesuit**). When the ambient atmospheric pressure is less than the fluid vapour pressure of the body, about 6.3 kPa (0.9 psi or 47 mmHg), the fluids vaporise and bubble through the body's mucous membranes: the eyes, mouth and other orifices.
[See also **decompression sickness, hypoxia**]

eccentricity A measurement of the ellipticity or non-circularity of an **orbit**, having a value between zero and one, such that a circular orbit has an eccentricity of zero; mathematically, the distance between the focal points of an ellipse divided by twice the length of the **major axis**.
[See also **elliptical orbit, orbital parameters**]

Echo The name given to a series of two spherical metallised balloons placed in **low Earth orbit** in August 1960 and January 1964. Echo 1 and 2, 30 m and 41 m in diameter respectively, were designed to reflect incident radio signals towards a ground-based **receiver**. Because they contained no **active** components, they were termed **passive** communications satellites.

eclipse The partial or total obscuration of one celestial body by another, or of a spacecraft by another body.

Eclipses of the Sun by the Earth are particularly important for spacecraft which derive their **power** from **solar cells**. During eclipse most spacecraft draw power from batteries which are recharged once sunlight is available again. Eclipses also affect the **thermal energy balance** of a spacecraft by removing a source of heat.

Spacecraft in **low Earth orbit** enter the Earth's shadow for a similar time on every orbit [see figure E2]: although it depends on their precise orbital

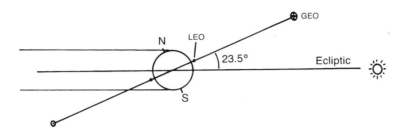

E2: A satellite in an equatorial **low Earth orbit** goes into **eclipse** on every revolution. A satellite in **geostationary orbit** (shown to scale) is eclipsed only around the **equinox**es [see figure E3] – this diagram shows the Earth at winter **solstice**.

period, they spend about 35 minutes of a 100-minute orbit in eclipse and may experience as many as 14 eclipses per day. For a spacecraft in **geostationary orbit**, eclipses occur on a less frequent but equally predictable basis – see **eclipse season**.

[See also **battery, thermal control subsystem, depth of discharge (DOD)**]

eclipse season The part of the year when a spacecraft in **geostationary orbit** (GEO) passes through the Earth's shadow and is unable to obtain **power** from the Sun via its **solar array**.

Since GEO is an **equatorial orbit** and the Earth's axis is tilted at 23.45° to the **ecliptic**, for most of the year the entire geostationary orbit is sunlit [see figure E2]. However, as the Earth approaches the vernal (spring) equinox (21 March) and the autumnal equinox (23 September), part of the orbit moves into the shadow cast by the Earth and the satellite becomes eclipsed for a small part of its circuit [figure E3]. The angle of **inclination** of the equatorial plane relative to the Sun governs the duration of the eclipse, which is greatest when the plane exactly bisects the Sun. The 'eclipse season' begins about 22 days before the equinox and ends about 22 days after it, the eclipse duration increasing to a maximum at equinox when it peaks at about 72 minutes (penumbral eclipse is 71.8 minutes and that of the umbra 67.5 minutes). In addition to the regular **eclipse** by the Earth, the Sun is eclipsed by the Moon, but on a much less regular basis.

Communications satellites generally have sufficient **battery** capacity to enable a service to continue throughout the eclipse, which occurs around midnight at the sub-satellite longitude. However, some satellites designed in the 1980s for **direct broadcasting by satellite (DBS)** could not carry the large number of batteries required to power their high power **transponders** and

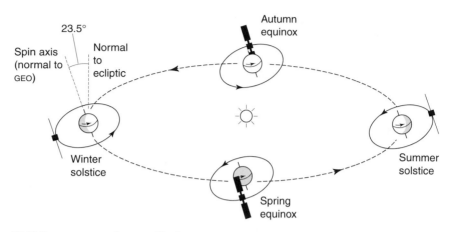

E3: **Eclipse** geometry for a satellite in **geostationary orbit** showing the **eclipse season**s at the **equinox**es.

were placed in **orbital position**s to the west of their respective **service area**s where they would be eclipsed in the early morning, when a 'closedown' was commercially less important.

[See also **solstice, solar array, sub-satellite point**]

eclipse-protected A characteristic of a **communications satellite** service whereby the satellite **transponder**s can remain powered during an **eclipse**, thereby providing a continuous service. The provision of 'eclipse protection' depends on the ability of the spacecraft's batteries to supply sufficient power for the **communications payload** and other **subsystem** equipment. High-powered **DBS** satellites are generally not eclipse-protected, because the mass of the batteries required would be too great.

[See also **eclipse season**]

ecliptic In astronomy, the apparent annual path of the Sun relative to the stars as seen from the Earth; the plane containing the **Sun** and the **Earth**: the ecliptic plane.

For historical reasons, the night sky visible from the Earth is known as 'the celestial sphere'; the projection of the Earth's equator on this sphere is called 'the celestial equator', and this and any similar projection is known as a 'great circle'. The projection of the Earth's orbital plane on the celestial sphere is a great circle – the ecliptic. Since the Earth's axis is tilted at an angle of 23.45° to its orbital plane, the ecliptic makes an angle of 23.45° with the celestial equator. It crosses the celestial equator at two opposing points called the **equinox**es; the two points on the ecliptic furthest from the equator are the **solstice**s.

[See also **eclipse, eclipse season, nodes**]

ECLSS – see **environmental control and life support system**

ECS An abbreviation for European Communications Satellite, a series of satellites developed by the European Space Agency (**ESA**) and operated by Eutelsat.

edge of coverage Strictly the edge of the **coverage area** of a **satellite communications** service, but used more widely in the vaguer sense of 'the locus of points furthest from the centre of a satellite **footprint** where a signal can still be received'. See **coverage area** and **service area**.

Edwards Air Force Base The site of the NASA Ames Dryden Flight Research Facility [see **NASA**], where much of the USA's advanced flight research has been conducted (location approximately 34.5° N, 117.5° W). It includes the 170 km² Rogers Dry Lake, an ideal natural surface on which to land high performance research aircraft [e.g. see **X-1, X-15**] and the Space Shuttle **orbiter** (e.g. during the 'Approach and Landing Tests' (ALT) [see **Enterprise**]).

EELV – see **Evolved Expendable Launch Vehicle**

EGNOS An acronym for European geostationary navigation overlay system, the first phase of **GNSS** in Europe.

egress An exit; the act of emerging. A word commonly used as a modifier (e.g. 'egress hatch', 'egress port'). The opposite of **ingress**.

EGSE An acronym for electrical ground support equipment. See **ground support equipment**.

EHF (extra high frequency) – see **frequency bands**

EHT An abbreviation for electrothermal **hydrazine** thruster or electrically heated thruster. See **hiphet thruster**.

EIRP – see **equivalent isotropic radiated power**

ejection velocity – see **exhaust velocity**

ELA A French abbreviation for Ensemble de Lancement Ariane or '**Ariane** launch site'. The individual **launch complex**es at the **Guiana Space Centre** are thus known as ELA-1, ELA-2 and ELA-3.

ELDO An acronym for European Launcher Development Organisation (also known by its French acronym CECLES), a body formed in April 1962 for the development of the **Europa** multistage **launch vehicle**. Founding countries were Australia, Belgium, France, Italy, Netherlands, the United Kingdom and West Germany. ELDO was disbanded in 1975 when it was amalgamated with

ESRO (European Space Research Organisation) to form the European Space Agency (**ESA**).

[See also **Blue Streak**]

electric propulsion (EP) A collective term for a form of rocket propulsion in which electrical energy is used to derive or augment the kinetic energy of the propulsive jet. There are three types of electric propulsion device:

  (i) electrothermal, in which electrical energy is used to heat a gaseous **propellant**, typically to increase the efficiency of a propulsive device [see **hiphet thruster**, **arcjet thruster**];

  (ii) electrostatic, in which electrical energy is used to ionise a gaseous propellant and an electrostatic field accelerates the positive ions to produce **thrust** (as in **ion** engines or ion thrusters) [see **ion engine**];

  (iii) electromagnetic, in which an electromagnetic field is used to accelerate a neutral, gaseous **plasma** to produce thrust (as in plasma engines, plasma rockets, plasma thrusters or magnetoplasmadynamic thrusters) [see **ion engine**].

[See also **chemical propulsion, nuclear propulsion, exhaust velocity, specific impulse**]

electrically heated thruster (EHT) – see **hiphet thruster**

electromagnetic propulsion – see **electric propulsion**

electromagnetic pulse (EMP) A short-duration burst of **electromagnetic radiation** which can damage electronic components that have not been 'hardened' against it. The term is used mainly in the context of the detonation of thermonuclear devices and their effect on electronics-based weapon systems, but can be extended to more natural sources such as lightning.

[See also **radiation hardening**]

electromagnetic radiation Energy propagated through space or material media as an advancing disturbance in the intrinsic electric and magnetic fields, which oscillate in orthogonal planes mutually perpendicular to the direction of propagation.

[See **electromagnetic spectrum**]

electromagnetic spectrum The entire range of **electromagnetic radiation**, from the lowest to the highest **frequency** or the shortest to the longest **wavelength** [see figure E4]. All such radiations travel at the velocity of **light** in **free space**. In increasing frequency (and decreasing wavelength), sections of the EM spectrum are commonly labelled radio, microwave, infrared (IR), visible,

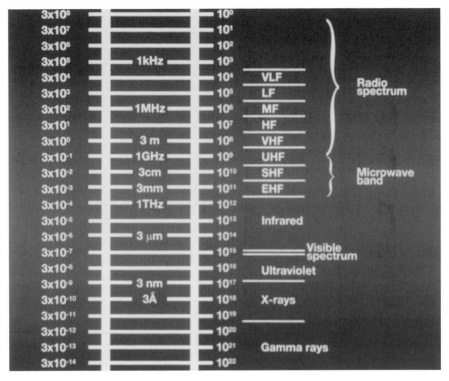

E4: The **electromagnetic spectrum**. [Mark Williamson/Willis Inspace]

ultraviolet (UV), X-rays and gamma-rays and cover the frequency ranges as follows:

- Radio 10^3–10^9 Hz (1 kHz–1 GHz)
- Microwave 10^9–10^{11} Hz (1 GHz–100 GHz)
- Infrared (IR) 10^{11}–10^{15} Hz (100 GHz–1000 THz)
- Visible 3.95×10^{14} Hz–7.69×10^{14} Hz (red–violet)
- Ultraviolet (UV) 10^{15}–10^{17} Hz
- X-rays 10^{17}–10^{19} Hz
- Gamma-rays 10^{19}–10^{23} Hz ...

NB All divisions and ranges are approximate. For example, the **microwave** band is more accurately defined as 1 GHz–300 GHz. **Satellite communications** frequencies are predominantly in the microwave band [see **frequency bands**].

[See also **radio frequency (RF)**, **waveband**]

electron An elementary particle which exists, in numbers equal to the atomic number, around the nucleus of every atom. The electron's rest mass is 9.11×10^{-31} kg (1/1837 of the proton's mass) and its charge is 1.60×10^{-19}C (symbol: e⁻ since the charge is negative).

The term 'electron' was suggested in 1891 by George Johnstone Stoney [1826–1911], an Irish physicist, as the name for a hypothetical particle believed to be a discrete unit of electricity. The electron was discovered experimentally at Cambridge University in 1897 by the English physicist Sir Joseph John Thomson [1856–1940].

[See also **electron beam, electron gun, ion**]

electron beam A collimated stream of electrons produced by an **electron gun**. Used for example:

(i) as the source of energy for RF amplification in a **travelling wave tube**: the beam of electrons passes through the TWT from electron gun to **collector** along the axis of the helical **slow-wave structure**, where it interacts with the **RF wave**.

(ii) in the **ion engine** for the neutralisation of the positively charged ion beam.

electron bombardment ion thruster – see **ion engine**

electron gun A device for producing a narrow beam of electrons from a heated cathode. Electrons liberated from the cathode are accelerated to high velocities by an electric field between the cathode and the anode. Although this definition covers equally the devices found in cathode ray oscilloscopes and domestic television receivers, the version used in space technology (e.g. within a **travelling wave tube (TWT)**) supplies an **electron beam** with a power density significantly higher than for the other two examples. The electron gun in a TWT injects a narrow beam of electrons into a **slow-wave structure** where they are used to amplify an RF **signal** [see figure T7].

electron tube An electrical device, sometimes termed a 'valve' or 'electron valve', in which a flow of **electrons** between electrodes takes place (see **electron gun**). Two main types of electron tube are used in spacecraft communications systems as high (RF) power amplifiers: the **travelling wave tube (TWT)** and the **klystron**.

electronic power conditioner (EPC) A device which supplies the operating voltages to the components of a **travelling wave tube (TWT)**: the **electron gun**, **slow-wave structure** and **collector**. A TWT is, in effect, useless without an EPC and is never found in an operational condition in isolation [see figure T6].

[See also **travelling wave tube amplifier**]

electronic steering A method for aligning or **pointing** an **antenna beam** used by a **phased array antenna**; an alternative to the mechanical steering of a reflector **antenna** – see **antenna pointing mechanism (APM)**.

electrostatic discharge (ESD) An instantaneous loss of static electrical charge; a phenomenon which constitutes a potential hazard to **spacecraft** operations, generally to electronic devices but particularly to astronomical **detectors** and the like. The main mechanisms are photoemission due to solar illumination, **micrometeoroid** impact and pressure increase (e.g. from thruster exhausts). Differential charging of the outside surface of the spacecraft can lead to a subsequent discharge which may, for instance, cause a communications system to switch off or severely disrupt an astronomical measurement.

    The severity of the hazard to a communications **repeater** is represented by the Rosen scale, where 0 is 'no hazard', 1 is a 'nuisance' or outage of one second or less, 5 an outage of a few hours and 10 a catastrophe. The risk can be reduced to less than '1' by discouraging the build-up of charge: parts of the satellite should be conductive and grounded to the structure, particular attention being paid to shaded areas, which are not discharged by photoemission. Electronic circuits can also be protected against ESD by shielding and the addition of protective devices, such as filters and diode-like limiters, in the input and output lines.

electrostatic propulsion – see **electric propulsion**

electrothermal hydrazine thruster (EHT) – see **hiphet thruster**

electrothermal propulsion – see **electric propulsion**

elevation (angle) The angle above the horizon measured in the horizon system of celestial coordinates. The word **altitude** is synonymous with elevation in this context. See also **azimuth**: knowledge of a **satellite**'s azimuth and elevation angle allow an **earth station** to be aligned with it. An alternative 'equatorial' or 'geographical' system uses **declination** and **hour-angle**.

elevon An aerodynamic control surface in the form of a hinged flap on the trailing edge of a delta-wing; a combined *elev*ator and ailer*on*. For example, the elevons on the Space Shuttle, which are controlled by hydraulic power provided by an **auxiliary power unit (APU)**.

elint An abbreviation for electronic intelligence.

    [See also **reconnaissance satellite, surveillance satellite**]

elliptical antenna Any **reflector antenna** with an elliptical aperture. On a spacecraft, an elliptical antenna, in its simplest form, provides an elliptical

footprint. However, the number, arrangement and detailed design of the **feedhorns** used with a particular antenna make a wide variety of footprints possible with the same reflector.

The **major axis** of a simple elliptical antenna produces a narrower beamwidth than the **minor axis**, meaning that the resultant footprint is oriented orthogonally to the antenna reflector – see **beamwidth**.

[See also **boresight**]

elliptical orbit A non-circular **orbit**; an orbit with a degree of **eccentricity** defined by a **major axis** and a **minor axis**. Owing to irregularities in the gravitational attraction of a **planetary body**, all orbits have some eccentricity: a 'circular orbit' is therefore an approximation. An elliptical orbit with a high degree of eccentricity is known as a highly elliptical orbit (HEO).

[See also **perturbations, orbital parameters**]

elliptical polarisation – see **polarisation**

ELT An abbreviation for emergency location transmitter. See **COSPAS-SARSAT**.

[See also **navigation satellite, Global Positioning System (GPS)**].

ELV – see **expendable launch vehicle**

EM An abbreviation for (i) electromagnetic and (ii) **engineering model**, the version of a spacecraft constructed to verify that the electrical and RF **performance** meets the **specification**.

EMC – an abbreviation for electromagnetic compatibility.

EMI – an abbreviation for electromagnetic interference.

EMP – see **electromagnetic pulse**

EMU – An abbreviation for **extra-vehicular mobility unit**.

encounter The moment in time or point in space at which a planetary exploration **spacecraft** on a fly-by **trajectory** is at its closest to the **planetary body** under investigation; the period over which the spacecraft is considered to be within close observing range of the body (the 'encounter period'). The term tends not to be used for spacecraft which enter an **orbit** about a planetary body.

encryption Coding. Most military and some commercial radio transmissions are 'encrypted' to make them unintelligible to unauthorised receivers.

[See also **scrambling, descrambling, decryption**]

Endeavour The name given to the fifth '**flight model**' of the Space Shuttle **orbiter** (Orbiter Vehicle OV-105), which was first launched on 7 May 1992 (STS-49). Endeavour was built to replace Challenger and bring the shuttle fleet back to four **operational** orbiters. It was named after a sailing ship commanded by British Captain James Cook in the late 1700s.

[See also **Space Shuttle**, **space transportation system (STS)**, **Atlantis**, **Challenger**, **Columbia**, **Discovery**, **Enterprise**]

end effector A device fixed to the end of the Space Shuttle's **remote manipulator system (RMS)**: effectively the 'hand' on the mechanical arm. The standard end effector can 'grapple' a **payload**, keep it rigidly attached as long as required and then release it. For specific tasks, specialised end effectors can be fitted as necessary.

[See also **grapple fixture**]

end of life (EOL) The end of a **spacecraft**'s **operational** lifetime. The term is used to specify the magnitude of a parameter at the end of a spacecraft's **lifetime**, as opposed to at **beginning of life (BOL)** when it may be different (e.g. the output power from a **solar array** [see **degradation**], the mass of **station-keeping** propellant, etc). Also refers to components – see **bathtub curve**.

end user The user of a communications system at the end of a link (e.g. the recipient of a **DBS, DTH** or **cable TV** service). A **cable** system operator, for example, could be classed as a user of a **satellite** since he rents a **transponder** for the distribution of his service. There is, however, a user of the system beyond the operator: the consumer, who is the end-user.

end-burning – see **propellant grain**

Energia A Russian **launch vehicle** developed as a **heavy lift vehicle (HLV)** to carry, among other things, the Soviet **space shuttle** [see **Buran**]. It had three stages and four **liquid propellant** strap-on boosters and had a **payload** capability of over 90 tonnes to **low Earth orbit (LEO)**. It was first launched, with an unmanned payload canister, in May 1987, then in November 1988 with Buran. It is no longer **operational**.

[See also **Proton**]

energy The capacity of a body or system to do work, measured in joules (J).

[See also **power**]

energy density – see **specific energy**

energy dispersal A process which involves adding a low frequency waveform to a **baseband** signal before **modulation**, to reduce the frequency-modulated signal's peak power per unit **bandwidth** and thus its **interference** potential. Particularly useful in avoiding the peaks of signal power which occur in a TV **signal**.

[See also **frequency modulation (FM)**]

engine – see **rocket engine**

engine bay The part of a **launch vehicle** or **spacecraft** which houses the **engines**, typically incorporating a **thrust frame** and bordered by engine **cowlings** and other **fairing**-components. Also called a **propulsion bay**.

engine cut-off – see **burn-out**

engine re-start (capability) The ability to resume **combustion** in a **rocket engine**. It is only possible to 're-light' engines using **liquid propellant**; once a **solid rocket motor** has been ignited, it burns until all the propellant is consumed. Typically only the uppermost **stage** of a **launch vehicle** has a re-start capability, usually for injecting its **payload** into a **transfer orbit**, or to inject separate payloads into different orbits [see **restartable upper stage**, **dispenser**]. Spacecraft liquid propulsion systems make continual use of this capability for **orbital control**.

[See also **combined (bipropellant) propulsion system, liquid apogee engine, parasitic station acquisition, hybrid rocket**]

engineering model A version of a **spacecraft** built for ground-based testing rather than flight [see **flight model**]. Depending on its specific function, it may alternatively be known as a **structure model** or **thermal model**, for example.

ENT (equivalent noise temperature) – see **noise temperature**

Enterprise The name given to the first-built Space Shuttle **orbiter** (Orbiter Vehicle OV-101), named after the *Starship Enterprise* from the TV series *Star Trek*.

Enterprise was used to determine the orbiter's aerodynamic characteristics in a series of approach and landing tests (ALT) conducted between February and November 1977 from a specially modified Boeing 747, now known as the Shuttle Carrier Aircraft (SCA). There were five 'captive flights' (with the orbiter fixed to the 747, inert and unmanned), three manned captive flights and five 'release-flights', in which the orbiter was allowed to glide to a landing at **Edwards Air Force Base**, California. In 1978 Enterprise was used for vibration tests, and in 1979 transported to **Kennedy Space Center** for mating with an **external tank** and **solid rocket boosters** for 'fit-checks'. Enterprise was never intended for **spaceflight** and is now part of the collection of the Smithsonian Air and Space Museum, Washington, DC.

[See also **Space Shuttle, space transportation system (STS), Atlantis, Challenger, Columbia, Discovery, Endeavour**]

entry interface The point in a space vehicle's **trajectory** on its return to Earth which defines the beginning of its **re-entry** (e.g. for the Space Shuttle it occurs about 30 minutes after the **de-orbit** 'burn' and about two minutes prior to **S-band blackout**).

[See also **re-entry corridor**]

entry-probe A planetary exploration **spacecraft**, colloquially termed a **probe**, which is designed to enter a planet's atmosphere. For example, the probe released by the **Galileo** spacecraft into Jupiter's atmosphere and the Huygens probe carried by the Cassini spacecraft to Saturn's largest moon, Titan.

[See also **Pioneer Venus**]

envelope A physical or conceptual boundary defining the limits of a parameter. For example:

  (i) the dimensional constraint on the volume available for a **payload** – see **payload envelope**;

 (ii) the operating limits of an **aerospace vehicle** (e.g. jargon: 'expanding the envelope').

environment (of space) – see **space environment**

environmental chamber A ground-based test facility which simulates some aspect(s) of the space or launch environment: e.g. temperature, pressure, humidity, noise and motion.

[See also **thermal-vacuum chamber**, **acoustic test chamber**, **vibration facility**, **anechoic chamber**]

environmental control and life support system (ECLSS) A system which maintains an air-conditioned environment (for crew and electronic equipment) and provides **life support** functions within the **cabin** of the Space Shuttle **orbiter**. The ECLSS comprises atmospheric control and purification equipment and thermal control, water and waste management systems.

The nominal atmosphere in the orbiter comprises 80% nitrogen and 20% oxygen at a pressure of 101.4 kPa (14.7 psi). Recirculated air is passed through replaceable canisters containing a mixture of activated charcoal to remove odours, and lithium hydroxide to remove carbon dioxide. Cabin temperature can be regulated between 16 °C and 32 °C; relative humidity between 35% and 55%. The air is passed through a **heat exchanger** and the excess heat is passed to a water **coolant** loop, from which it is transferred to a freon coolant loop and thence to **radiator** panels mounted inside the **payload bay** doors. Water for food preparation, drinking and personal hygiene is produced by the **fuel cell**s, as a by-product of energy generation. The 'waste collection system' collects and processes liquid and solid 'biowaste', washing water, and condensed water from the cabin heat exchanger.

[See also **portable life support system (PLSS), carbon dioxide absorber, spacesuit, cabin pressure**]

EOL – an abbreviation for **end of life**

EP – see **electric propulsion**

EPC – see **electronic power conditioner**

ephemeris A table of data providing position information for a **planetary body**, comet or **spacecraft** for a specified period. Plural: ephemerides.

[See also **orbital parameters**]

EPIRB An acronym for emergency position indicating radio beacon. See **COSPAS-SARSAT**.

[See also **navigation satellite, Global Positioning System (GPS)**]

EPROM An acronym for erasable programmable read only memory. See **ROM**.

EPS An abbreviation for electrical **power** system or subsystem. A more common term is **power subsystem**, since all **spacecraft** power is electrical.

equatorial mount A structure for the support and guidance of an astronomical telescope which has its axis aligned with the Earth's rotational axis, with the result that the apparent motion of the stars can be 'cancelled out' by driving the telescope about one axis only (cf. **altitude-azimuth mount**). When used for **satellite communications** it is called a **polar mount**.

equatorial orbit An **orbit** in the same plane as a planet's equator, e.g. **geostationary orbit (GEO)**.

[See also **circular orbit, elliptical orbit**]

equilibrium point One of four points on the **geostationary orbit** produced by the variation in the gravitational force around the Earth's equator, which is roughly elliptical in cross-section. There are two 'stable equilibrium' points aligned with the minor axis and two 'unstable equilibrium' points aligned with the major axis. The approximate **orbital position**s of the stable points are 75° E and 105° W, while those of the unstable points are 165° E and 15° W (a 90-degree spacing around the orbit) [see figure E5].

Although satellites stationed at the stable equilibrium points are less likely to **drift**, other **perturbations** of the orbit make **orbital control** necessary: correction of the drift around the orbit is known as east–west station keeping. For a satellite stationed at an arbitrary position between the equilibrium points, the drift mechanism operates in the following manner. If the effective component of the gravitational force is in the same direction as

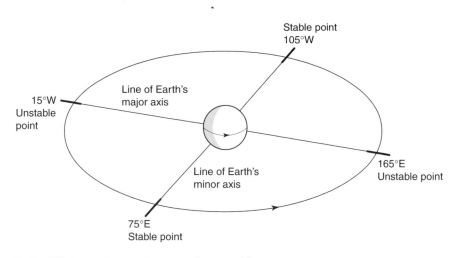

E5: **Equilibrium points** on the **geostationary orbit**.

the satellite's motion (west to east), energy is added to the system and the orbital height increases. The satellite decelerates, its orbital period increases and it drifts back towards the equilibrium point on the minor axis. Passing the equilibrium point, the satellite experiences a force in opposition to its direction of motion, which reduces its energy. The orbital height decreases and the satellite accelerates, reducing its orbital period. It moves back towards the equilibrium point and the cycle continues. During its eastward drift, the satellite is up to 34 km below the nominal height of geostationary orbit, and during the westward drift it is above it to the same degree. The period of the oscillation is about 840 days, and the maximum rate is 0.4 degrees per day.

An **operational** satellite is not allowed to perform this oscillation, since it is required to remain within its station-keeping 'box'. The necessary velocity increment (or change in velocity, ΔV) required to correct the motion is a maximum of about 2 m s^{-1} per year depending on the **nominal orbital position**.

[See also **triaxiality, libration, libration period**]

equinox Either of two annual occasions when day and night are of equal length: the vernal (spring) equinox on 21 March and the autumnal equinox on 23 September.

[See also **solstice, eclipse, solar array**]

equipment bay – see **instrument unit**

equivalent isotropic radiated power (EIRP) In **telecommunications**, the level of **transmitter** power, P, available at an **antenna** multiplied by the antenna gain, G (or added if working in decibels: EIRP(dBW) = P(dBW) + G(dB)). From the viewpoint of the **satellite**, it indicates the power (measured in dBW) of a particular combination of **transponder** and antenna (for an **earth station**, the antenna and **high power amplifier (HPA)**), less any associated losses.

The 'benchmark' for **radiated power** is the **isotropic antenna**, which radiates equally in all directions and therefore has no gain. The EIRP gives the equivalent power which would have to be radiated from an isotropic antenna to provide the same **power flux density**. Increasing the **directivity** of the antenna, and thereby the gain, decreases the transmitter power required to provide an equivalent radiated power. (For example, suppose an EIRP of 50 dBW is needed: to provide this with an isotropic antenna [0 dB gain] would require a transmitter power of 100,000 W, which is about 400 times greater than is available at present from satellite transponders. However, if an antenna of readily attainable 30 dB gain is substituted, the required transmitter power is reduced to 20 dBW or 100 W. The example demonstrates the utility of the antenna.)

[See also **gain**, **decibel (dB)**]

equivalent noise temperature An alternative term for **noise temperature**.

ESA An acronym for European Space Agency, an organisation which provides for and promotes cooperation among member states in space research and technology and its **application**s. ESA was established in May 1975 by the member countries of ESRO (Belgium, Denmark, France, Italy, Netherlands, Spain, Sweden, Switzerland, the United Kingdom and West Germany) and by Ireland (Eire). It became a legal entity in October 1980 when its convention was ratified, and gained two further members (Austria and Norway) in January 1987. Finland joined ESA in 1995 and Portugal became a member in 1999. In addition, Canada takes part in certain programmes under a cooperation agreement.

ESA combined the respective activities of **ESRO** and **ELDO**, in **satellite** and **launch vehicle** development, in a single organisation. Its main establishments are: headquarters (Paris); ESTEC (European Space Research and Technology Centre), Noordwijk, Netherlands (design, development and testing of spacecraft); ESOC (European Space Operations Centre), Darmstadt, Germany (satellite operations – **telemetry, tracking and command**); ESRIN (previously European Space Research Institute), Frascati, Italy (acquisition and distribution of satellite remote sensing data, development and operation of ESA's information systems).

[See also **Guiana Space Centre**]

escape tower A structure mounted on top of a manned **expendable launch vehicle,** containing **rocket motor**s which can be used to pull the spacecraft **capsule** clear of the launch vehicle in case of emergency. Otherwise called a 'launch escape tower' or 'launch escape system'. Escape towers were used with **Mercury, Apollo** and **Soyuz** spacecraft; **Gemini** and **Vostok** had ejection seats.

[See also **Voskhod**]

escape velocity The minimum velocity necessary for a body to escape from the **gravitational field** of a **celestial body.**

Three so-called 'space velocities' can be defined. The 'first space velocity' is that required for a body to become a **satellite**: at the Earth's surface, for example, it is about 7.9 km s^{-1} (disregarding air resistance). The 'second space velocity' is that required to escape the gravitational pull of the celestial body: about 11.2 km s^{-1} from the Earth's surface. The 'third space velocity' is that required to escape from the solar system: about 16.7 km s^{-1} from **Earth.**

ESD – see **electrostatic discharge**

ESOC – see **ESA**

ESRIN – see **ESA**

ESRO An acronym for European Space Research Organisation, a body formed in June 1962 to promote, and provide the facilities for, collaboration among West European countries in space research and technology (also known by its French acronym CERS: Conseil Européen de Recherches Spatiales). Founding countries were Belgium, Denmark, France, Italy, Netherlands, Spain, Sweden, Switzerland, the United Kingdom and West Germany. ESRO was disbanded in 1975 when it was amalgamated with ELDO (European Launcher Development Organisation) to form the European Space Agency (**ESA**).

ESTEC – see **ESA**

ET – an abbreviation for the Space Shuttle's **external tank.**

Eumetsat An abbreviation for the European Meteorological Satellite Organisation, an intergovernmental organisation established in 1986 to maintain and exploit Europe's operational **meteorological satellite**s (initially the **Meteosat** and later the Eumetsat Polar System (EPS) series), which were developed by the European Space Agency (**ESA**). Eumetsat took formal responsibility for the Meteosat system in January 1987.

[See also **NOAA**]

Europa A European three-**stage** launch vehicle proposed in the 1960s and cancelled in the early 1970s. See **Blue Streak**.

European Space Agency – see **ESA**

Eurovision The system of regular television programme and news exchange in Europe set up in 1954 by the European Broadcasting Union (**EBU**). The first Eurovision-by-satellite tests performed using **ESA**'s Orbital Test Satellite (OTS) were followed by a regular service via the **Eutelsat** spacecraft.

Eutelsat An abbreviation for the European Telecommunications Satellite Organisation, a body established provisionally in June 1977, and on a permanent basis in 1982, by the members of CEPT (European Conference of Postal and Telecommunications Administrations) to oversee and operate the **space segment** of European **telecommunications** embodied in the Eutelsat series of satellites. The first generation of Eutelsat satellites were developed by the European Space Agency (**ESA**) under the European Communications Satellite (ECS) programme.

EVA – see **extra-vehicular activity**

Evolved Expendable Launch Vehicle (EELV) The generic name for a US military launch vehicle designed to replace the **Titan** IV. Individual vehicles, from different suppliers, were named **Delta** IV and **Atlas** V.

exhaust

  (i) The **combustion** products of a **rocket motor** or **rocket engine** ejected from an exhaust **nozzle** to produce **thrust**.

  (ii) The waste products of a non-propulsive combustion process in a rocket engine, ejected from an 'exhaust pipe' (e.g. a **turbopump** turbine exhaust).

[See also **exhaust velocity, plume, plume impingement**]

exhaust nozzle – see **nozzle**

exhaust plume – see **plume**

exhaust velocity The average axial velocity of the **combustion** gases at the exit of an exhaust **nozzle**. Also called 'ejection velocity'. Exhaust velocity (measured in m s^{-1}) is equivalent to specific impulse (N s kg^{-1}) [see **specific impulse**]. **Liquid propellants** have higher exhaust velocities than **solid propellants**. [See also **characteristic velocity**]

exit cone The divergent, bell-shaped part of a rocket exhaust nozzle which converts the thermal energy of the **combustion** gases to the kinetic energy of the exhaust **plume** and controls the expansion of the plume. See **nozzle**. [See also **expansion ratio**]

exosphere – see **atmosphere**

expansion nozzle – see **nozzle**

expansion ratio

  (i) The ratio of the area of an exhaust **nozzle** exit (A_s) to the **throat** area (A_t); otherwise called the 'area expansion ratio', $\varepsilon = A_s/A_t$.

  (ii) The ratio of the pressure in a **combustion chamber** (p_o) to that at the exit of an exhaust nozzle (p_a); otherwise called the 'pressure expansion ratio', $\varepsilon = p_o/p_a$. When the mean pressure in the exit area is equal to the ambient atmospheric pressure, the expansion ratio and the nozzle with that ratio are defined as 'rated'. A nozzle with an expansion ratio lower than rated is termed 'underexpanded' and one with a ratio greater than rated is 'overexpanded'. If a **launch vehicle** first stage nozzle is rated at sea level, it will be underexpanded at altitude; a second **stage** nozzle will typically have a higher expansion ratio than a first stage nozzle.

  [See also **flow separation, plug nozzle**]

expendable launch vehicle (ELV) A **launch vehicle** which is only used once. The trajectories of ELVs tend to ensure that the discarded **stages** and components fall into a sea or an ocean upon **re-entry**, although the lower stages of some Russian vehicles fall on uninhabited areas of land. Some upper stages 'burn up' in the atmosphere to an extent, while their remains tend to fall into the ocean.

  [See also **reusable launch vehicle**]

experimental Generally, 'appertaining to experiment'. When used in the context of space-based **hardware**, the term tends to be confined to scientific **payload**s and smaller scientific **spacecraft**. An advanced technology payload carried by a **communications satellite** (e.g. a **communications payload** operating in a non-operational **frequency** band) is typically termed 'a demonstration payload' or 'technology demonstrator', rather than 'experimental', because the word 'experiment' implies a lack of knowledge of the outcome; by contrast, the intended outcome of a technology demonstration is closely specified.

expert system – see **artificial intelligence (AI)**

Explorer The name given to a series of American **unmanned spacecraft** launched in the late 1950s and early 1960s. Explorer 1, America's first satellite, was launched by a Juno I **launch vehicle** (a converted **Jupiter** C) on 31 January 1958. Its Geiger-counter **payload** detected a concentration of charged

E6: Space Shuttle **external tank** at **Kennedy Space Center**, Florida. [NASA]

particles around the **Earth** which was later named after the instrument's principal investigator, James Van Allen.

[See also **Van Allen belts**, **Vanguard**, **Sputnik**]

extension module – see **SYLDA**

external tank (ET) The **propellant tank** which carries the **liquid hydrogen** and **liquid oxygen** propellants for the **Space Shuttle main engines** (SSMEs), so-called because it is mounted externally to the Space Shuttle **orbiter** [see figure E6]. The orbiter and two **solid rocket booster**s (SRBs) are attached to the tank in the launch configuration [see figure S9]. The ET also acts as a structural 'backbone' of the Shuttle-combination by absorbing the thrust **load**s from the SSMEs and SRBs.

The ET is 47 m long and 8.4 m in diameter and contains two individual propellant tanks, one at the aft end containing about 1.5 million litres of liquid hydrogen (LH_2) and the other over 0.5 million litres of liquid oxygen (LO_2 or 'LOX') in the forward compartment. Subsequent layers of cork/epoxy and polyurethane-like foam insulation are sprayed onto the ET's exterior surface. They insulate the propellants against the higher external temperatures, protect the tank from the effects of **aerodynamic heating** during launch and prevent the formation of ice prior to launch. Apart from increasing the vehicle's weight, the ice could vibrate loose during lift-off and damage the orbiter's **thermal protection system**. Tanks for the first two missions were painted white, a cosmetic coating which was eliminated for subsequent flights thus reducing the launch-mass by about 270 kg. An even lighter version, known as the super light weight tank (SLWT) was developed in the late 1990s.

The ET is the only expendable part of the re-usable **space transportation system**, being discarded after its tanks are empty some 10 to 15 seconds after **main engine cut-off (MECO)**. The philosophy of **staging** therefore remains, with parts of the launcher being jettisoned throughout the launch-phase to reduce the mass carried into **orbit**. However, proposals have been made to carry ETs into **low Earth orbit**, where they could be converted for use as a type of orbiting **space station**.

[See also **cryogenic propellant**, **beanie cap**, **tumbling**]

extranet A limited-access adjunct to an **intranet** which typically provides secure, restricted access to that intranet (e.g. for a company's clients or subcontractors). An extranet operates between companies as opposed to within a company.

[See also **Internet**]

extra-solar planet A planet outside the solar system.

[See also **planetary body**]

extra-vehicular activity (EVA) Any activity which takes place outside a **spacecraft**, whether in space or on the surface of a **planetary body**; the opposite of **intra-vehicular activity (IVA)**. An EVA in space is colloquially termed a 'spacewalk' [see figure E7]. The first EVA was performed by the Russian **cosmonaut** Alexei Leonov on 18 March 1965 from the **Voskhod** 2 spacecraft, the second by American **astronaut** Ed White on 3 June 1965 from **Gemini** 4. The first EVA on another planetary body was performed by Neil Armstrong at 2:56 am GMT on 21 July 1969 from **Apollo** 11's lunar module.

[See also **spacesuit**]

extra-vehicular mobility unit (EMU) An alternative term for a **spacesuit**; more generally extended to include the **manned manoeuvring unit (MMU)** and emergency **life support** and rescue equipment (e.g. the **personal rescue sphere**).

extra-vehicular pressure garment – see **spacesuit**

E7: Space Shuttle astronauts attach a **grapple fixture** to an Intelsat VI **communications satellite** during an **extra-vehicular activity** (**EVA**) to repair the spacecraft. Two astronauts are secured by foot **restraints**, the other is attached to the end of the **remote manipulator system (RMS)**. [NASA]

F

f/D ratio The ratio between the focal length, f, and the diameter, D, of an **antenna**, or a lens or mirror of an optical system.

face-skin A thin sheet of material bonded to the surface of a spacecraft structural panel, **antenna** reflector, etc.; an integral part of a **honeycomb panel** or the conductive surface of a reflector, for example. Typically made from an alloy of **aluminium**, **titanium**, **carbon composite**, etc.
[See also **materials**]

facsimile A telegraphic system whereby textual or pictorial documents are scanned photoelectrically, producing a signal which can be transmitted via **transmission lines**, **microwave link**s, **satellite communications** links, etc., and reproduced at a distant receiver. Often abbreviated to 'fax'.

failure rate – see **bathtub curve**

fairing Any component part of a **launch vehicle** or **aerospace vehicle** designed to reduce **drag** by smoothing the flow of air around the vehicle. A section of a vehicle covered by a fairing is said to be 'faired in'.
[See also **payload fairing**, **skin**, **skirt**, **shroud**, **ogive**, **airframe**, **inter-stage**, **aerodynamics**]

false colour image A colloquial term for an **image** produced by an **infrared** imaging sensor or film, which, for example, indicates vegetation in red tones as opposed to green. The intensity of the colour is related to the health of the vegetation in that bright red indicates vigorous growth. Colours as perceived by the average human eye, by contrast, are termed 'true colour' or 'natural colour'.

far infrared – see **waveband**

fatigue
  (i) The weakening or deterioration of a material under load (e.g. 'metal fatigue'), especially when subjected to cyclic stresses such as vibration.
  (ii) Physical or mental exhaustion due to exertion (e.g. 'pilot fatigue').
  [See also **fracture mechanics**, **load path**, **vibration table**, **stress-corrosion cracking**]

fax A colloquial abbreviation for **facsimile**.

FCC An abbreviation for Federal Communications Commission, the regulatory body for **telecommunications** within the United States and between the USA and foreign **carriers**.

[See also **ITU**]

FDM – see **frequency division multiplexing**

FDMA – see **frequency division multiple access**

feed A device designed to illuminate the surface of an antenna reflector when transmitting an RF **signal**, and to collect radiation reflected from the antenna when receiving. The most common type is the **waveguide** termination known as a **feedhorn**, but various types of 'wire antenna' are also used [see **antenna**].

feeder link An alternative term for **uplink**.

feed(er) losses RF losses in a **feed** system – see *I²R loss*.

[See also **radio frequency (RF)**]

feedhorn A flared and open-ended termination to a **waveguide** designed to illuminate the surface of an **antenna** reflector when transmitting an RF signal, and to collect radiation reflected from the antenna when receiving. Horns can be rectangular in cross-section ('E-plane' or 'H-plane' horns), square-ended to accommodate both E- and H-plane waves ('pyramidal'), or circular. Feedhorn is often abbreviated to 'feed', although this term is sometimes used to describe the feedhorn and the length of waveguide to which it is attached; it also includes other types of **feed**.

FET An acronym for **field effect transistor**.

FGB A **module** of the **International Space Station (ISS)**, otherwise known as the functional cargo block or 'Zarya'.

fibre optics A technology which uses optical fibre for the transmission of **signals** modulated onto a **laser** beam; a communications technique using optical fibre; the field of study involving optical fibre. See **optical fibre**. [Note: functions as singular, e.g. 'fibre optics is a competitor to satellite communications'.]

field effect transistor (FET) A voltage-controlled, unipolar transistor in which the transverse application of an electric field produces amplification. Its good thermal stability makes it useful for space applications; in contrast, an ordinary junction transistor is a current-controlled bipolar device and can be thermally unstable.

[See also **GaAsFET, MMIC, solid state power amplifier (SSPA)**]

figure of merit (*G/T*) A parameter which defines the quality of a receiving installation (**antenna** and **receiver**), commonly called G/T ("G over T"), where G is the **gain** (dB) of the antenna and T is the receive system **noise temperature** (dBK). Both quantities must be referenced to the same point in the **receive chain** and are typically referred to the **feed** flange or **LNA** input. G/T is a useful parameter in that it is independent of the reference point chosen, even though the gain and system noise temperature, individually, are different at different points. The units of G/T are dB/K.

filament-winding A method of manufacturing low-density spacecraft structures, which involves winding a continuous length of resin-impregnated carbon-fibre, or similar material, around a former to produce a hollow body, such as a payload support structure or propellant tank [see figures F1, C1, R7]. Filament-winding is a finer version of '**tape-wrapping**'.

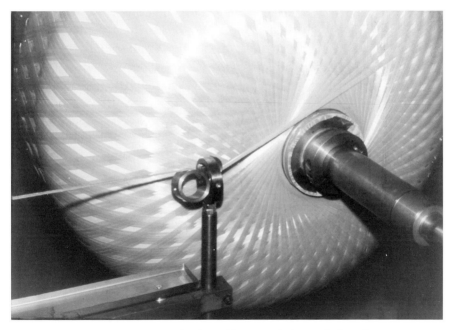

F1: Manufacturing a **propellant tank** using a **filament-winding** technique. [MAN Technologie]

filter An electrical or **microwave** device designed to allow a selected range of signal frequencies to pass, while obstructing those outside the range. The wanted signals are said to be in the 'passband'; the unwanted signals are in the 'stopband'. In addition to confining the **bandwidth** of the signals entering a communications system, a filter reduces the possibility of interference between transmitted signals (but see **mixing products** and **spurious signal**).

[See also **low-pass filter**, **high-pass filter**, **band-pass filter**, **channel filter**, **input filter**, **output filter**]

fin

(i) An aerodynamic appendage fixed to the body of a **launch vehicle** (or **aerospace vehicle**), typically at the base of the first stage (or rear of the fuselage), to provide stability during flight.

[See also **airframe**]

(ii) A projecting rib or plate which increases the effective surface area of a **radiator** and enhances heat dissipation.

[See also **heat rejection**, **heat sink**]

finite-element modelling A computer-based technique for predicting the behaviour of a structure under mechanical loads. Otherwise known as finite-element analysis. The structure is divided into a finite number of separate parts, or 'elements', small enough to be assigned realistic values for mechanical loading, etc., but not so numerous that the integration process becomes unwieldy. The technique allows the behaviour of a structure under stress to be accurately assessed at the design concept stage, which can lead to improvements in the final manufactured structure.

firing range – see **range** (iv)

firing room – see **launch control centre (LCC)**

firmware A type of **software** that is not designed to be altered or updated. This is a particularly important consideration for **spacecraft**, since firmware cannot be changed without access to the **hardware**.

first space velocity – see **escape velocity**

fixed conductance heat pipe (FCHP) – see **heat pipe**

fixed service structure – see **service structure**

fixed-satellite service (FSS) ITU terminology for a **satellite communications** system which uses **earth stations** at specified fixed points. This service is restricted to a relatively small set of user-organisations and is not directly accessible to the general public. The earth stations at one end of the link tend to be those with very large **antennas**, trunk stations or **gateway stations**, while those at the **user** level can be of any size down to about 2 m in diameter (depending on the specific parameters of the system). In the formal International Telecommunication Union (**ITU**) definition, FSS may include satellite-to-satellite links, or **inter-satellite link**s, which are included in the inter-satellite service (ISS).

[See also **broadcasting-satellite service (BSS)**, **mobile-satellite service (MSS)**]

flame bucket – see **flame deflector**

flame deflector An obstruction designed to intercept a **launch vehicle**'s exhaust gases and deflect them away from the structure of the **launch pad**, the ground and the vehicle itself. Designs vary in relation to the size and **thrust** of the vehicle, from relatively small devices fixed to the top surface of the pad to very large deflectors, sometimes called 'flame buckets', beneath the pad. The larger devices are typically deluged with water during a launch to constrain the temperature of the structure and avoid damage.

[See also **flame trench**, **sound suppression system**]

flame inhibitor A material placed between the **motor case** and the **propellant** in a solid rocket motor which protects parts of the case that might otherwise be damaged (even burned through) before propellant combustion is complete. Also known as a 'propellant liner' – see **rocket motor**.

flame trench A channel or gully beneath a **launch pad** which carries a **launch vehicle**'s exhaust products away from the pad [see figures F2, L3]. It usually contains a **flame deflector**.

flat absorber – see **surface coatings**

flat reflector – see **surface coatings**

flight

   (i) Of a **spacecraft**, any journey in space: a **spaceflight**.

  (ii) Of a **launch vehicle**, the time interval between **lift-off** and the completion of its mission – when its **payload** has been injected into an **orbit** or **trajectory**.

 (iii) Of an **aerospace vehicle**, any journey in space or the portion of its journey which takes place in the atmosphere.

[See also **launch**]

flight controller A person in a **launch control centre** with responsibility for the **flight** of a **launch vehicle** or **aerospace vehicle**.

flight deck The part of an aircraft or **manned spacecraft** from which the **flight** is controlled by a pilot and/or co-pilot. In the case of the Space Shuttle **orbiter**, the 'aft flight deck' also allows a view into the **payload bay** and houses the controls for the **remote manipulator system (RMS)**.

[See also **mid-deck**]

F2: **Flame trench** beneath Space
Shuttle **launch pad**. Also note
beanie cap above **external tank**,
white room against Shuttle orbiter,
tail service masts on **mobile launch
platform** and **rotating service
structure** at left. [NASA]

flight hardware Any equipment which leaves the surface of the Earth, or another **planetary body**, in the course of a **flight**; a **launch vehicle, aerospace vehicle, spacecraft** or any of their constituent **subsystem**s or equipment.
[See also **ground support equipment (GSE), hardware**]

flight model The version of a spacecraft built specifically for flight, as opposed to some phase of ground testing (e.g. **structure model, thermal model, engineering model**). The flight model undergoes tests slightly less severe than the other models (e.g. narrower temperature range).
[See also **thermal-vacuum chamber**]

flight path The course of a **spacecraft** or **launch vehicle** through a planetary atmosphere; also called a **trajectory**. The term 'trajectory' is also used for a body moving outside an atmosphere, whereas 'flight path' tends to be confined to atmospheric flight.
[See also **flight, glide path**]

flight readiness firing (FRF) The final test-firing of a **launch vehicle**'s main **propulsion system** prior to an actual **launch**, particularly of the **Space Shuttle main engine**s.

Such tests, which are conducted with the vehicle held down on the **launch pad**, concern only **liquid propellant** engines since they can be tested for a short period compared with the flight **burn**-duration, shut down, and then refuelled on the pad; **solid propellant** motors, once ignited, must burn for the full duration, until their propellant is exhausted, and cannot be refuelled on the pad.
[See also **static firing, test stand, hold-down arm**]

flight termination system (FTS) A **launch vehicle** system used to terminate its mission, by **command**, if there is a serious malfunction or deviation from the planned **flight path**. It typically involves destroying the launch vehicle by pyrotechnic means and is also known as a **propellant dispersal system**.

flotation bag An inflatable device on board a spacecraft re-entry **capsule** (e.g. in the nose cone of the **Apollo** spacecraft), designed to inflate on **splashdown**, right the spacecraft if it lands 'upside down' and keep it afloat until **recovery** services arrive.

flotation collar An inflatable collar attached by divers to **Gemini** and **Apollo** spacecraft after **splashdown** to keep the spacecraft upright and to stop it sinking.
[See also **flotation bag**]

flow separation In general, the detachment of the boundary of any moving fluid from the surface by which its flow is guided (e.g. the separation of the air flow from an aerodynamic surface). In **rocket** propulsion, the separation of the gas flow from the inside of an exhaust **nozzle**. If a nozzle is designed for optimum efficiency at high altitude (i.e. in vacuum or near-vacuum conditions), air pressure at lower altitudes will compress the exhaust **plume** causing flow separation, thus reducing the **expansion ratio** and the efficiency of the device. Conversely, a nozzle designed for optimum efficiency at low altitude performs less efficiently at high altitude. The **exit cone** of a **launch vehicle** nozzle is therefore specifically designed to operate with optimum efficiency over the range of altitude flown by its parent **stage**. [See also **plug nozzle**]

fly-by An alternative term for **swing-by**.

FM An abbreviation for (i) **frequency modulation** and (ii) **flight model**.

focal plane The plane perpendicular to the axis of an optical or **radio frequency (RF)** device, such as a **telescope** or **antenna** subsystem, and passing through the focal point (or focus).

focal plane assembly (FPA) A group of components positioned in the **focal plane** of an optical or **radio frequency (RF)** device.

foot restraint – see **restraint**

footprint

   (i) The area within which a spacecraft is expected to land. Usually depicted on a map as an ellipse with its major axis aligned with the direction of travel.

   (ii) The projection of a satellite's antenna **beam** on the Earth's surface. Usually depicted as a contour map of **radiated power** received at the surface, with the **peak** power at the centre of the footprint [see figure F3]. The footprint or 'beam shape' may be circular, elliptical or more complex depending on the design of the originating antenna [see **beam-forming network**]. 'Footprint' is a descriptive and inexact term and bears no mathematical relationship to the more closely defined **service area** and **coverage area**.

    [See also **beamwidth, multiple beam, equivalent isotropic radiated power (EIRP), power flux density (PFD)**]

forward contamination The contamination of a **planetary body** other than the Earth, which may occur as a result of landing a **spacecraft** on that body. Early planetary spacecraft were sterilised by heating to high temperatures to avoid

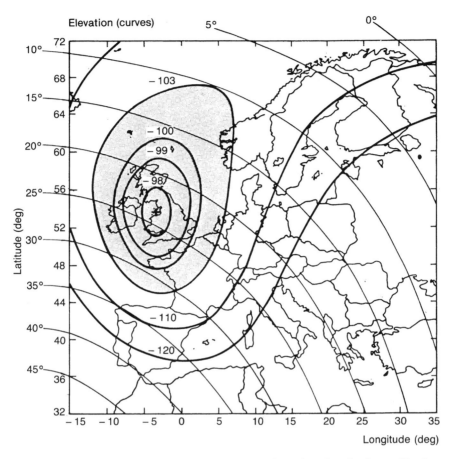

F3: **Footprint** devised at the 1977 **WARC** conference on **direct broadcasting by satellite** for the United Kingdom. The contours indicate levels of **power flux density** (**peak** is at -97.7dBW m^{-2}; **coverage area** is within the -103 dBW m^{-2} contour). [British Aerospace]

forward contamination; unfortunately, this sometimes rendered their electronic circuitry inoperative and the practice ceased.

[See also **back-contamination**, **sample return**, **quarantine facility**]

forward error correction (FEC) A process for detecting and correcting errors in a communications link without retransmission. The information is divided into blocks and labelled with 'checking information' which is recalculated after transmission and compared with the original. Any errors are corrected on this 'forward' path rather than requesting retransmission on the return path.

FOV An abbreviation for field of view (e.g. of a **sensor**).

FPA An abbreviation for **focal plane assembly**.

fracture mechanics The science of predicting the mechanisms of material failure and designing against them.

[See also **fatigue, stress (engineering)**]

frame of reference A known arrangement of planes or curves used to determine the position of a point in space. The most common frame of reference is that of Cartesian geometry, characterised by three mutually perpendicular (orthogonal) planes.

[See also **spacecraft axes, degrees of freedom**]

free fall – see **weightlessness**

free space A term used to describe a volume of 'space' which is free of anything which might perturb or affect the system under discussion: matter, electromagnetic radiation, gravitational fields, etc. Free space is therefore something of an approximation. For radio communications between an **earth station** and a **satellite** in orbit, the distance of free space is used in the calculation of signal strength (see **free space loss**).

free space loss A loss in power density of a radiated telecommunications signal due simply to the distance of **free space** between **transmitter** and **receiver**. This major source of loss is inherent in a spacecraft communications system owing to the remote nature of the spacecraft.

For example, for a simple **antenna** with some degree of **directivity**, radiated power is spread into a cone in accordance with the antenna **beamwidth** (like a huge searchlight beam). In the case of an earth station with an antenna beamwidth of 0.25 degrees, the **beam** is over 150 km in diameter at the height of **geostationary orbit** (36,000 km). The power level at the satellite is reduced by the same ratio as that between the area of the beam at that height and the surface area of the ground antenna, which is around 10^{10}:1. In these terms space loss is otherwise known as 'spreading loss', but since beamwidth is frequency-dependent, it is common (and convenient) to include a wavelength term in the calculation of free space loss.

Thus the loss in signal power in a communications link between an **earth station** and a **satellite**, due to the signal's passage through **free space**, may be calculated using the following 'rule-of-thumb' expression:

$$20 \log_{10} (\lambda / 4 \pi R)$$

where λ is the **wavelength** and R is the distance, both in metres (m), separating the satellite and the earth station (the depth of the atmosphere is usually ignored in this calculation but is taken into account in a calculation of **atmospheric attenuation**).

Freedom A name originally given to the **International Space Station (ISS)**, before it was redesigned and renamed 'Alpha', and later ISS.

free-drift strategy A satellite **station keeping** method which increases the north–south **dead-band**, allowing the satellite to drift north and south of the equatorial plane; also known as **inclined orbit** operation [see figure I1]. It reduces the **propellant** requirement for north–south station keeping.

This strategy can be applied if the **earth station**s can accommodate a large excursion in latitude by **tracking** the satellite [see **box** and figure B6]. Under the right conditions, a satellite can be given an **orbital inclination** at **beginning of life** which causes it to describe a 'figure-8' completely filling its box. The perturbing effects of **luni-solar gravity** decrease the inclination, reducing the size of the figure-8 to a minimum about halfway through the satellite **lifetime**, and then increase it until a maximum is reached at **end of life**. The **operational** variations between satellites, which may call for smaller boxes or longer lifetimes, often mean that a combination of free-drift and north–south station keeping is required.

free-flying pallet – see **pallet**

free-return trajectory A **trajectory** which takes a spacecraft out to, around and back from a **planetary body** using only the gravitational field of that body, i.e. with no energy input from the spacecraft. This trajectory was used, for example, during the **Apollo** lunar programme, for both the Apollo 8 and 13 missions.

[See also **trans-lunar trajectory**, **ballistic trajectory**]

frequency The number of oscillations per second of an electromagnetic wave. Frequency is measured in hertz (Hz) or multiples of hertz. Satellite communications frequencies are predominantly in the GHz (gigahertz: 10^9 Hz) and MHz (megahertz: 10^6 Hz) bands. See also **frequency bands**.

Relationship between frequency and **wavelength**: $f = c\lambda$, where f is frequency (Hz), λ is wavelength (m) and c is the speed of light (3×10^8 m s^{-1}).

frequency allocation The process by which the International Telecommunication Union (**ITU**) allocates either exclusive or shared **bandwidth** to telecommunications services and all other users of the **radio frequency** spectrum. For frequency allocation, the ITU has divided the world into three regions: Region 1 includes Europe, Africa, the former USSR and Mongolia; Region 2 the Americas and Greenland; and Region 3 Asia, Australasia and the Pacific [see figure I5]. Frequency allocation is also known as frequency assignment or allotment.

Although frequency allocation is a complex subject, in general particular **frequency bands** are dominated by certain services: commercial communications satellites operate at C-band, Ku-band, K-band and Ka-band; X-band is largely, though not exclusively, reserved for military satellite communications; L-band is used for the **mobile-satellite service**; and S-band is used, among other things, for satellite **telemetry** and **telecommand**. Frequency combinations such as 4/6 GHz, 12/14 GHz, 12/18 GHz, 20/30 GHz are commonly used: these refer to approximate **uplink** and **downlink** frequency bands for various satellite services. The uplink is almost always the higher figure, since the greater losses (due to **atmospheric attenuation**) at higher frequencies can be more readily overcome by higher-power earth station transmitters (cf. increasing the spacecraft **HPA** power).

frequency allotment – see **frequency allocation**

frequency assignment – see **frequency allocation**

frequency band A continuous range of frequencies between two limits; an alternative description of this range is embodied in the term **waveband**, which is based on **wavelength** as opposed to **frequency**. The **frequency bands** used by all communications services, and those protected for the benefit of astronomical observations, are allocated by the International Telecommunication Union (**ITU**) at its World Radiocommunication Conference (**WRC**) meetings.
[See also **frequency allocation**]

frequency bands The **radio spectrum** has been divided into the following bands by the International Telecommunication Union (**ITU**):

Symbol †	Frequency range †† (lower limit exclusive, upper limit inclusive)	Corresponding metric subdivision [note: not widely used]
VLF	3–30 kHz	Myriametric waves
LF	30–300 kHz	Kilometric waves
MF	300–3000 kHz	Hectometric waves
HF	3–30 MHz	Decametric waves
VHF	30–300 MHz	Metric waves
UHF	300–3000 MHz	Decimetric waves
SHF	3–30 GHz	Centimetric waves
EHF	30–300 GHz	Millimetric waves
*	300–3000 GHz	Decimillimetric waves

† Key: V, very; U, ultra; S, super; E, extra; L,low; M, medium; H, high; F, frequency.

†† prefixes are those assigned by the ITU:

k = kilo (10^3): >3 kHz, ≤ 3000 kHz;

M = mega (10^6): >3 MHz, ≤ 3000 MHz;

G = giga (10^9): >3 GHz, ≤ 3000 GHz;

T = tera (10^{12}): >3000 GHz (i.e. centimillimetric, micrometric and decimicrometric waves).

* Note that 100 GHz–1000 THz (10^{11}–10^{15} Hz) is also defined as infrared (IR); the boundary between radio and far IR is indistinct.

Satellite communications frequencies are mainly in the SHF band, but UHF and EHF are also used. For many communications applications the sub-bands are given letter-designations, for example:

P-band: 0.23–1 GHz

L-band : 1–2 GHz

S-band : 2–4 GHz

C-band : 4–8 GHz

X-band : 8–12 GHz (in USA, 8–12.5)

Ku-band: 12–18 GHz (in USA, 12.5–18)

K-band : 18–27 GHz (in USA, 18–26.5)

Ka-band: 27–40 GHz (in USA, 26.5–40)

O-band: 40–50 GHz

V-band: 50–75 GHz

These designations are intended as a guide to common usage in the satellite communications industry and should not be taken as definitive for all applications. Over the years, several frequency band standards have evolved for use in terrestrial radio, radar and spacecraft communications, often differing according to which side of the Atlantic they were devised. The above compilation is broadly based on the IEEE (Institute of Electrical and Electronic Engineers) Radar Standard 521, which also defines 40–100 GHz as the millimetre-waveband (mm-wave) and 0.3–1 GHz as UHF (at variance with the ITU definition). Readers may also come across, among others, R-band (1.7–2.6 GHz), H-band (3.95–5.85 GHz), and designations which differ in that Ku-band is shown as J-band and Ka-band as Q-band (although the range is 33–50 GHz). The designations Ku-band and Ka-band are derived from their position 'under' and 'above' K-band, respectively. Still other designations refer to U-band as 40–60 GHz and W-band as 75–110 GHz, so band designations are far from standardised throughout the world. It is also common for the 1–300 GHz band to be defined as the **microwave** band, which is part of the 'radio frequency' (RF) spectrum, itself defined as covering $3 \times 10^3 - 3 \times 10^{11}$ Hz (3 kHz–300 GHz).

[See also **frequency allocation**]

frequency coordination The process which ensures the simultaneous operation of both **satellite** and terrestrial communications systems without **interference**. The case for the frequency coordination of a satellite or earth station is usually made to the radiocommunication sector of the International Telecommunication Union (**ITU**) via the national radio regulatory administration, the local coordinating body.

The **frequency** coordination of a satellite earth station, for example, begins when a new site is proposed. The procedure entails the submission of details of **channel** frequencies, **orbital position** of the target satellite, geographical location of the intended station, a polar diagram of the antenna **radiation pattern** and other parameters such as **transmitter** power and data format. An **interference analysis** is performed to decide whether the earth station will interfere with existing communications links, and vice versa. A similar range of criteria is applied to the frequency coordination of a satellite.

[See also **frequency allocation, frequency bands, paper satellite**]

frequency division multiple access (FDMA) A coding method for information transmission between a potentially large number of users, whereby each user is allocated a relatively narrow section of the total transponder **bandwidth** (i.e. at slightly different frequencies) to which they have continual access. The satellite **transponder** can thus be accessed by a relatively large number of users, thereby making efficient use of the total available bandwidth. FDMA was the first multiple access scheme to be developed and was first used on **Intelsat** satellites. It can be used for either analogue or digital **signals**.

[See also **time division multiple access**]

frequency division multiplexing (FDM) In **telecommunications**, the process of placing more than one **signal** on a **carrier**, whereby each signal is given its own **subcarrier** frequency. Each subcarrier is modulated independently, then each modulates the main carrier [see **modulation**]. Thus the main carrier conveys all the information signals (or **channels**) at the **frequency** of operation of the **transmitter**. Frequency multiplexing allows continuous **data** transmission from all channels, unlike **time division multiplexing**.

[See also **frequency modulation**]

frequency modulation (FM) A transmission method using a modulated **carrier** wave, whereby the **frequency** of the carrier is varied, around its nominal or 'centre frequency', in accordance with the **amplitude** of the lower frequency input signal. The amplitude of the carrier remains unchanged.

FM, which came into practical use just before the Second World War, was developed to improve on **amplitude modulation (AM)** which suffers from **noise**. Practically all natural and man-made radio noise consists of amplitude disturbances which, in the radio receiver, are indistinguishable from the modulation of AM. Raising the transmission power improves the **signal-to-noise ratio**, but is costly and becomes ineffective over long distances. Frequency modulation, by contrast, is unperturbed by amplitude noise since the amplitude remains nominally constant; any amplitude modulation can be removed by means of a 'limiter' in the receiver.

Terrestrial AM broadcasting was also characterised by a lack of fidelity, but this was mainly due to the historical allocation of channel **bandwidths** too narrow to encompass the complete range of frequencies detectable by the human ear. FM removed this limitation by reproducing the full audio band (from 20 Hz to 15 kHz) at VHF where **frequency-space** for a wider bandwidth is available.

[See also **modulation, phase modulation (PM), pulse code modulation (PCM), delta modulation (DM), QPSK (quadrature phase shift keying)**]

frequency re-use The practice of using the same **radio frequency** more than once in the same satellite **beam** or **coverage area**. The aim of this practice is to make the most efficient use of the limited resource of **frequency-space**. In one form it involves the use of opposite **polarisations** (polarisation frequency re-use) [see **dual-gridded reflector**]. In another form (spatial frequency re-use) the same frequencies are used in more than one **spot beam**. A more advanced form utilises **antenna**s which transmit three or four different frequencies into adjacent spot beams arranged in a pattern repeated many times across the coverage area. A practical system of this type requires a complex switch network on the satellite to direct the communications **traffic** into the correct beam via the correct **feedhorn**.

frequency separation The allocation of different radio frequencies to communications services which, owing to their physical proximity, would otherwise interfere with each other. Essentially the same as **channel separation**. See **interference**.

[See also **radio frequency (RF)**]

frequency source – see **local oscillator**

frequency spectrum An alternative name for the **electromagnetic spectrum**; the spectrum in terms of **frequency** as opposed to **wavelength**. The frequency spectrum contains, by definition, the whole range of frequencies; the **radio frequency (RF)** part of the frequency spectrum is more confined (3 kHz–300 GHz) and is also known as the radio frequency spectrum or radio spectrum.

[See also **frequency bands, ITU**]

frequency-hopping The practice of switching a radio transmission between different radio frequencies in order to deter unauthorised reception, used mainly in military communications systems.

frequency-space In theoretical terms, the collective term for a conceptual commodity known as '**frequency**', a usable but not unlimited asset described by the **frequency spectrum**. On a more practical level the term refers to a range of frequencies (if continuous, a **band** of frequencies) in the sense of 'space in the **frequency band**'.

frisbee ejection system A method of releasing spacecraft from the **payload bay** of the Space Shuttle specifically developed for the **spin-stabilised** Leasat and Intelsat VI satellite programmes (named after the 'Frisbee' recreational product, a light plastic disc thrown with a spinning motion). The spacecraft is mounted in a U-shaped **cradle**, with its spin axis parallel to the Shuttle's longitudinal axis. An attachment point on one side of the bay acts as a pivot point and an ejection spring on the other side, when released, causes the spacecraft to rotate about the pivot. This creates a simultaneous translation and rotation about the spacecraft's centre of mass (typical values: 0.36 m s^{-1} and 2 rpm), and the spacecraft appears to roll up an imaginary ramp out of the payload bay.

front end – see **receiver**

FSS – see **fixed-satellite service**

FTS – see **flight termination system**

fuel The component of a rocket **propellant** which is burned or 'oxidised' in a chemical reaction with an **oxidiser** to produce **thrust** [e.g. **liquid hydrogen, kerosene, polybutadiene, hydrazine, MMH, UDMH**].

[See also **solid propellant, liquid propellant**]

fuel budget – see **propellant budget**

fuel cell A spacecraft power-source which generates energy by the electrochemical combination of two fluids, usually hydrogen and oxygen. The conversion of chemical energy to electrical energy is the opposite to the electrolysis of water, itself a by-product of the fuel cell reaction. Otherwise known as the 'Grove cell' after the British physicist Sir William Robert Grove [1811–1896] who invented the fuel cell in 1839.

F4: Russian-built **functional cargo block** (FGB/Zarya) from Space Shuttle STS-88, which docked US Node 1 (Unity) to the FGB, beginning the in-orbit assembly of the **International Space Station (ISS)**. [NASA]

The fuel cell has been used in American manned spacecraft since the **Gemini** programme, but the requirement for consumables limits the duration of the mission and makes the fuel cell highly unsuitable for unmanned spacecraft with **design lifetime**s of several years.

[See also **power**, **solar cell**, **radioisotope thermoelectric generator**]

fuel estimation – see **propellant budget**

functional cargo block A **module** of the **International Space Station (ISS)**, otherwise known as FGB or 'Zarya' (Russian for 'daybreak') [see figure F4]. Launched by a **Proton** on 20 November 1998, it was the first module of the ISS to reach **orbit**.

G

G/T – see **figure of merit**

GaAs The chemical designation for gallium arsenide, a **semiconductor** material used for **solar cell**s.

GaAsFET An abbreviation for gallium arsenide **field effect transistor**, a semiconductor device widely used in **satellite communications** equipment (e.g. the **solid state power amplifier (SSPA)**).

Gagarin Centre Russia's main facility for **cosmonaut** training, at Star City (northeast of Moscow). Founded in 1960, the centre received its current name in 1968.

gain The ratio of **output** power to **input** power of an **amplifier**, usually measured in decibels (dB). See **antenna gain**.

[See also **decibel (dB)**]

gaiter A flexible sealing-collar, fixed between a **rocket motor** or **rocket engine** and the surrounding structure of a **propulsion bay**, which protects the bay from the effects of heat and **exhaust** gases while allowing the **nozzle** to **gimbal** for **thrust vector** control.

Galileo An American planetary exploration **probe** launched on 18 October 1989; the first **spacecraft** to **orbit** the planet Jupiter (from 1995). On 13 July 1995, it released an **entry-probe** which entered the Jovian atmosphere on 7 December 1995.

[See also **GNSS**]

gallium arsenide A **semiconductor** material used for **solar cell**s; chemical designation GaAs.

gamma-ray spectrum – see **electromagnetic spectrum**

gantry A tower framework which allows access to a **launch vehicle** on its **launch pad**. See **service structure**.

GAS An acronym for **Getaway Special**.

gas generator A device used to drive a turbine (e.g. in the **turbopump** of a **rocket engine** or in an **auxiliary power unit (APU)**). In the **Space Shuttle** APU, for example, **hydrazine** is decomposed into a gas by passing it over a bed of a granular catalyst (similar to that of the **hydrazine thruster**). The gas is then expanded to decrease its pressure, accelerated to a high velocity and directed by a nozzle at a set of turbine blades to drive an hydraulic pump. In a rocket engine, the turbine drives a turbopump which delivers **propellant** to the engine(s).

gas jet A colloquial term for **reaction control thruster**.

gaseous oxygen arm A **swing-arm** on the fixed **service structure** at the Space Shuttle **launch pad** which supports a device to collect gas vented from the **liquid oxygen** tank within the Shuttle's **external tank**. See **beanie cap**.

[See also **vent and relief valve**]

gateway station A major **earth station** which handles incoming and outgoing international telecommunications traffic – analogous to a 'gateway airport'. Originally the definition was limited to a few **Intelsat** earth stations, characterised by their very large **antennas** (30 m in diameter or larger). Nowadays, it refers to earth stations as small as 9 m in diameter operated by any authorised **common carrier**, or those installed by commercial international satellite network operators.

[See also **Intelsat standard earth stations**]

Gaussian distribution – see **probability distribution**

Gemini A series of ten American **manned spacecraft** launched in the mid-1960s to continue research into manned orbital flight after the **Mercury** programme and prepare for the **Apollo** lunar programme. Gemini achieved this mainly by practising **rendezvous** and **dock**ing techniques, using the **Agena** target vehicle, and proving that **astronauts** could survive in space for up to two weeks as future Apollo crews would have to. The first manned mission was Gemini 3, launched 23 March 1965 (mission duration 4 h 53 m); Gemini 6 rendezvoused with Gemini 7 (15 December 1965) and the latter remained in orbit nearly 2 weeks.

The Gemini spacecraft comprised two sections: a **re-entry** module (the capsule itself) and an adapter module. Unlike Mercury, the two-man Gemini **capsule** had two ejection seats instead of an **escape tower**. The capsule was depressurised for EVA (extra-vehicular activity). The adapter module, which was tapered to fit the capsule to the **Titan** II launcher, carried service and **propulsion** subsystems. Power was supplied by **fuel cells** and **attitude control** was accomplished by **thrusters** using **MMH** and **nitrogen tetroxide**. The adapter module was jettisoned prior to re-entry to expose a **heat shield** based on a glass-fibre honeycomb and an ablative organic compound.

[See also **extra-vehicular activity (EVA)**]

generation A modifier indicating a specific level of development of **hardware** or **software** (e.g. first-generation **spacecraft**, second-generation **launch vehicle**). [See also **block**]

GEO A common abbreviation for **geostationary orbit**. It can be thought of as an acronym for 'Geostationary Earth Orbit', but this is tautologous since the prefix 'geo' refers to Earth. It is sometimes used as an acronym for Geostationary Equatorial Orbit.
[See also **geosynchronous orbit**]

Geographic Information System (GIS) An organised collection of computer hardware, software and digital geographic data designed to provide, and allow the manipulation of, geographically referenced information.

geolocation A satellite **application** which involves the accurate definition of positions on the Earth's surface.
[See also **Global Positioning System (GPS)**]

geomagnetic storm A storm in the Earth's **magnetosphere** caused by an increase in solar activity [see **solar flare**]. This type of storm has been known to cause equipment malfunctions on board **satellite**s orbiting the **Earth**.
[See also **single event effect, space radiation, space weather**]

geometric resolution – see **resolution**

geostationary arc The portion of the **geostationary orbit** 'visible' from a particular place on Earth. The observer's latitude and longitude define the length of the arc, and therefore the number of **orbital position**s visible, and its **declination**, which is important in that it governs the degree of **atmospheric attenuation**.

geostationary orbit (GEO) A **circular orbit**, of radius 42,164 km (26,200 miles) and height (above the Earth's surface) 35,786 km (22,237 miles), in the same plane as the Earth's equator [see figures G1, O1]. Commonly abbreviated to GEO [see **GEO**]; sometimes GSO (for Geostationary Satellite Orbit or simply GeoStationary Orbit). The term 'geosynchronous' is often heard in place of 'geostationary', but the latter is a sub-set of the former [see **geosynchronous orbit**].

A satellite in geostationary **orbit** has an orbital period equal to the Earth's **sidereal period** of rotation (23 h 56 m 4 s); the Earth's **synodic period** is 24 h. The satellite therefore appears stationary with respect to the Earth – hence 'geostationary' – which means it can be given an **orbital position** related to the line of longitude above which it is stationed [see figure O2].

GEO is the orbit used for most **communications satellite**s, partly because

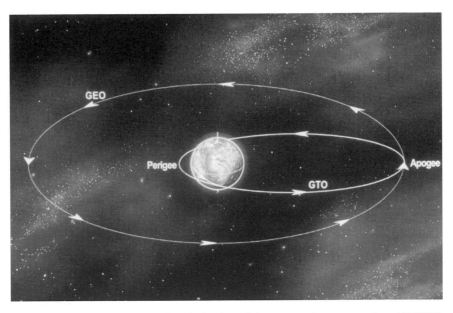

G1: Route to **geostationary orbit (GEO)** via a launch into **geostationary transfer orbit (GTO)**. Note that GEO is circular and GTO is elliptical, with an **apogee** and **perigee**. [Mark Williamson/Willis Inspace]

small **earth terminal**s, with their relatively wide **beamwidth**s, do not need to 'track' the satellite, which decreases the cost of the **ground segment**. In addition, a system of three equally spaced satellites can provide full coverage of the Earth, except for the polar regions where the **elevation angle** is very low. Although an Austrian engineer, H. Noordwig, had observed that a satellite at an altitude of 35,786 km in the equatorial plane would appear stationary, the possible application of the orbit was first brought to public attention in the October 1945 issue of '*Wireless World*' by the writer Arthur C. Clarke (in an article entitled 'Extra-Terrestrial Relays'). Geostationary orbit is thus sometimes known as the 'Clarke Orbit' or, even more colloquially, the 'Clarke belt'.

[See also **perturbations**, **Syncom**, **Early Bird**]

geostationary transfer orbit (GTO) An elliptical Earth orbit used to transfer a spacecraft from a low altitude **orbit** or flight **trajectory** to **geostationary orbit (GEO)** [see figure G1]. The orbit has its **apogee** at geostationary height (36,000 km); if a spacecraft carries a solid propellant **apogee kick motor**, it is fired when the spacecraft reaches this point to inject it into GEO. If the spacecraft utilises a **liquid apogee engine**, it is likely to be fired several times to achieve GEO [see **parasitic station acquisition**].

Expendable launch vehicles (ELVs) usually inject their **payload**s directly into GTO using their third (or fourth) **stage**, although some ELVs are specified to deliver to GEO. The Space Shuttle delivers its payloads to a **low Earth orbit** or **parking orbit**, from where an additional rocket system (a **perigee kick motor** or **upper stage**) boosts it to geostationary orbit (via a GTO). The geostationary transfer orbit is a derivative of the minimum energy **Hohmann transfer orbit**.

geosynchronous orbit An **orbit** whose period of rotation is some multiple or submultiple of the Earth's rotational period; an orbit in which the orbiting body is synchronised with the Earth. A **satellite** in such an orbit will pass over the same point on the Earth at a given time (or times) each day. For example, a satellite in an **equatorial orbit** with a period of 12 hours is synchronous in that it passes over the same point twice each day. Geostationary orbit is a special type of geosynchronous orbit [see **geostationary orbit (GEO)**].
[See also **supersynchronous orbit, sun-synchronous orbit**]

Getaway Special (GAS) A colloquial term for the small self-contained payload (SSCP), a **payload** designed to be enclosed in a cylindrical canister mounted in the **payload bay** of the Space Shuttle. The Getaway Special offers relatively inexpensive access to the space environment for non-commercial payloads, typically from school and university groups.

g-force An abbreviation for 'gravitational force', used particularly with regard to the force experienced by an **astronaut** in an accelerating or decelerating spacecraft, and by aircraft test-pilots, etc. See **gravity**.

GHz An abbreviation for gigahertz – see **hertz, frequency bands**.

gimbal

    (i) A device used to support a **gyroscope** and give its spin axis a **degree of freedom**. See **inertial platform**.
    [See also **strapdown gyro system, control moment gyro, gyrodyne**]

    (ii) A device on which the **nozzle** of a **rocket motor** or **rocket engine** (or the entire engine) can be mounted to provide angular movement in one or two dimensions to vary the direction of the **thrust vector**, thereby steering the vehicle [see figure G2].

    (NB also used as a verb: 'to gimbal')
    [See also **vector steering**]

Giotto A European Space Agency (**ESA**) spacecraft designed to investigate Halley's

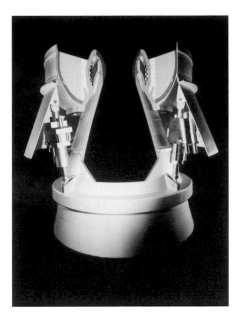

G2: Cutaway of **nozzle** of **Ariane** 5 **solid rocket booster** showing hydraulic rams which **gimbal** the nozzle for steering. [SEP]

Comet. The **encounter** occurred 13 March 1986, when the spacecraft made a close approach of 596 km $\pm$ 2 km at a relative velocity of 68 km s^{-1}. Under the Giotto Extended Mission (GEM) the spacecraft was retargeted towards Comet Grigg–Skjellerup, which it encountered on 10 July 1992, passing within 200 km of the nucleus. Giotto was returned to hibernation on 23 July 1992.

[See also **Whipple shield**]

GIS – see **Geographic Information System**

glass cockpit A cockpit of an aircraft or **spacecraft**, such as the **Space Shuttle** orbiter, which uses cathode ray tubes or other types of screen to display vehicle parameters and other data (as opposed to displaying data on individual, dedicated instruments).

Glenn Research Center – see **NASA**

glide path The course of a **spacecraft**, particularly an unpowered winged vehicle (e.g. the Space Shuttle), which generates **lift** in a planet's atmosphere to enable a controlled landing to be made. Since the path invariably subtends an angle to the planet's surface, it is also known as a 'glide slope'.

[See also **flight path**, **trajectory**, **lifting surface**, **aerospace vehicle**]

glide slope – see **glide path**

glitch Jargon: any spurious pulse experienced in the operation of an electrical circuit. The term has been extended to include any unexpected occurrence in any pre-planned sequence of events (e.g. 'a glitch in the **countdown sequence**').

[See also **spurious signal**]

global beam A **beam** formed by a **satellite** communications **antenna** which provides coverage to all parts of the globe 'visible' from the satellite. A satellite in **geostationary orbit** can communicate effectively with about one third of the Earth's surface, hence global communications satellite networks feature at least three satellites – usually stationed over the Atlantic, Pacific and Indian Oceans.

[See also **beamwidth, hemispherical beam, zone beam, spot beam, multiple beam**]

Global Positioning System (GPS) A **constellation** of **navigation satellite**s designed to determine the position of ground- and satellite-based receivers, their relative velocity, and time, with great accuracy. Although operated by the US Department of Defense and originally designed for use by the US military services, GPS has subsequently been applied to a wide range of civil applications, such as business aviation and geological surveying. Following publicity gained during the 1990 Gulf War, commercial GPS **receivers** became widely available as handheld units and were integrated into commercial and passenger vehicles. Some **spacecraft**, chiefly in **low Earth orbit (LEO)**, use GPS receivers for position fixing and **attitude control**.

The GPS constellation originally comprised 18 **Navstar** satellites (plus three **in-orbit spares**) in six different 19,150-km-altitude, 12-hour orbits inclined at 55 degrees to the equatorial plane. Eleven Block I **pre-operational** satellites were launched between 1978 and 1985, followed by nine Block II **operational** satellites (1989–1990) and 19 Block IIA spacecraft (1990–1997). GPS was declared fully operational in April 1995, once the constellation had reached 21 primary satellites and three in-orbit spares. Block IIR (replacement) satellites were launched from the late 1990s, when a Block IIF (follow-on) system was ordered.

The positioning error of a GPS receiver depends primarily on the type of signal being processed, P-code or C/A-code (precision or coarse acquisition), whether or not the signal accuracy has been degraded (known as selective availability*), and whether or not differential navigation techniques are used.

---

* Selective availability was deactivated in May 2000.

The military P-code signal gives an average accuracy of 7.5 m, the civilian C/A-code signal gives 15 m, and the degraded C/A-code signal gives an average of 50 m. With differential navigation, where GPS is used to define a receiver's position with respect to a known, fixed base station, an average of 2.5 m is achievable (with even greater accuracy – down to 0.5 cm – for receivers that remain stationary for short periods).

[See also **Glonass, GNSS, Transit**]

Glonass An acronym for global navigation satellite system; a system of **navigation satellite**s designed to determine the position of ground- and satellite-based receivers, their relative velocity, and time, with great accuracy. Although operated by and originally designed for the use of former-Soviet military forces, the system is broadly compatible with the US Global Positioning System (GPS) and can be accessed by civilian receivers with similar performance characteristics (i.e. providing accuracies corresponding to the GPS C/A-code: 50–100 m). See **Global Positioning System (GPS)**.

GMDSS An abbreviation for global maritime distress and safety system, a public service obligation of **Inmarsat**.

GMPCS An abbreviation for global mobile personal communications system, a term introduced in the mid-1990s to categorise the satellite **constellation** systems designed to offer worldwide **telecommunications** coverage to individual mobile telephone handsets. The broader term, **personal communications system**, includes services provided by conventional terrestrial means.

GNC An abbreviation for guidance, navigation and control.

[See also **inertial platform, Global Positioning System (GPS), Glonass, GNSS, control system**]

GNSS An abbreviation for global navigation satellite system, a system of **navigation satellite**s designed to determine the position of ground- and satellite-based receivers, their relative velocity, and time, with great accuracy; a civilian successor to the US military-operated **Global Positioning System (GPS)** named Galileo. The first phase of the GNSS in Europe was called EGNOS (European geostationary navigation overlay system).

[See also **WAAS**]

Goddard Space Flight Center (GSFC) – see **NASA**

Gorizont The name given to a series of Soviet/Russian geostationary **communications satellite**s.

GPS – see **Global Positioning System**

GPS rollover A potential problem for **Global Positioning System (GPS)** receivers related to the design of system software which counts weeks in modules of 1024, because it uses 10 bits of binary code for its calendar function (1024 is two to the power of ten (2^{10})). The possibility was that, when counters reset to '0000', older receivers would reset to an invalid date and the systems they supported would fail. Since the GPS system was activated at midnight on 5 January 1980, the first 'week 1023 rollover' occurred at midnight on 21 August 1999 (universal time). Most receivers had been designed to cope with the rollover, but some commercial receivers required the replacement of microprocessors. The second week 1023 rollover is due on 25 May 2019.

graceful degradation Jargon: a gradual decrease in the performance or **capacity** of a **system**, as opposed to a sudden failure.
[See also **degradation**, **soft fail**]

grain – see **propellant grain**

graphite epoxy – see **carbon composite**

grapple fixture A device attached to a spacecraft which allows it to be handled by a remote manipulator system. For example, grapple fixtures designed for use with the Space Shuttle's **remote manipulator system (RMS)** typically include a shaft with a broadened tip, which can be 'grappled' by three 'snare wires' within the manipulator arm's **end effector** [see figure R4]. They also include an arrangement of three 'guide ramps' which mate with compatible 'key-ways' in the end effector to secure the spacecraft to the RMS.

graveyard orbit A colloquial term for an **orbit** above or below **geostationary orbit** to which satellites are relocated at the end of their lives, in order to leave room for their replacements. It is common to allocate an amount of thruster **propellant**, which would otherwise be used for **station keeping**, for this activity. Spacecraft in these orbits are not controlled and therefore constitute a potential source of future **space debris**.
[See also **end of life**, **propellant budget**]

gravitational anomaly A local irregularity in the gravitational field of a **planetary body** – see **mascon**.
[See also **perturbations**]

gravitational attraction – see **gravity**

gravitational boost A technique in which the **gravitational field** of a **planetary body** is used to accelerate and re-direct a **spacecraft** on an interplanetary **trajectory**; also known as **gravity propulsion**.

gravitational constant – see **gravity**

gravitational field – see **gravity**

gravitational perturbation A disturbance (of a spacecraft **orbit** or **trajectory**) caused by the gravitational attraction of the planetary bodies. See **perturbations**.

gravity The force of attraction which exists between any two bodies. Also called 'gravitation'.

The gravitational force, F, between two bodies is proportional to the product of their masses (M and m) and inversely proportional to the square of the distance, r, between them. It is given by the expression

$$F = \frac{GMm}{r^2}$$

where G is the gravitational constant, 6.67×10^{-11} N m^2 kg^{-2}.

Gravity is an inherent property of matter. Any conglomeration of matter, however small, creates a gravitational attraction which accelerates any other body towards it – this is the acceleration due to gravity, which on **Earth** is approximately 9.8 m s^{-2} ('1g'). The conceptual 'field-of-influence' of the gravitational force is termed the 'gravitational field'.
[See also **microgravity**, **perturbations**, **gravitational anomaly**, **gravity gradient stabilisation**]

gravity gradient stabilisation A method of stabilising the attitude of a spacecraft, used predominantly in **low Earth orbit**, which makes use of the decrease in the Earth's gravitational field-strength with distance from its centre. The difference in gravitational attraction between the two ends acts to maintain the body's orientation.

Its use requires that the spacecraft body should be as long as possible with its mass concentrated about an axis aligned with the Earth's centre (the **yaw axis**). Although the body can rotate about the yaw axis, its orientation is restricted and this method of **attitude control** is useful predominantly for the simplest satellites [see **tethered satellite**]. The method can, however, be used for larger **space platforms**.

gravity propulsion A method of propulsion in which the **gravitational field** of a **planetary body** is used to accelerate and re-direct a **spacecraft** on an interplanetary **trajectory**; also known as a 'gravitational boost' or 'gravitational sling-shot' technique [see figure G3]. For example, the planet Jupiter was used to boost the **Pioneer** and **Voyager** spacecraft on their trajectories to the outer solar system.

Gregorian reflector A style of dual reflector **antenna** using a concave ellipsoidal **subreflector** and paraboloidal main reflector. The **feedhorn**, subreflector and

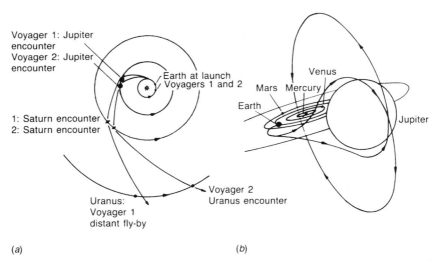

(a)   (b)

G3: **Gravity propulsion** technique used to supplement the kinetic energy of a spacecraft on an interplanetary **trajectory**: (a) **Voyager** 1 and 2 trajectories to Saturn and beyond; (b) Double sling-shot manoeuvre originally planned for two International Solar Polar Mission (ISPM) spacecraft, later reduced to one **ESA** spacecraft renamed Ulysses.

main reflector are coaxial [see figure A3]. This geometry was first described for optical astronomical telescopes in 1663 by James Gregory, a Scottish mathematician and astronomer, but owing to the difficulty of grinding glass into accurate curves, the first practical example was constructed several years later by Robert Hooke who presented it to the Royal Society.
[See also **Cassegrain reflector, Newtonian telescope, Schmidt telescope, Coudé focus, space telescope**]

ground segment The terrestrial part of a **satellite**-based communications system or any communications link with a spacecraft. The term is currently synonymous with **earth segment**, but spacecraft communications installations on any planetary surface could be termed 'ground segment'.
[See also **space segment**]

ground spare A **satellite** stored on Earth which is available to replace a defective satellite in **orbit** or enhance an existing satellite system. The value of such a spare depends on the capability to launch the satellite in time to replace a defective spacecraft.
[See also **in-orbit spare, operational, pre-operational**]

ground station An installation on the Earth comprising all the equipment necessary for communications with a **spacecraft**. See **earth station**.
[See also **ground segment, space segment**]

G4: Polar platform **payload module** for **ESA**'s Envisat **Earth observation** satellite mounted on its integration stand, part of the **ground support equipment**. The stand allows the **spacecraft** to be rotated, tilted and translated throughout the **cleanroom**. [Matra Marconi Space]

ground support equipment (GSE) Any ground-based equipment designed to support a spacecraft mission. It includes non-flight **hardware** for ground-handling, servicing, inspecting and testing **flight hardware** [see figure G4], and **software** for launch checkout and in-flight support. The definition does not include land or buildings but may include physical equipment-supports.

GSE is often divided into MGSE and EGSE (pronounced 'megsy' and 'egsey') for mechanical, and electrical ground support equipment. Other similar terms are also used (e.g. PGSE for payload GSE).

ground track The projection of the **flight path** of a vehicle on the surface of a planet, usually depicted as a line on a map. If the vehicle is in **orbit**, it is also known as an **orbital track** [see figures G5, M5].

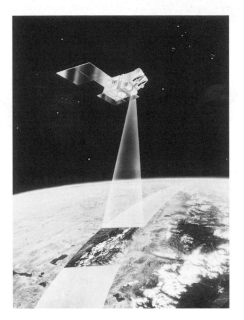

G5: **Swath** for the Vegetation **payload** of the SPOT **Earth observation** satellite, which follows the satellite's **ground track**. [Aerospatiale]

GSE – see **ground support equipment**

GSLV An abbreviation for Geosynchronous Satellite Launch Vehicle, an Indian **launch vehicle**, designed mainly to inject satellites into **geostationary transfer orbit (GTO)**. Its **payload capability** is about 2500 kg to GTO. [See also **ASLV, PSLV, Sriharikota**]

GSO – see **geostationary orbit**

GTO – see **geostationary transfer orbit**

guardband A band of frequencies separating two **channel**s to guard against **interference**. [See also **channel isolation**]

Guiana Space Centre The **launch site** of the European **Ariane** launch vehicle (at approximately 5.2° N, 52.8° W) near Kourou, French Guiana; also known by the French abbreviation CSG (Centre Spatiale Guyanais) [figure G6]. The centre was established in 1964 by the French space agency, **CNES**, to launch **sounding rocket**s and conducted its first satellite launch on 10 March 1970. It

G6: **Ariane** 4 launch from the **Guiana Space Centre**. [Arianespace]

was also used by **ELDO** (European Launcher Development Organisation) for its **Europa** launch vehicle and in 1976 was chosen by the European Space Agency (**ESA**) for Ariane.

[See also **ELA**]

guidance (inertial) – see **inertial platform**

gyro An abbreviation for **gyroscope**.

gyrodyne A magnetically suspended **control moment gyro** used on the Russian **Salyut** and **Mir** space stations as part of an inertial **attitude control** system; a type of **momentum wheel**. For example, Mir's Kvant 1 **module** incorporated six gyrodynes, mounted in **redundant** pairs on each **orthogonal axis**. Spun at 10,000rpm, they were used to orient the station without consuming **propellant** and to damp out perturbations resulting from crew activities.

gyroscope A device containing a spinning disk or 'fly-wheel' mounted on a system of **gimbals** such that it maintains a fixed orientation irrespective of the motion of the supporting structure or vehicle. The gyroscope remains fixed in **inertial space**; its **frame of reference** is the 'celestial sphere'. Gyroscopes are used for the detection of rotation relative to inertial space in both **launch vehicle**s and **spacecraft**. See **inertial platform**.

[See also **control moment gyro, gyrodyne, strapdown gyro system**]

gyroscopic stiffness The type of stability provided by a **gyroscope**, which tends to maintain the orientation of its spin axis. This property is used, among other things, to stabilise **three-axis stabilised** spacecraft – see **momentum wheel**. [See also **inertial platform**]

H

H (launch vehicle) A series of **launch vehicles** developed under the auspices of the National Space Development Agency of Japan (**NASDA**). The three-stage H-I variant, which first flew in August 1986, used **liquid oxygen (LOX)** and **kerosene** in its first stage, LOX/**liquid hydrogen (LH$_2$)** in its second and a **solid propellant (HTPB)** in its third. Its first stage **thrust** was augmented by nine solid (**CTPB**) **strap-on** boosters (the same stage and strap-ons were used for the N-II booster [see **N**]). The H-I's **payload capability** to **geostationary transfer orbit (GTO)**, the most common delivery orbit for commercial satellites, was 1100 kg, equivalent to 550 kg to **geostationary orbit (GEO)**. A second variant, the H-II (using LOX and LH$_2$ for its two stages and two large HTPB/AP/aluminium **solid rocket booster**s), was first flown in February 1994. It has a payload capability of 4000 kg to GTO and 2000 kg to GEO. An upgraded commercial variant, the H-IIA, is designed to deliver 4000 kg to GEO.
[See also **M-5**]

habitable module A part of a modular **manned spacecraft** [e.g. a **space station**] designed for human habitation, as opposed to **propellant** storage, **power** generation, etc.; a spacecraft **module** which incorporates a **life support system**.
[See also **Spacelab**, **Spacehab**]

HA-DEC mount – see **polar mount**

half-power beamwidth The width of the beam of radiation formed by a communications **antenna**, measured in degrees, between the points where the radiated **power** is half of its **peak** value. See **beamwidth**.

half-transponder In **satellite communications**, half of the frequency **bandwidth** available to the user in a **transponder**. Service providers (e.g. **Intelsat**, **Eutelsat**, etc.) sometimes lease half-transponders, since not all users require the bandwidth available in a whole transponder. For example, two analogue television signals can be transmitted through a satellite transponder by reducing the deviation and power allocated to each; half-transponder TV carriers operate typically 4–7 dB below single carrier saturation power. Where satellite systems use **digital compression**, users can lease smaller fractions of a transponder.
[See also **saturated output power**]

Hall effect thruster – see **ion engine**

halo orbit An **orbit** centred on a point in space rather than a **celestial body**. See **Lagrange point**.

hard dock – see **dock**

hard vacuum Jargon: a 'near-perfect' **vacuum**, as found in **space** (as opposed to a 'partial vacuum' produced in a laboratory).

[See also **vacuum chamber**]

hardening (of spacecraft) – see **radiation hardening**

hardware Any physical equipment. For example, **flight hardware**, **ground support equipment (GSE)**, the physical equipment comprising a computer system (cf. **software**).

harness An **assembly** of electrical wiring bundles, more fully known as a wiring harness.

head-end

 (i) The central distribution point for a **cable TV** system.

 (ii) An alternative term for **head-end unit**.

head-end unit An **amplifier** and/or associated electronics mounted on an **earth station** antenna, at the **prime focus** of a simple **paraboloid** or behind the dish in a Cassegrain or other dual reflector system [see figure L7]. An alternative term for LNA, LNB, LNC or 'outdoor unit'.

[See also **low noise amplifier (LNA)**, **low noise converter (LNC)**, **integrated receiver decoder**, **Cassegrain reflector**]

heat exchanger A device designed to transfer thermal energy from one fluid (liquid or gas) to another without allowing the two to mix; used, for example, in a **rocket engine**.

[See also **cooling system**, **heat sink**, **heat pipe**, **radiator**]

heat pipe A device designed to transfer thermal energy from one point to another; part of a **thermal control subsystem**. On board a spacecraft, heat pipes are used to maintain the temperature of heat-producing components (e.g. a **travelling wave tube (TWT)**, **multiplexer** or **battery**) and are often an integral part of the structural panels on which the items are mounted.

Although more complex than a simple **heat sink**, **heat spreader-plate** or **radiator**, the heat pipe is more efficient in transferring heat and removing 'hot-spots'. It contains a fluid which is vaporised by the applied heat at one end (the evaporator) and condensed at the other end where it relinquishes its heat [see figure H1]. The condensed liquid returns to the evaporator through a porous wick or via a system of axial grooves by means of capillary action.

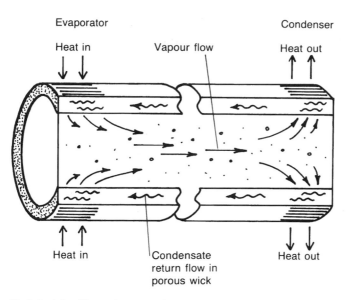

Evaporator

Condenser

Heat in   Vapour flow   Heat out

Heat in   Condensate return flow in porous wick   Heat out

H1: Principle of **heat pipe** operation.

This fixed or constant conductance heat pipe (FCHP or CCHP) has a disadvantage in that it transports heat even when it would be useful to retain it – during an **eclipse** for example. The variable conductance heat pipe (VCHP) uses a non-condensable gas (such as nitrogen) to close off part of the condenser when the heat load is reduced, thereby limiting the heat transfer area.

[See also **heat rejection**]

heat rejection The removal of thermal energy. For a spacecraft this is limited to **radiation**, since convection and conduction are not possible in a vacuum. Heat loss by evaporation is technically possible but would require excessive amounts of coolant for the long **lifetime**s expected of current spacecraft. Evaporation to the local environment would also degrade the satellite's optical surfaces and **solar cells**. However, 'closed-loop' evaporation, as used in the operation of the **heat pipe**, is non-polluting.

heat shield A coating or surface which offers protection from excessive heat, such as that experienced by a spacecraft on **re-entry** or during **aerocapture**. For example, the 'ablative material' used on the **Apollo** command module, which was subjected to temperatures of about 2700 °C, was a phenolic epoxy resin, a type of reinforced plastic [see **composite**]. The heat shield of the European Atmospheric Re-entry Demonstrator (ARD), a **technology**

demonstrator for a crew transport vehicle (CTV) launched in October 1998 by an **Ariane** 5, was a silica-based material impregnated with phenolic resin.

[See also **ablation, thermal protection system**]

heat sink The opposite of a heat source; a **passive** device for the removal of thermal energy.

The typical spacecraft heat sink is commonly known as a 'heat spreader-plate', because its primary function is to spread heat over a wide enough area for it to be radiated away. Although a heat sink may resemble a **radiator**, and they can be combined, there is a subtle difference: a radiator is primarily a **heat rejection** device, whereas a heat sink temporarily stores and then disperses its heat, tending to reject that heat by conduction (perhaps to a radiator) rather than radiation. It can act, therefore, as a kind of 'thermal capacitor': as the electrical capacitor stores and later releases electrical energy, the heat sink absorbs thermal energy, distributes it throughout its volume and eventually dissipates it into other structures. Heat sinks are, of course, particularly useful for equipment which generates large amounts of heat (e.g. some **radio frequency** components in a satellite **communications payload**). In general, metals which are good conductors are also good heat sinks, which is why, coupled with its relatively low weight, **aluminium** is widely used.

[See also **thermal control subsystem**]

heat spreader-plate An item of spacecraft hardware with the properties of a **heat sink**.

[See also **thermal control subsystem**]

heater A device which uses the heat developed when an electric current passes through an electrical resistance to maintain the temperature of an item of **spacecraft** equipment. Dependent on requirements, a heater may be on continuously, cycled on and off between a maximum and minimum temperature, controlled by a thermostat to maintain a constant temperature, or ground-controlled. This flexibility, the power of the heater and its position are all parameters in the thermal design of the spacecraft.

[See also **line heater, thermal control subsystem**]

heat-soak The phenomenon whereby the temperature of a **rocket engine** and surrounding or connected components continues to increase after the engine has ceased firing. Sometimes called 'heat soakback'. The effect is due to the 'thermal conduction inertia' of structures which have no active cooling system and is one of the factors which must be considered in the vehicle's thermal design.

[See also **cold-soak**, **thermal control subsystem**, **heat rejection**]

heavy-lift launch vehicle (HLLV) – see **heavy lift vehicle (HLV)**

heavy-lift vehicle (HLV) A **launch vehicle** capable of lifting relatively large and heavy **payload**s into **orbit**; otherwise called a heavy-lift launch vehicle (HLLV). The term is imprecise, but a vehicle that delivers a payload of five tonnes or more to **geostationary transfer orbit (GTO)** could be considered an HLV – see **payload capability**.

heliopause – see **heliosphere**

heliosphere A region of **space** which is influenced by the charged particle flux from the **Sun**. The boundary between the heliosphere and 'interstellar space' is known as the 'heliopause'. Theory states that beyond the heliopause the **solar wind** ceases, and the cosmic ray intensity is constant and greater than it is inside since it is no longer affected by the solar wind. When the interplanetary spacecraft **Voyager** 1 overtook Pioneer 10 on 17 February 1998, it was about 70 AU (**astronomical unit**s) from the Sun (some 10.4 billion kilometres from **Earth**), and estimated to be some 10 years from crossing the heliopause.

[See also **magnetosphere**]

heliosynchronous orbit A type of **polar orbit** in which the **sub-satellite point** remains approximately fixed at the same local time on Earth; an **orbit** which is synchronous with the Sun (also called 'sun-synchronous orbit') [see figures H2, O1]. This type of orbit has an **altitude** of between about 600 and 800 km and is used for Earth observation or solar study.

[See also **remote sensing**, **astronomical satellite**, **low Earth orbit**, **geostationary orbit**]

helix

(i) A type of **antenna**.

(ii) A type of **slow-wave structure**; a component of a **travelling wave tube**.

hemispherical beam A **beam** formed by a **satellite** communications **antenna** which provides coverage of a hemisphere. Although this should mean coverage of half the Earth, the term is invariably used more loosely to include a smaller but significant portion of a hemisphere. Moreover, the hemispheres in **satellite communications** are usually divided by lines of longitude rather than the equator, so that a hemispherical beam may provide coverage of North *and* South America, for example.

[See also **global beam**, **beamwidth**, **zone beam**, **spot beam**, **multiple beam**]

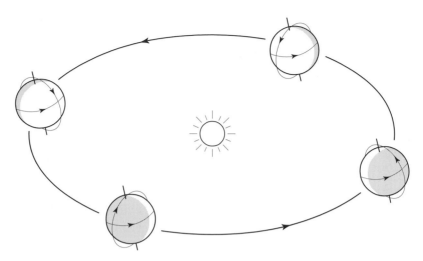

H2: A **heliosynchronous orbit**. The angle between the orbital plane and the Sun-direction is constant throughout the year.

HEO An acronym for:
  (i) **high Earth orbit**;
 (ii) **highly elliptical orbit**.

Hermes A European manned **space shuttle**, proposed by the French space agency **CNES** in the mid-1980s, but later cancelled owing to lack of funds. The shuttle was to be built and operated under the auspices of the European Space Agency (**ESA**) and launched by the **Ariane** 5 launch vehicle.

hertz (Hz) The SI unit of **frequency**; the number of cycles or oscillations per second. The most common multiples used in **satellite communications** are MHz (megahertz, 10^6 Hz) and GHz (gigahertz, 10^9 Hz); kHz (kilohertz, 10^3 Hz) is more common in terrestrial communications and when referring to **baseband** signal characteristics.

    The unit is named after Heinrich Rudolf Hertz, the German physicist [1857–1894], in honour of his experimental confirmation of Maxwell's equations, which predicted that an oscillating electrical circuit should radiate an electromagnetic wave. The waves produced became known as 'Hertzian waves'.

    [See also **frequency bands**]

heterodyning The mixing or combination of two input **signals** to produce two output signals with frequencies corresponding to the sum and the difference of the input signals. See **mixer**, **upconverter**, **downconverter**.

heterogeneous propellant – see **solid propellant**

HF (high frequency) – see **frequency bands**

HGA An abbreviation for high gain **antenna**.

high Earth orbit (HEO) A class of orbits with an altitude of at least 20,000 km; this places them above the two **Van Allen belts** where **radiation** exposure to **spacecraft** can be minimised. The most important example of a high Earth orbit is **geostationary orbit (GEO)**.

[See also **highly elliptical orbit, orbit**]

high power amplifier (HPA) A generic term for a device that amplifies electronic signals to a high power level, usually for transmission over long distances, e.g. between the Earth and a **spacecraft**. The term is sometimes used to refer to a satellite **travelling wave tube amplifier (TWTA)**, but is more often applied to an amplifier in an **earth station**.

[See also **solid state power amplifier (SSPA)**]

highly elliptical orbit (HEO) An orbit with a high eccentricity – see **elliptical orbit**.

high-pass filter A **filter** with a low-**frequency** cut-off which allows only high-frequency signals to pass.

[See also **band-pass filter**]

high-power DBS – see **direct broadcasting by satellite**

hiphet – see **hiphet thruster**

hiphet thruster A propulsive device that uses electrical heating to increase the **exhaust velocity** of the standard **hydrazine thruster**; a type of **reaction control thruster**. It is otherwise known as an electrically heated thruster (EHT), electrothermal hydrazine thruster (EHT) or power-augmented hydrazine thruster (PAHT). Between the **combustion chamber** and the **nozzle**, the combustion products pass through an electrically heated tube, which increases the temperature of the gases, and in turn increases their velocity. Although this improves performance [see **specific impulse**], the use of hiphets has an impact on a spacecraft **power budget**.

Note: the term 'hiphet' is derived from *high performance electrothermal hydrazine thruster (hipeht)*, but is more often called a hiphet because 'high-fet' is easier to pronounce.

[See also **electric propulsion, arcjet thruster**]

hi-rel A colloquial abbreviation for high **reliability**. Usage e.g. 'the **specification** of hi-rel parts'.

HLV – see **heavy lift vehicle**

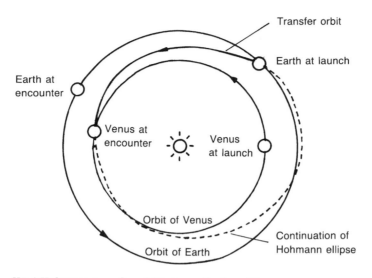

H3: A **Hohmann transfer orbit** between Earth and Venus.

Hohmann transfer orbit A **trajectory** linking the **orbits** of two planets, which can be traversed by a spacecraft using a minimum amount of **propellant** (and carrying the maximum **payload**); a 'minimum energy satellite transfer orbit', e.g. **geostationary transfer orbit (GTO)**.

The transfer ellipse, named after the German scientist and engineer Walter Hohmann, who published the theory in 1925, is also called a 'Hohmann ellipse' or 'semi-ellipse', since the vehicle follows half of an elliptical path between the two orbits [see figure H3]. The path is tangential to both orbits and has the Sun or a **planetary body** at one focus. For interplanetary trajectories, if the spacecraft is launched from Earth to a planet further from the Sun, **launch** is at **perihelion** and **encounter** is at **aphelion**; for planets closer to the Sun, launch is at aphelion, encounter at perihelion. The main drawback of the 'Hohmann transfer' is the relatively long journey-time: from Earth to Mars, for example, would take about 260 days; Saturn would be six years away. However, many planetary exploration **probes** have used the trajectory for at least the first leg of their journey, and then used **gravity propulsion**.

[See also **escape velocity, launch window**]

hold (in countdown) A period during which the **countdown** to a **launch** is halted, due either to an unforeseen problem or, with more complex vehicles, as a pre-planned measure to allow time for any procedures which have fallen behind to 'catch up', and for final checks (e.g. on the weather) to be made. Such contingency measures are known as 'built-in holds'.

[See also **recycle countdown**]

hold-down arm One of a set of devices which holds a **launch vehicle** on the **launch pad** until its **thrust** has increased to a level sufficient to ensure a successful **lift-off**. Also termed 'launcher release gear' and colloquially termed a 'clamp' [see **clamp release**]. Typically, the arms are counterbalanced to ensure that they rotate away from the launch vehicle 'under gravity', as soon as the **pyrotechnic** release charges are fired.

homogeneous propellant – see **solid propellant**

honeycomb panel A type of structural panel commonly used on spacecraft to reduce the overall mass. It is generally a 'sandwich' of two **face-skins** bonded to a cellular 'honeycombed' core with epoxy-film adhesive. The core is typically an **aluminium** or similar light-weight alloy, while the face-skins are aluminium or **titanium** alloys, or some form of **carbon composite**. Typical applications: spacecraft body panels, **antenna** reflector cores.

[See also **materials**]

horizon scanner – see **earth sensor**

horizon sensor – see **earth sensor**

horizontal polarisation – see **polarisation**

horn – see **feedhorn**

horn antenna – see **antenna**

Hotol A **single-stage-to-orbit** unmanned **aerospace vehicle** proposed in the 1980s by the UK as a next-generation satellite-launcher. Hotol is an acronym for *ho*rizontal *t*ake *o*ff and *l*anding, derived from a design requirement to take off and land on a runway like a conventional aircraft. Its **propulsion system** comprised an air-breathing engine which could also be used as a **rocket engine**, using **liquid oxygen** and **liquid hydrogen** propellants, once above the atmosphere.

[See also **Sanger, reusable launch vehicle (RLV)**]

hour-angle The angle between the observer's meridian and the meridian of the subject under observation, measured in the equatorial or geographical system of celestial coordinates. Hour-angle is measured towards the west, from 0 to 360 degrees; in astronomy, from where the term originates, it is usually expressed in units of time, where $15°$ equals one hour. See also **declination**: knowledge of a **satellite's** declination and hour-angle allows an

earth station to be aligned with it. An alternative 'horizon system' uses
altitude (or elevation) and azimuth.

[See also polar mount, altitude-azimuth mount]

housekeeping The function of a number of subsystems, on both manned and
unmanned spacecraft, involved in maintaining the spacecraft in an
operational condition (e.g. telemetry, a certain degree of attitude control
and thermal control and, in a manned spacecraft, life support).

HPA – see high power amplifier

HPBW – see half-power beamwidth

HST – see Hubble Space Telescope

HTPB An abbreviation for the solid propellant hydroxyl-terminated
polybutadiene. See polybutadiene.

hub An earth station at the centre of a 'star network' VSAT system.

Hubble Space Telescope (HST) A space-based telescope named after American
astronomer Edwin Powell Hubble [1889–1953]. Designed to be launched and
retrieved from low Earth orbit by the Space-Shuttle, the HST is an optical
telescope of the Cassegrain type, with a 2.4-m-diameter primary mirror [see
figures H4, H5, N1].

H4: Hubble Space Telescope in
cleanroom. Note mirror door at top
and folded communications
antenna beneath. [ESA]

H5: **Hubble Space Telescope** deployment from the Space Shuttle in April 1990. Note partially unfurled **solar array** and deployed communications **antennas**. [NASA]

Following the HST's launch in April 1990, a defect in its primary mirror was discovered: it had been ground incorrectly as a result of a measurement error. A servicing mission to correct the spherical aberration of the mirror was conducted in December 1993, replacing one of the HST's five main **payload**s with an optical instrument known as COSTAR (corrective optics space telescope axial replacement), which deployed five pairs of small corrective mirrors between the primary mirror and three of the remaining payloads. The HST went on to make many successful observations, including the derivation of a new value for the Hubble Constant (the constant of proportionality between the recession speeds of galaxies and their distances from each other) and thus an improved estimate of the age of the Universe.

hybrid A **microwave** device used to connect more than one **input** path to more than one **output** path, in a satellite **transponder** for example. A hybrid is generally located in a similar manner to a switch: the difference is that a switch position has to be 'selected' (by ground **command**), while a hybrid is a **passive** device which links several input and output paths 'simultaneously'.

hybrid rocket A rocket which uses one **propellant** in the solid phase and another in the liquid phase [see **solid propellant, liquid propellant**]. Although hybrids have not been used extensively, the most common combination is a solid **fuel** and a liquid **oxidiser** (e.g. **polybutadiene** and **liquid oxygen**), the latter being sprayed onto the former to enable **combustion**. The hybrid generally has a higher **performance** than a solid motor and is less complex than a liquid engine, although it retains the cut-off/re-start capability of the latter [see **engine re-start**].

hybrid satellite A largely archaic term (used mainly in the United States) for a **communications satellite** with separate **communications payload**s operating in different **frequency bands** (e.g. C-band and Ku-band). Not to be confused with the multi-purpose satellite, which carries payloads for different applications (e.g. **communications** and meteorology).

hydrazine (N_2H_4) A storable liquid **fuel** used in **rocket engines**. See **liquid propellant**.
[See also **hydrazine thruster, hiphet thruster, monopropellant, MMH, UDMH**]

hydrazine thruster A propulsive device that uses the chemical compound **hydrazine (N_2H_4)** as a **propellant**; a type of **reaction control thruster**. The propellant is stored as a liquid and usually fed to the thruster by a **pressurant** such as gaseous nitrogen or helium. It is decomposed to ammonia, nitrogen and hydrogen with the aid of a catalyst in a **combustion chamber** and the hot gases are ejected from the **nozzle** in a tiny jet (hence the colloquial term for reaction thrusters, 'gas jets'). The hydrazine thruster can deliver its **thrust** in short or long bursts, or in pulses as short as a few milliseconds.

[See also **pulsed thrust, hiphet thruster, arcjet thruster, cold-gas thruster, bipropellant thruster, gas generator**]

hydrogen peroxide (H_2O_2) A storable liquid **oxidiser** used in **rocket engines**. See **liquid propellant**.

hyperbola A conic section formed by a plane that cuts a cone at a steeper angle to its base than its side; strictly speaking, the curve consists of two branches asymptotic to two intersecting fixed lines.

[See also **conic sections**]

hyperboloid A geometric surface consisting of one sheet, or of two sheets separated by a finite distance, whose sections parallel to the three coordinate planes are **hyperbolas** or ellipses.

[See also **conic sections**]

hypergolic An attribute of a **bipropellant** combination (**fuel** and **oxidiser**) which ignites spontaneously when mixed (e.g. **monomethyl hydrazine** (fuel) and **nitrogen tetroxide** (oxidiser)). Propulsion systems using hypergolic propellants are simplified by the lack of an **ignition system**.

[See also **pyrophoric fuel, NHMF**]

hypersonic flow – see **Mach number**

hyperspectral Composed of many parts of the spectrum (typically more than ten); images composed of ten or less spectral bands are known as **multispectral**. In **remote sensing**, the term is used to describe the response of a film or **sensor** that produces an **image** derived from distinct and closely specified parts of the spectrum. Hyperspectral images can be used to identify characteristics of surfaces, vegetation, etc., to a more accurate degree than multispectral images.

[See also **waveband, electromagnetic spectrum, spectral resolution**]

hypoxia A deficiency in the amount of oxygen in the blood and body tissues. Hypoxia becomes evident at an altitude of about 3.5 km above the surface of the Earth, when the alveoli of the lungs experience difficulty in absorbing oxygen at the ambient pressure. A complete lack of oxygen, experienced at about 16 km, is called 'anoxia': this occurs when the ambient pressure drops to about 11.6 kPa (1.7 psi or 87 mmHg), the pressure maintained within the alveoli, at which point there can be no transfer of oxygen across the alveoli membrane.

[See also **decompression sickness, ebullism**]

I

I^2R **loss** The loss of **power** in a **transmission line** due to heating. The mathematical symbology (*I* stands for current and *R* for resistance) derives from Joule's Law for the production of heat in a conductor, otherwise known as 'the Joule effect' or 'Joule heating'. The effect is particularly important for long transmission lines: e.g. the **waveguide** runs in large **earth station**s, between the **feedhorn** and the **high power amplifier** (HPA), are generally kept to a minimum by mounting the feedhorn and HPA close together.

IAA An abbreviation for the International Academy of Astronautics, a scientific institution founded by the **IAF** in 1960 to foster the development of astronautics for peaceful purposes. Its four sections – basic sciences, engineering sciences, life sciences and social sciences – organise specialist symposia at IAF and **COSPAR** congresses. Among other things, the Academy publishes the journal *Acta Astronautica* and has developed a multilingual space dictionary.

IAF An abbreviation for the International Astronautical Federation, a body founded in 1949 to develop **astronautics** for peaceful purposes by means of public education, the dissemination of technical information, and the organisation of congresses and scientific meetings. In 1960 it created the cooperative but autonomous International Academy of Astronautics (**IAA**) and International Institute of Space Law (**IISL**).

ICBM An abbreviation for intercontinental ballistic missile, sometimes defined as a missile with a **range** greater than about 5000 km. Smaller missiles (e.g. **Blue Streak**), having a range between about 500 km and 5000 km, may otherwise be termed intermediate range ballistic missiles (IRBMs), though the distinction is inexact. Many early American ICBM boosters were adapted for use in the space programme (e.g. the **Atlas** booster used to launch the **Mercury** spacecraft and the **Titan** II used for **Gemini**). In the late 1980s the Atlas and the Titan III became commercial launchers of satellite payloads.

Disarmament negotiations conducted between the superpowers in the 1990s, resulting in the Strategic Arms Reduction Treaty (START), led to a number of programmes for the conversion of ICBMs into commercial launchers. See, for example, **Rockot, Dnepr, Start, Cosmos**.

[See also **ballistic trajectory**]

ICM An abbreviation for Interim Control Module, a **module** of the **International Space Station (ISS)**.

ICO An acronym for **intermediate circular orbit**.

iconoscope – see **vidicon**

IF – see **intermediate frequency**

IF amplifier A device which amplifies a signal at an **intermediate frequency** (**IF**).

IFOV An abbreviation for instantaneous field of view (e.g. of a **sensor**).

IFRB An abbreviation for the International Frequency Registration Board, a body established within the International Telecommunication Union (ITU) in 1947 to manage the **radio spectrum** and record assignments of the limited resources of radio frequencies and **orbital position**s (for use by **communications satellites**, etc.). On 1 March 1993, the ITU's radiocommunication sector took responsibility for the functions of the IFRB and other radiocommunications matters [see **ITU**].

igloo – see **Spacelab**

igniter A device in a **rocket engine** or **rocket motor** which initiates **propellant** combustion. See **ignition system**.
 [See also **percussion primer**]

ignition The act or process of initiating **combustion** by igniting the **propellant**(s) in a **rocket motor** or **rocket engine**.
 [See also **ignition system**]

ignition system The part of a propulsion system which initiates **combustion** in a **rocket engine** or **rocket motor**. Ignition systems may utilise **pyrotechnics**, spark-ignition or electrically heated 'hot-wire' devices. For the typical **solid propellant** motor, a **power supply** sends a current pulse to a pyrotechnic cartridge which ignites a small sample of the propellant in a steel or glass-fibre housing. This produces a controlled amount of hot gas which, in a similar way to a detonator in an explosive device, ignites the main motor.
 In the case of a **liquid propellant** engine, the propellants [**fuel** and **oxidiser**] are supplied either by a pressurising gas [**pressurant**] or by mechanical **turbopumps** and are injected into a **combustion chamber**, atomised and mixed. Initially a small amount of fuel meets the oxidiser and a 'pilot flame', produced by the ignition system, heats the liquids to vaporisation point. If the propellants are not self-igniting [**hypergolic**], one of the above devices is used to produce the heat required for evaporation, ignition and final combustion (if they are hypergolic, there is no need for an ignition system). After ignition the supply valves are opened fully and

combustion temperatures and pressures increase rapidly to produce a high-velocity exhaust.

[See also **nozzle, exhaust velocity, percussion primer**]

IGY An abbreviation for International Geophysical Year, an international scientific programme conducted jointly by 54 nations between July 1957 and December 1958. Its principal task was the comprehensive study of the influence of solar activity on the various phenomena observed in the **atmosphere, ionosphere** and near-**space**. Thus the time selected for the IGY corresponded to a period of maximum solar activity. Particularly useful were the first artificial **satellites** carrying scientific instruments: e.g. analysis of **data** returned from America's first satellite, **Explorer** 1, led to the identification of the **Van Allen belts**. IGY studies continued into 1958–59 were referred to as the International Geophysical Cooperation Year (IGCY). A similar programme to the IGY, known as the International Quiet Sun Year (IQSY), was carried out during the 1964–65 solar-minimum period.

IISL An abbreviation for the International Institute of Space Law, a body founded by the **IAF** in 1960 to replace the former Committee on Space Law, itself established by the IAF in 1958. The IISL organises an annual colloquium on the law of outer space at the IAF congress, which addresses a wide variety of legal aspects concerning the peaceful uses of outer space.

image A reproduction of a physical object or scene formed by a lens, mirror, or some other focusing system, projected onto an imaging device and displayed using a medium such as photographic film or **television**. The 'reproduction' is referred to as an image at all stages of the process: it may be the image formed by a **telescope** sensitive to any **wavelength** of radiation (X-ray, radio, etc., as well as optical); it may be the image on the photosensitive imaging tube of an electronic camera or the retina of the eye; or it may be the image on a screen, a piece of paper, or any other 'output device'. The process of obtaining 'an image' is generally known as 'imaging'.

[See also **pixel, resolution, imaging team, false colour image**]

image enhancement The process concerned with increasing the 'clarity' of an **image**; the process of making the information contained within an image more readily accessible, thereby increasing the image's intelligibility. At its simplest, this can be engineered by increasing the contrast (by converting the lowest grey-values to black and the highest values to white), or by enhancing the **signal-to-noise ratio** (which allows the signal to stand out against the **background noise**). Advanced forms of the technique, often referred to as computer enhancement, are used to improve the **raw data** received from spacecraft imaging systems to produce more intelligible images.

[See also **resolution**]

image intensification The process concerned with increasing the intensity of an **image**, using an image intensifier.

[See also **resolution**]

imaging – see **image**

imaging radar A **radar** used to produce an **image** as opposed to simply detecting an object and providing **range** information. See **synthetic aperture radar (SAR)**.

[See also **side-looking radar, scatterometer, radar altimeter**]

imaging team The collective term commonly used for a group of scientists and engineers concerned with the reception and interpretation of **images** received from a planetary exploration **probe** or other similar spacecraft.

impulse A product of the force (or **thrust**) acting upon a body and the time for which it acts (measured in N s). Used in connection with **propulsion system**s: e.g. 'the impulse of a **rocket engine**'.

[See also **specific impulse**]

Imux A colloquial abbreviation for input multiplexer. See **multiplexer**.

inclination – see **orbital inclination**

[See also **declination, elevation (angle)**]

inclined orbit In general, any orbit whose plane is inclined to the equatorial plane of the orbited body. Specifically, geostationary satellites which can no longer perform north–south station keeping manoeuvres are said to be operating in 'inclined orbit'; this method of operation is also known as a **free-drift strategy**. A satellite in an inclined **geosynchronous orbit** describes a 'figure-8' as seen by a ground-based observer [see figure I1].

[See also **orbital control, geostationary orbit (GEO)**]

indoor unit A colloquial term for a package of **receiver** electronics connected between a **head-end** or **outdoor unit** and a **television** set in a direct broadcasting system.

[See also **direct broadcasting by satellite (DBS), direct to home (DTH)**]

inertia wheel – see **momentum wheel** and **reaction wheel**

inertial frame In physics, an unaccelerated reference frame (e.g. the 'fixed stars'). The Foucault Pendulum experiment, first performed publicly in 1851 using a mass on a wire nearly 70 m long, demonstrates that the Earth is a rotating, non-inertial frame: the plane of motion of the pendulum, which is free to

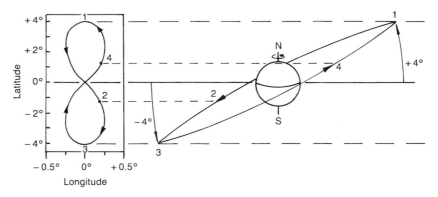

I1: Apparent ('figure-8') daily motion of a **satellite** in an **inclined orbit** as seen by a ground-based observer and its relationship to a 4°-inclination **geosynchronous orbit**.

rotate in any plane, remains fixed in an inertial frame while the Earth rotates beneath it.

[See also **inertial platform, frame of reference**]

inertial guidance – see **inertial platform**

inertial platform A device which uses a set of **gyroscopes** to define a **frame of reference** in a moving vehicle (e.g. a **launch vehicle**), with respect to which the vehicle's **attitude** can be determined and controlled; a platform stable with respect to inertial space [see **inertial frame**]. Sometimes called an inertial reference unit (IRU).

   An inertial platform typically comprises three gyros and three accelerometers mounted on gimbals which isolate them from the vehicle. The gyros are set spinning with their axes in the three orthogonal directions which, however the vehicle moves, will remain fixed in inertial space. Any movement of the gimbal frames around the gyros will be detected and the signal thus generated used to drive the gimbal motors to hold the gyro cluster in a fixed orientation. The stable gyro-platform provides the accelerometers with a reference. Any acceleration imposed upon the platform is detected by the accelerometers, the outputs from which are processed to calculate the vehicle's velocity and position. The actual **trajectory** is compared with a theoretical trajectory and the vehicle **autopilot** corrects any errors. This is known as inertial guidance or inertial navigation.

inertial reference unit (IRU) Another name for an **inertial platform**.

Inertial Upper Stage (IUS) A particular type of **upper stage**, used on the Space Shuttle and **Titan** expendable launch vehicle (ELV), which combines the functions of **perigee kick motor** and **apogee kick motor**. The IUS comprises two **solid propellant** stages, both of which separate from the **spacecraft** after firing (i.e. a spacecraft launched using an IUS has no integral apogee motor). The IUS also carries its own guidance and control **avionics**: it is **three-axis stabilised** and inertially navigated [see **inertial platform**]. Its **payload** capability when used on a Titan 34D was 1870 kg to **geostationary orbit (GEO)**, 2270 kg when used with the Shuttle. Its first flights on Titan and Shuttle were in October 1982 and April 1983, respectively.

[See also **payload assist module (PAM), Centaur**]

infrared – see **waveband**

infrared sensor A device used to establish the **attitude** or **pointing** of a spacecraft or **launch vehicle** relative to the Earth by detection of the Earth's infrared **radiation**; also called an **earth sensor**.

[See also **sensor**]

ingress An entrance; the act of entering. A word commonly used as a modifier (e.g. 'ingress hatch', 'ingress port'). The opposite of **egress**.

inhibitor – see **flame inhibitor**

injection

(i) The process of placing a spacecraft in **orbit** – orbital injection or insertion.

(ii) The time following **launch** when a vehicle is following a **ballistic trajectory**.

(iii) The introduction of **propellant**, etc., into an engine, **combustion chamber**, etc.

Inmarsat An organisation, formerly known as the International Maritime Satellite Organisation, established to provide satellite services to mobile receivers. In April 1999, it was restructured to form a two-tier private company (Inmarsat Holdings and Inmarsat Ltd.), and an intergovernmental body called the International Mobile Satellite Organisation (IMSO) to oversee the organisation's public service obligations, including the global maritime distress and safety system (GMDSS).

The original Inmarsat was established in July 1979 (following the adoption of its convention in September 1976) to develop and promote a global maritime **satellite communications** system. It began operations in February 1982 when it took control of satellites previously operated by 'Marisat' (an American joint venture) and expanded the network by leasing satellites from the European Space Agency (**ESA**) and **Intelsat**. Even before privatisation,

Inmarsat was known as the International Mobile Satellite Organisation in recognition of the fact that its operations had expanded beyond the original maritime remit.

in-orbit delivery – see **delivery in orbit**

in-orbit mass The mass of a **satellite** once it has reached its **operational** orbit. A satellite's in-orbit mass is generally less than its **launch mass**, because an amount of on-board **propellant** may be used to attain the final **orbit**. [See also **dry mass, mass budget, combined (bipropellant) propulsion system**]

in-orbit spare A **satellite** in **orbit** which is available to replace a defective satellite or enhance an existing satellite system. It usually remains inactive until needed, but it may be incorporated into the system on a part-time or part-**capacity** basis. The provision of an in-orbit spare can change the status of a satellite system from **pre-operational** to **operational**, since it offers a guarantee that the service will continue if the prime satellite fails. [See also **ground spare**]

input

(i) The entry point or path through which information or energy is applied to a device or system (refers to the hardware).

(ii) The information or energy (a **signal, carrier, RF wave**, etc.) fed into an electronic or **microwave** circuit or component.

[See also **input section, input filter, data**]

input filter A **filter** at the input end of a **receive chain** in a communications **transponder** which confines the signal to its appointed **bandwidth**. This type of device is known generically as a **band-pass filter** and is often installed as part of an input **multiplexer** (Imux). [See also **output filter**]

input multiplexer – see **multiplexer**

input section The section of a spacecraft **communications payload** from the point where the input signal enters from the receive antenna to the point prior to its entry to the **travelling wave tube amplifier**. The division of payload components into an input section and an **output section** is an alternative to their classification as part of a **receive chain** or a **transmit chain**.

insertion – see **injection**

insolation The quantity of solar **radiation** incident on a **spacecraft** or **planetary body** (measured in $W\,m^{-2}$). Outside a planetary **atmosphere** it is equal to the **solar constant**.

instrument unit The part of a **launch vehicle** containing the guidance and control equipment, **telemetry** transmitter and other electronic devices. Also called instrument bay, vehicle equipment bay, etc. On contemporary three-stage vehicles, for example, the instrument unit is an integral part of the third stage, the last stage to be ignited.

[See also **autopilot**]

insulation – see **thermal insulation**

integrated receiver decoder (IRD) An electronics unit, containing both **receiver** and decoder, connected to an **earth terminal** or **VSAT** installation. It receives a modulated **carrier**, demodulates the carrier, performs error correction as required and demodulates the **data** streams if applicable.

[See also **decryption, descrambling, descrambler, modulation, forward error correction (FEC), head-end unit**]

integrated services digital network (ISDN) An international network (operated by the **PTTs**) for the provision of telecommunications services. The internationally agreed definition of ISDN is 'a network evolved from the telephony IDN (integrated digital network) that provides end-to-end digital connectivity to support a wide range of services, including voice and non-voice services, to which the users have access by a limited set of standard multipurpose customer interfaces'. From the customer's point of view, ISDN offers access to a modern, digital network (System X in the UK) capable of handling both **data** and telephony transmissions.

[See also **digital compression, bit rate**]

integration – see **spacecraft integration, payload integration**

Intelsat The International Telecommunications Satellite organisation, created in August 1964 (originally as 'The Provisional Committee for Satellite Telecommunications') to develop the first worldwide satellite communications network. It offers **telecommunications** services, to all parts of the world except the polar regions, using **satellite**s in **geostationary orbit**. Its first satellite, Intelsat I, was nicknamed **Early Bird**.

[See also **Intelsat standard earth stations, Intersputnik, Inmarsat, space insurance**]

Intelsat standard earth stations Earth stations authorised by **Intelsat** to operate within the Intelsat system. There are three standards for international **gateway station**s:

- Standard A: originally **antenna**s 30 m in diameter or larger, operating in C-band (later 15–18 m owing to increased satellite power);

- Standard B: antennas 11–13 m in diameter, operating in C-band (for lower **traffic** demands than Standard A);
- Standard C: originally antennas 14–19 m (later 11–14 m) in diameter, operating in Ku-band.

There are also a number of standards for smaller **earth station**s:

- Standard D1: 4.5–6 m; C-band (for Vista low-density **SCPC** telephony service)
- Standard D2: 11 m; C-band and Ku-band (for Vista)
- Standard E1: 3.5–4.5 m; Ku-band (for Intelsat Business Service (IBS), etc.)
- Standard E2: 5.5–7 m; Ku-band (for IBS, etc.)
- Standard E3: 8–10 m; Ku-band (for IBS, etc.)
- Standard F1: 4.5–7 m; C-band (for IBS)
- Standard F2: 7–8 m; C-band (for IBS, etc.)
- Standard F3: 9–10 m; C-band (for international voice and data, including IBS)
- Standard G: all sizes; C- and Ku-band
- Standard Gx: <4.5 m C-band; <3.7 m Ku-band

(G and Gx are for international and domestic lease service).

[See also **frequency bands**]

intentional ignition A term used in **space insurance** to define the moment of **ignition** of a **launch vehicle**'s first-stage **rocket engine**(s) or **rocket motor**(s), as designated by the vehicle's operator. For example, this may be the moment an electronic **signal** is transmitted to command the opening of the first-stage propellant **valve**s.

intercontinental ballistic missile – see **ICBM**

Intercosmos A system for collaboration in **space science** and **astronautics** between the former Soviet Union and other nations which developed following the launch of the first **Sputnik** satellites in 1957. Agreements made under the Intercosmos banner have included manned spaceflights by members of several Eastern Bloc countries, and by French **cosmonaut**s, following a Franco-Soviet agreement on cooperation in space science signed in June 1966.

interface filler A material placed between two surfaces to improve the thermal conductivity between them. Also called 'thermal grease'.

interference In **telecommunications**, a term applied to the undesirable superposition of two or more **radio frequency (RF)** signals in a **receiver**. In a satellite **communications system**, there are three main factors used to reduce interference: the allocation of different frequencies (**frequency**

separation); the use of opposite **polarisation**s; and the angular separation between satellites in different **orbital positions** on the **uplink** and that between **earth station**s in different geographical areas on the **downlink**. [See also **multipath, frequency coordination, interference analysis, protection ratio**]

interference analysis A comparison of radio communication parameters, such as **carrier** powers at the **input** to a **receiver**, which concludes whether or not a particular service will suffer radio **interference** from other services. If the power of the wanted carrier is sufficiently greater than that of the unwanted carriers, there will be no interference.
[See also **isolation**]

Interim Control Module (ICM) A **module** of the **International Space Station (ISS)**.

interlayer A layer of a material interposed between two others. For example, **Nomex** and **Dacron** are used as interlayers in spacecraft **multi-layer insulation (MLI)**.
[See also **thermal insulation**]

intermediate circular orbit (ICO) A specific type of **medium Earth orbit (MEO)**, a class of orbits with an altitude of approximately 10,000 km; often used as a synonym for MEO, but the MEO classification is not restricted to circular orbits.
[See also **orbit**]

intermediate frequency (IF) A **frequency** used in a communications **transponder**, or other receive/transmit device, between the input and the output, or other specified points. In a satellite communications link the **uplink** and **downlink** frequencies must be different to obviate interference: the uplink frequency is generally the higher of the two, since the greater (**atmospheric attenuation**) losses at higher frequencies can be more readily overcome by higher-power earth station transmitters (cf. increasing the spacecraft power). In the so-called 'dual-conversion' transponder, the intermediate frequency is the **output** of the **downconverter** and the **input** to the **upconverter**. A typical example for C-band might be: uplink 6 GHz; IF 750 MHz; downlink 4 GHz.

It is possible to convert directly from the received uplink frequency to the transmitted downlink frequency (so-called 'single-conversion'), but a signal at IF is easier to handle than the high-power, high-frequency **carrier** wave required for long-distance communication. The technology for amplification, filtering and switching is simpler at the lower frequencies, and the hardware is easier to produce and therefore less expensive. In addition, the lower frequencies can be conducted by **coaxial cable** which, since it is lighter than **waveguide**, is an important factor for the mass-limited spacecraft **payload**.

International Space Station (ISS) A **space station** based on an international collaboration between the USA, Europe, Japan, Canada and Russia, all of which are expected to provide **modules** and/or equipment required to operate the station for at least 15 years. Although it is being assembled at an orbital **altitude** of about 380 km, its **operational** orbit, attained late in the assembly sequence, is expected to have an altitude of 426 km and an inclination of 51.6°. The station's mass at completion is expected to be about 420 tonnes, offering a total pressurised volume of some 900 m³, and its overall dimensions will be approximately 110 m by 60 m [see figure I2]. The power generated by the station's **solar arrays** will total some 110 kW at completion.

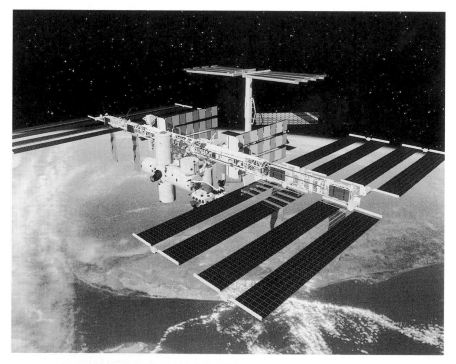

I2: **International Space Station (ISS)** as expected to appear at completion. Note **solar arrays** at either end of main truss (and at top) and **deployable radiator** panels at right angles to arrays. [NASA]

US President Reagan directed **NASA** to build a space station in 1984, but redesigns and cost overruns delayed the beginning of orbital assembly until November 1998. Over that period, its original name, 'Freedom', was changed first to 'Alpha' and then to ISS. The first module to reach orbit, on 20 November 1998, was 'Zarya', formally known as the **functional cargo block** (or, from its Russian name, FGB) [see figure F4]. It is a self-contained **spacecraft** – a **power**, propulsion and **orbital control** module which is pressurised and thus provides habitable accommodation. The second module, a connection node called 'Node 1' or 'Unity', was docked to Zarya on 7 December 1998.

Other modules include:

- the Russian-built Service Module, 'Zvezda', a habitable **command** and control centre [see figure Z1], docked to the station on 26 July 2000
- the US-supplied Interim Control Module (ICM), built initially as a back-up for the Service Module
- the **ESA**-supplied Columbus Orbital Facility (COF), a general-purpose science and technology laboratory
- the US laboratory module
- the US habitation module
- the Japanese Experiment Module (JEM) supplied by **NASDA** [see figure J1]
- the Multipurpose Pressurised Logistics Module (MPLM), supplied by the Italian Space Agency (ASI).
- the ESA-supplied Cupola observation module
- the Canadian **remote manipulator system**.

[See also **nodes**, **CTV**, **CRV**]

Internet A global network of **computer** systems and personal computers interconnected by the global telephone network and used, in general, for **data** transfer and **communications**. Internet services became an important **satellite communications** application in the late 1990s.

[See also **IP**, **TCP**, **ISP**, **intranet**, **extranet**, **multicasting**, **WWW**]

inter-orbit link (IOL) A **communications** link between spacecraft in different **orbit**s, e.g. between a communications or data-relay satellite in **geostationary orbit** and a **remote sensing** spacecraft or **space station** in **low Earth orbit** [see figures I3, T4]. Sometimes called (less formally) a crosslink, although this term does not differentiate between inter-orbit links and inter-satellite links.

[See also **inter-satellite link (ISL)**]

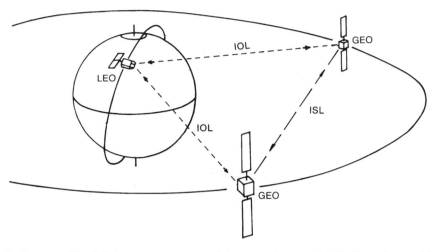

I3: **Inter-satellite link** between two spacecraft in **geostationary orbit (GEO)** and **inter-orbit links** between GEO and **low Earth orbit (LEO)**.

interplanetary Between the planets. Generally used to refer to **spacecraft** which travel between the planets (colloquially termed interplanetary **probes**), or to the space between the planets (e.g. interplanetary space, interplanetary medium).

inter-satellite link (ISL) A **communications** link between spacecraft in **orbit**, sometimes called (less formally) a crosslink, although this term does not differentiate between inter-satellite links and inter-orbit links. The term inter-satellite link is generally reserved for spacecraft in the same orbit. When the link is between two spacecraft in different orbits, say geostationary orbit and a **low Earth orbit**, it is termed an **inter-orbit link (IOL)** [see figure I3].

ISLs utilise different radio frequencies from the fixed, mobile and broadcasting services and may use laser links. The link obviates the need for double-hopping, the transmission of a signal up to a satellite and down to an **earth station** twice in succession in order to span a large distance around the Earth [see **double hop**]. Alternatively ISLs can be used to link two or more small satellites in relatively closely-spaced **orbital position**s (a 'local ISL'), so that the cluster acts as a large-capacity satellite. ISLs can also reduce the number of earth stations needed, by allowing the user to point a single earth station at any one satellite and communicate with the others through the ISLs.

[See also **fixed-satellite service, mobile-satellite service, broadcasting-satellite service**]

inter-satellite service (ISS) A **satellite communications** system providing links between spacecraft orbiting the Earth – **inter-satellite links** (ISLs).
[See also **fixed-satellite service, mobile-satellite service, broadcasting-satellite service**]

Intersputnik An Eastern Bloc telecommunications satellite organisation established in 1972 (draft agreements were signed by nine countries in May 1968; the final Intergovernmental Agreement was signed in November 1971 and entered into force on 12 July 1972). The American company Lockheed Martin and Intersputnik formed a joint venture, known as Lockheed Martin Intersputnik (LMI), in June 1997.
[See also **Intelsat**]

interstage A section of a **launch vehicle** between two **stages**, which surrounds the **engine**(s) of the uppermost stage. It may be an open latticework structure, or 'faired in' with the rest of the vehicle to provide aerodynamic continuity between stages. It is usually jettisoned during **staging**, before **ignition** of the uppermost stage.
[See also **fairing, upper stage**]

intertank structure In a **launch vehicle**, a supporting structure between two **propellant** tanks which enhances the strength of the **stage** and provides continuity of the **load path**. It also provides a protected site for instruments, etc.

intranet A limited-access, small-scale version of the **Internet** intended to be used by a closed community (e.g. an office or a company). An intranet allows a company to link its offices together electronically, in order to share information (**data**) which is not accessible from the outside. Note: whereas the prefix 'inter' means 'between', 'intra' means 'within' (contrast: **extranet**).

intra-vehicular activity (IVA) Any activity which takes place inside a **spacecraft**; the opposite of **extra-vehicular activity (EVA)**.

intra-vehicular pressure garment – see **spacesuit**

Invar An alloy composed of iron, nickel and carbon which has a low coefficient of expansion; a trademark: an abbreviation of 'invariable'. Used to manufacture **radio frequency** components (e.g. **filters**, input and output **multiplexers**), etc.

IOC An abbreviation for in-orbit check-out, an initial series of measurements made on a **spacecraft**'s systems once it has reached **orbit**; also called in-orbit test (IOT).

IOL – see **inter-orbit link**

ion An electrically charged atom or group of atoms formed by the loss or gain of one or more **electron**s. See **ionisation**.

ion engine A propulsive device in which electrical energy is used to ionise a gaseous **propellant** and an electrostatic (or electromagnetic) field accelerates the positive **ion**s to produce **thrust**. Also called an ion thruster [see figures I4, S2].

In one device (the Kaufman electron bombardment ion thruster), electrons bombard the atoms of the propellant vapour forming a cloud of positive ions, which drifts towards a pair of perforated electrode plates (known as the 'screen grid' and 'accelerator grid') which together form the 'exit port' of the discharge chamber. The ions are accelerated by the high potential gradient across the grids and ejected as a high velocity exhaust. A flux of **electron**s is emitted from a neutraliser to prevent ions being attracted back towards the thruster and to avoid a build up of positive charge in the vicinity of the spacecraft.

In the Hall effect ion thruster, ions and electrons are accelerated together by crossed electric and magnetic fields, the axial electric field being produced by an electron plasma in the thrust chamber and the radial magnetic field by a set of coils. The Hall effect states that when a conductor carrying a current is placed in a magnetic field, a potential difference is produced across the conductor (perpendicular to both electric and magnetic fields). An example of a Hall effect thruster is the stationary plasma thruster (SPT), originally developed in Russia.

As an alternative to electron bombardment, ionisation can be induced in a quartz discharge chamber by a **radio frequency** (typically 1 MHz) generator coil. This type of device is generally known as a radio frequency ionisation thruster (RIT).

The **specific impulse** (I_{sp}) of ion engines is typically an order of magnitude greater than those of chemical propulsion systems. For example, RIT devices using xenon as the propellant have demonstrated an I_{sp} of around 3500 s with a thrust of about 10 mN. A commercial version, developed by US spacecraft manufacturer Hughes, is known as the **xenon-ion** propulsion system (XIPS). Although the low thrust available from ion engines has usually suggested applications on interplanetary spacecraft, which could benefit from low thrust over a long period, this type of engine is also used on geostationary **communications satellite**s, particularly for north–south **station-keeping**. Other applications include fine **attitude control**, **orbital relocation** and

I4: UK-10 **ion engine** with 10 cm accelerator grid (visible at right). [Matra Marconi Space]

'**graveyard orbit**' transfers for satellites based in **geostationary orbit**, and **drag compensation** for those in **low Earth orbit**.

[See also **electric propulsion (EP)**]

ion propulsion – see **electric propulsion**

ion thruster – see **ion engine**

ionisation The formation of **ions**. In chemistry, ions can be positively or negatively charged (cations or anions), but in physics and engineering they are generally positively charged atoms. Ionisation may be the result of chemical reaction, heat, electrical discharge or radiation bombardment, all of which cause **electrons** to be liberated from or 'stripped off' the neutral atom, leaving a net positive charge. The resultant 'ionised gas' of ions and electrons is called a plasma.

[See also **ionosphere**, **plasma sheath**]

ionosphere A region of the Earth's atmosphere, in which ionising radiation (chiefly ultraviolet and X-ray emission from the Sun) causes the liberation of **electrons** [see **ionisation**]. The resultant 'ionised gas' of **ions** and electrons is called a plasma.

The depth of the ionosphere varies with the extraterrestrial particle flux, and therefore with the local time of day: during the daylight hours its lower reaches extend to an altitude of about 50 km, but at night it lifts to about 80 km. The top of the ionosphere is about 1000 km above the surface, but this depends on the value of plasma density used in the definition since the ionosphere can be thought of as 'thinning into' and contiguous with the interplanetary plasma.

Since a plasma contains 'free electrons', capable of independent movement relative to the surrounding ions, the layers of ionised gas have properties similar to those of a metallic conductor, in that they can reflect radio waves. It was this property which led to the experimental determination of the ionosphere's structure, by the English physicist Sir Edward Victor Appleton [1892–1965], in the 1920s. This knowledge has since been usefully exploited for certain types of terrestrial communication.

Especially at times of high particle flux, for instance following a **solar flare**, particles entering the **ionosphere** lead to the displays known as the Aurora Borealis and Aurora Australis.

[See also **ionisation**, **magnetosphere**]

IOT An abbreviation for in-orbit test, a series of measurements made on a **spacecraft**'s systems once it has reached **orbit**; also called in-orbit check-out (IOC).

IP An abbreviation for **Internet** protocol, a **software** procedure concerned with the operation of the Internet.

IRBM An abbreviation for intermediate **range** ballistic missile. See **ICBM**.

[See also **ballistic trajectory**]

IRD – see **integrated receiver decoder**

IRU An abbreviation for inertial reference unit, another name for an **inertial platform**.

ISAS An acronym for Institute of Space and Astronautical Science, a body responsible for space science in Japan. It was established in 1981, by reorganising the former Institute of Space and Aeronautical Science of Tokyo University, and developed Japan's first satellite, Ohsumi, which was launched in 1970. Its HQ is in Sagamihara City and it operates the **Kagoshima** Space Center (**launch** and **telemetry, tracking and command (TT&C)** facilities).

[See also **NASDA, M-5**]

ISDN – see **integrated services digital network**

ISL – see **inter-satellite link**

isochronous In **telecommunications**, a property of a transmission where the signals are of equal duration and are sent in a continuous sequence.

isoflux contours Contours on a satellite antenna **footprint** which represent the same level of incident flux, or **power flux density** (e.g. 'the -3 dB contour', which is '3 dB down' from the **peak**) [see figure F3].

[See also **antenna, beamwidth, coverage area, decibel**]

isolation In **telecommunications**, a measure of separation, in terms of **RF power**, between the **carrier** waves of two potentially interfering communications systems. The isolation, measured in dB, may be realised by transmitting on different **polarisations** or in different directions (e.g. into separate satellite antenna **beams**). In an **earth station**, isolation entails preventing high transmission powers from entering the receiver, which is achieved largely by filtering.

[See also **interference analysis, channel isolation, decibel**]

isotropic Having uniform physical properties in all directions.

isotropic antenna A hypothetical **omnidirectional**, point-source **antenna** used as a reference for **antenna gain** measurements. Such an antenna radiates equally in all directions and can therefore be defined as having 'no gain' in any direction, i.e. no **directivity**. (To be precise, 'no gain' here means a gain value of '1' numerically, or a gain of 0 dB)

[See also **effective isotropic radiated power (EIRP), decibel**]

ISP An abbreviation for **Internet** service provider.

ISRO An acronym for the Indian Space Research Organisation.

ISS An abbreviation for:

  (i) **International Space Station**;

 (ii) **inter-satellite service**.

ISY An abbreviation for International Space Year, a programme to promote multinational cooperation in space which was held in 1992. It was modelled on the International Geophysical Year [see **IGY**] and selected in honour of the 500th anniversary of Columbus's discovery of America and the 75th anniversary of the Russian revolution.

ITT An abbreviation for invitation to tender, industrial terminology for a document which invites bids for a contract. Similar to a request for proposals (RFP) or request for quotations (RFQ).

ITU An abbreviation for International Telecommunication Union, a body responsible for the planning and regulation of international **telecommunications** services [see **fixed-satellite service, mobile-satellite service, broadcasting-satellite service**]. For administrative purposes, the ITU divides the world into three regions: Region 1 includes Europe, Africa, the

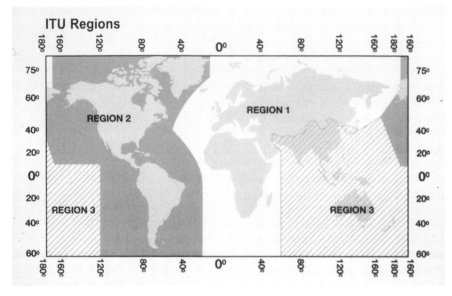

I5: Administrative regions as defined by the **ITU**. [ITU/Mark Williamson/Willis Inspace]

former USSR and Mongolia; Region 2 the Americas and Greenland; and Region 3 Asia, Australasia and the Pacific [see figure I5].

The ITU was formed, as the International Telegraph Union, on 17 May 1865 and was renamed the International Telecommunication Union, on 1 January 1934, in recognition of its broader responsibilities. It became a specialised agency of the United Nations on 15 October 1947. On 1 March 1993, the ITU was reorganised into three sectors: ITU-R for radiocommunication; ITU-T for telecommunication standardisation; and ITU-D for telecommunication development. The bureaux operating in each sector – the Radiocommunication Bureau (RB), Telecommunication Standardisation Bureau (BST) and Telecommunication Development Bureau (BDT) – took responsibility for the functions of bodies such as the **IFRB**, **CCIR** and **CCITT** (the bureau abbreviations reflect their French titles). ITU-R includes, among others, the Radio Regulations Board (RRB) and Radiocommunication Advisory Group (RAG), and is responsible for the organisation of the World Radiocommunication Conference (**WRC**).

[See also **WARC, RARC**]

IUS – see **inertial upper stage**

IVA – see **intra-vehicular activity**

J

jamming The intentional transmission of interfering radio signals, which prevent the clear reception of the 'jammed' **signal**. Used also as a verb (to jam). Its countermeasures are referred to as 'anti-jamming' or 'anti-jam'. [See also **radio frequency (RF)**, **spread spectrum**, **encryption**, **scrambling**, **scrambler**]

JEM An acronym for Japanese Experiment Module, a **module** of the **International Space Station (ISS)** supplied by **NASDA** [see figure J1].

Jet Propulsion Laboratory (JPL) – see **NASA**

jettison To cast off or release. The word is typically used for items which are being discarded, no longer of any use (e.g. spent rocket **stages** [see figure J2], **heat**

J1: Japanese Experiment Module (**JEM**) attached to the **International Space Station (ISS)**. Note the external experiment **pallets**, **remote manipulator system** and **cupola** (windowed module at centre). [NASA]

J2: **Jettison** of **Ariane 5 solid rocket
boosters** at **burn-out**. [Arianespace]

shields, aerodynamic **fairing**s, refuse containers, etc.); in cases where the
item is not being discarded, terms such as 'release' or 'eject' are applied (e.g. a
satellite released from a **launch vehicle**).

Jiuquan A Chinese **launch site**, in the Gansu Province on the edge of the Gobi
Desert (at approximately 41° N, 100° E), formerly known in the West as Shuang
Chen Tse (or Shuang-ch'eng-tzu). It conducted its first launch, of DFH-1, on 24
April 1970 and, until 1987, launched most of China's satellites. Although it is
still used for remote sensing and recoverable **reconnaissance satellite**s,
launches into **geostationary orbit** are now made from the **Xichang** site.
[See also **Taiyuan**]

Johnson noise – see **thermal noise**

Johnson Space Center (JSC) – see NASA

Joule–Thomson effect The decrease in temperature when a gas expands; a
phenomenon which proved useful in the production of **cryogenic
propellant**, among other things. (Note: under certain conditions a reverse
effect can occur, i.e. temperature increases). Also called the Joule–Kelvin
effect (since the Scottish mathematician and physicist William Thomson
[1824–1907] became Lord Kelvin in 1892).
[See also **kelvin (K)**]

JPL An abbreviation for **Jet Propulsion Laboratory**. See **NASA**.

Jupiter An American single-**stage** liquid **propellant** intermediate range ballistic missile (IRBM) based on the US Army **Redstone** and developed in the late 1950s as a **satellite** launcher (propellant: **liquid oxygen/kerosene**). No longer **operational**.

The Jupiter C (C-for-composite **re-entry** test vehicle) was a more extensively modified Redstone with two upper stages, developed to test the ablative **nose cone** of the Jupiter's warhead. The second and third stages were formed from clusters of 11 and three Sergeant guided missiles, respectively. The Jupiter C was converted into a satellite launcher, called Juno I, by adding a fourth stage in the shape of a single Sergeant: it launched America's second satellite, **Vanguard** 1, in March 1958.

[See also **Thor, ICBM, ablation**]

K

Ka-band: 26.5–40 GHz – see **frequency bands**

Kagoshima The location of a Japanese **launch site** (at approximately 31° N, 131° E), which is operated by the Institute of Space and Astronautical Science (**ISAS**) and conducted its first launch, of the Ohsumi satellite, on 11 February 1970.

[See also **Tanegashima**]

Kakuda Propulsion Center – see NASDA

Kapton A plastic material used for spacecraft **thermal insulation** (a trademark). Aluminised and used in **multi-layer insulation (MLI)**, it is typically employed for the outer layers since it is more rugged than **Mylar** and provides protection during installation and handling. Moreover, Kapton is a preferred outer layer since Mylar disintegrates under prolonged **ultraviolet** exposure.

Kapustin Yar The location of a Russian **launch site** (at approximately 48.4° N, 45.8° E), which conducted its first launch on 16 March 1962.

[See also **Baikonur Cosmodrome**, **Northern Cosmodrome**, **Svobodny**]

K-band: 18–26.5 GHz – see **frequency bands**

kelvin (K) The SI unit of temperature, named after the Scottish mathematician and physicist William Thomson, Baron Kelvin of Largs [1824–1907]; the unit of **absolute temperature**. Thomson suggested that -273 °C should be considered the 'absolute zero' of temperature, at which the kinetic energy of molecules would be zero, and that the degrees on the 'absolute scale' should be the same as those on the Celsius scale (devised in 1742 by the Swedish astronomer Anders Celsius [1701–1744]). The modern figure for absolute zero is about -273.15 °C. Although it is more or less obligatory to use the (°) symbol for °C, it is more common, by convention, to write 273 K than 273 °K.

[See also **rankine (°R)**]

Kennedy Space Center (KSC) NASA's main **launch** facility (located at approximately 28.5° N, 80.6° W), known more fully as the John F. Kennedy Space Center, after the past president of the USA; also known colloquially as 'Cape Canaveral', after the part of Florida upon which it is built [see figure K1]. When KSC was established in July 1962 the land itself was renamed 'Cape Kennedy', but public confusion between NASA and US Air Force facilities and

K1: **Launch complex** 39A (foreground with Space Shuttle **Columbia**) and 39B (background with **Discovery**) at **Kennedy Space Center**. [NASA]

local opposition to the name-change led to the reinstatement of the original title in 1973–74, when the USAF facility became Cape Canaveral Air Force Station (it was later renamed Cape Canaveral Air Station). The USAF site is also known as the Eastern Test Range (ETR), and prior to May 1964 was known as the 'Atlantic Missile Range'. KSC also manages NASA launches conducted at the 'Western Test Range' in California [see **Vandenberg Air Force Base (VAFB)**]. The first successful space-related launch from Cape Canaveral was that of **Explorer** 1 on 1 February 1958.

[See also **vehicle assembly building (VAB), orbiter processing facility (OPF), Shuttle landing facility (SLF), launch control centre (LCC), launch complex, Wallops Flight Facility**]

kerosene A storable liquid **fuel** used in **rocket engine**s. Also referred to as RP-1. See **liquid propellant**.

Kevlar composite A material composed of a polymer matrix reinforced by
threads or fibres of Kevlar (a form of the organic polymer
polyparabenzamide; a proprietary material developed by Du Pont).
Otherwise known as Kevlar reinforced plastic (KRP). Kevlar exhibits a typical
density only 55% that of aluminium with the additional advantage of
improved impact resistance (which is why it is also used in bullet-proof vests).
Typical spacecraft applications: **face-skin**s, **antenna** reflector surfaces,
filament-wound motor cases and structural components.

[See also **composites**, **carbon composite**, **materials**, **Whipple shield**]

kHz An abbreviation for kilohertz – see **hertz**, **frequency bands**.

kinetheodolite A **tracking** camera, used to track and film rocket and missile
launches, and to track and photograph **satellites** in **orbit**. It comprises a
telescope mounted on an **altitude-azimuth mount**, which allows it to be
pointed in any direction, and is usually operated by two observers following
the object through smaller, finder-telescopes, one observer controlling
pointing in elevation, the other in azimuth.

kinetic heating – see **aerodynamic heating**

klystron A type of 'electron tube' used as a high (RF) power amplifier in radar
systems and satellite **earth station**s; also known as a klystron power amplifier
(KPA) [see figure K2]. The klystron is a narrow-**bandwidth**, high output-**power**
device, typically delivering 45 MHz and 3 kW at **C-band**, and 80 MHz and
2 kW at **Ku-band**. Another type of electron tube, the **travelling wave tube**
(TWT), is used in both earth stations and spacecraft.

K2: 1500 W/18 GHz earth station
klystron for satellite **uplink**
transmissions. [Thomson-CSF]

The klystron operates by accelerating **electrons** in an **electron beam** through a positive potential difference. The electrons enter a series of cavities, typically five, which are tuned around the operating **frequency** of the device and connected by cylindrical 'drift tubes'. In the input cavity the electrons are velocity-modulated by the time-varying electromagnetic field produced by the input **radio frequency (RF)** signal. Faster electrons catch up with slower ones to form bunches, with optimum bunching occurring in the output cavity. RF currents are generated in the cavity wall by the density-modulated beam, thereby generating an amplified output signal. The remaining energy of the electron beam is dissipated as heat in a **collector**. Intermediate cavities are used to optimise power, **gain** and bandwidth characteristics.

[See also **high power amplifier**, **solid state power amplifier (SSPA)**]

Kosmos The name given to a series of Soviet/Russian spacecraft and launch vehicles; alternative spelling Cosmos. More than 2000 spacecraft launched by the former Soviet Union were designated Kosmos as part of a policy of restricting information on their **application**.

Kourou The location of the **Guiana Space Centre** (at approximately 5.2° N, 52.8° W), the **launch site** of the European **Ariane** launch vehicle.

KPA An abbreviation for klystron power amplifier. See **klystron**.

KRP An abbreviation for Kevlar reinforced plastic. See **Kevlar composite**.

Krystall A module of the **Mir** space station.

KSC – see **Kennedy Space Center**

Ku-band: 12.5–18 GHz – see **frequency bands**

Kvant A module of the **Mir** space station.

L

LAE – see **liquid apogee engine**

Lagrange point One of a number of points in space where gravitational forces are balanced so that a body at that point remains nominally stationary. Also called Lagrangian points or libration points. The existence of these points was calculated by the Italian–French astronomer and mathematician Joseph Louis Lagrange [1736–1813], who predicted that three astronomical bodies could move in a stable configuration, with one at each of the vertices of an equilateral triangle, if one of the bodies was small. The discovery of Jupiter's 'Trojan asteroids' in the early twentieth century proved the theory: they orbit in two groups, 60° ahead and 60° behind the planet, in its orbit about the Sun.

For a given system of two celestial bodies orbiting about their common **centre of mass**, five Lagrangian points (L_1–L_5) can be defined, all in the same orbital plane. L_1, the inner Lagrangian point, is located between the two bodies where the gravitational acceleration due to the smaller mass (augmented by **centrifugal force**s) is balanced by that of the greater mass. L_2 and L_3, the outer Lagrangian points, are located at either end of a line passing through the two bodies (and L_1), where the combined **gravity** of the masses is balanced by centrifugal forces. L_4 and L_5, the equilateral points, are located on the orbital track of the smaller body around the larger, 60° ahead and 60° behind the smaller body.

The L_1 point between the Earth and the Sun, approximately 1.5 million kilometres from Earth, is a recognised position for a solar observatory, since it is never eclipsed by the Earth or any other **planetary body**. The first spacecraft to be stationed at L_1, in a so-called 'halo orbit', was NASA's International Sun–Earth Explorer 3 (ISEE-3), which, in 1978, became the first spacecraft to orbit a point in space rather than a **celestial body**. L_1 was also chosen for the Solar and Heliospheric Observatory (SOHO), part of **ESA**'s Solar–Terrestrial Science Programme (STSP), and has been suggested for **Earth observation** spacecraft. L_2, on the opposite side of the Earth, has been suggested for **deep space** observatory spacecraft. For the Earth–Sun system, L_3 is on the opposite side of the Sun from Earth, L_4 is 60° ahead of the Earth in its orbit and L_5 is 60° behind. L_4 and L_5 have been suggested as positions for space stations and 'space colonies'.

[See also **libration, libration period**]

Laika The name of the first living creature to be launched into **space**. The dog Laika was one of several **payloads** on **Sputnik** 2, the world's second **satellite**, which was launched on 3 November 1957.

land earth station An **earth station** installed on land; a term sometimes used to differentiate a land-based station from a **ship earth station**. Although 'land earth' may, at first sight, seem tautologous, 'land' refers to ground and 'earth' refers to the planet itself. The term is not widely used.

[See also **coast earth station**]

Landsat A series of American **remote sensing** satellites used predominantly for land-use studies. Landsat 1, launched in July 1972 and formerly known as the Earth Resources Technology Satellite (ERTS), was heavily based on NASA's Nimbus **technology demonstrator** series. The early Landsats carried two main **payloads**: a return beam **vidicon** (RBV) camera, which produced 4500-line images of the Earth, and a multispectral scanner (MSS), which used photomultiplier tubes to produce images in three spectral bands (visible, near IR and thermal IR).

[See also **SPOT, waveband**]

Langley Research Center – see NASA

laser An acronym for light amplification by stimulated emission of radiation. A device for producing a narrow, monochromatic beam of coherent light (i.e. of a single **frequency** and constant relative phase). The laser was a later result of the development of the first **maser** (**microwave** amplification . . . etc.), in 1953, by Charles Hard Townes, an American physicist. In 1958, Townes and his brother-in-law published a paper on the possibility of an 'optical maser', for the amplification of infrared or visible light instead of microwaves. The first practical laser was constructed by Theodore Harold Maiman, another American physicist, in 1960.

In terrestrial **communications**, the laser is used in **optical fibre** links. In space, it is used in **inter-satellite link**s and **inter-orbit link**s, and can also be used for the measurement of distance – see **laser ranging retroreflector**.

laser gyro A type of **gyroscope** which uses a **laser**, instead of a spinning mass, to detect rotation (which is used as an input to a **launch vehicle** guidance system). Also known as a 'laser ring gyro' or 'ring laser gyro' owing to the typical arrangement of three mirrors in a triangle, with one laser beam

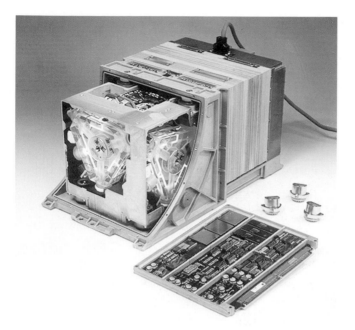

L1: **Laser gyro** inertial control unit for **Ariane**. [Sextant Avionique]

propagating in a clockwise direction and another travelling anti-clockwise [see figure L1]. The operation of the laser gyro requires a standing wave to be maintained between the mirrors, which implies the existence of a node in the wave pattern at each of the mirrors. However, if the device rotates, a given wavefront on one beam has a little further to go before it reaches a mirror (conversely the other beam reaches the mirror sooner). Automatic adjustments mean that the wavelength of the first beam is increased to maintain the standing wave, and that of the other beam is reduced. The light from the two beams is incident on a photodetector where the frequency difference generates a beat-frequency. This frequency is proportional to the rotation rate of the device. A complete **inertial platform** uses three such laser rings, one for each axis, 'strapped down' to the vehicle rather than mounted in a set of **gimbal**s. See **strapdown gyro system**.

laser ranging retroreflector A type of reflector used in the measurement of distance, which returns incident **laser** radiation to the point of its origin, i.e. it ensures that the outgoing rays are parallel to the incoming rays.

Retroreflectors mounted on a spacecraft allow its distance from Earth (its range) to be accurately calculated, using a laser beam directed towards it [see figure L2]. For example, the **Apollo** programme left a number of retroreflector

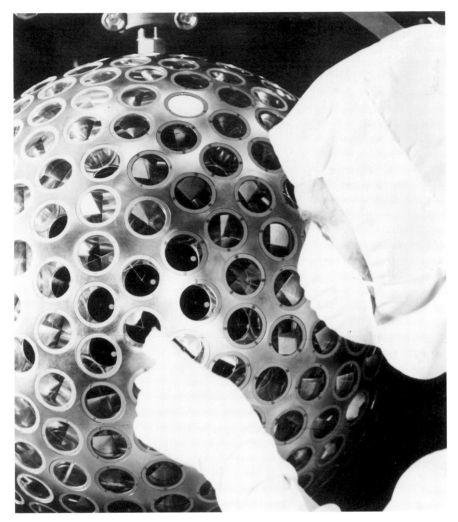

L2: Laser geodynamics satellite, Lageos 1, designed for experiments in accurate range determination using the **laser ranging retroreflector**s which cover its surface [Aeritalia].

packages on the **Moon** which have been used to determine its distance, and the variations in that distance, to accuracies of a few tens of centimetres (the Moon's distance from Earth varies between about 348,294 km and 398,581 km, surface-to-surface; 356,410 km and 406,697 km, centre-to-centre).

When retroreflectors take the form of three equilateral triangles joined to make an open pyramidal shape, they are called corner-cube reflectors. (Radar retroreflectors of this design can be mounted on the masts of boats to assist in their detection.)

laser ring gyro – see **laser gyro**

last mile Jargon (chiefly American) for the communications link between a customer's premises and a provider's national or international network. See **wireless local loop**.

latch-up A 'change of state' in a **computer** or **microprocessor** (which typically causes it to become inactive or unresponsive) caused either by a **software** problem or an external influence such as a **single event effect (SEE)**.

latency Jargon (chiefly American) for 'delay'. Typically used to describe the inherent **signal delay** in a transmission via satellite, which results from its distance from Earth and the finite propagation velocity of **electromagnetic radiation**.
[See also **light (velocity of)**]

launch The procedure by which a **spacecraft** or **launch vehicle** is propelled into an **orbit** or **trajectory**; the moment of **lift-off** (also called 'blast-off'). Most launches are made from a **launch platform** on the surface of the Earth, but the term can be used in connection with any **planetary body**. Spacecraft are also 'launched' from the **payload bay** of the Space Shuttle, once it has reached **low Earth orbit**.

   In **space insurance** terms, the moment of 'launch' is carefully defined to avoid a gap in insurance coverage between a '**pre-launch** policy' and a 'launch policy'. For example, launch may be defined as 'the time after **intentional ignition** at which the launch vehicle separates from the **launch pad**' or, alternatively, 'the moment of **clamp release**'.
[See also **flight**]

launch campaign The period during which a **launch site** and a **launch vehicle** are prepared for a **launch**, typically including the delivery of the vehicle **stages** to the launch site, their assembly, **payload integration** and all other activities leading to the **countdown** and **lift-off**.
[See also **mission**]

launch capability An alternative term for **payload capability**, a measure of a **launch vehicle**'s ability to lift **payload** to a given **orbit**; also known as payload capacity.

launch commit criteria The key factors which influence a launch controller's decision regarding whether or not to approve a **launch**. They may include aspects of a **launch vehicle**'s **operational** status, local weather conditions, etc.
[See also **triggered lightning**]

L3: **Space Shuttle** Columbia, carried by a **Crawler** transporter, arrives at the **launch pad** at **launch complex** 39A. The opening of the **flame trench** and the **flame deflector** are visible beyond the Shuttle/Crawler combination. [NASA]

launch complex The general area surrounding a **launch pad**; a self-contained facility for launching **launch vehicles**. Sometimes called a **launch site**, but there is usually more than one launch complex (and more than one launch pad) on a launch site [see figures K1, L3].

launch control centre (LCC) A building from which the **launch** of a **launch vehicle** is controlled and monitored. When rocket launches were simpler affairs, the equivalent building would have been termed a 'firing room' or 'blockhouse', especially in a military context.

Most LCCs are known colloquially as 'mission control', since the **mission** of the typical launch vehicle is completed with the injection of its **payload** into a particular **trajectory**. In the case of a manned **NASA** launch from **Kennedy Space Center**, a distinction is made between LCC and MCC (**mission control centre**): once the launch vehicle has cleared the tower, responsibility for the mission is handed from the LCC at KSC [see figure V2] to the MCC in Houston. [See also **Sea Launch**]

launch escape tower – see **escape tower**

launch mass The mass of a **satellite** or **spacecraft** at launch. The **launch mass** is generally greater than the **in-orbit mass**, because an amount of on-board **propellant** may be used to attain the final **orbit**.
[See also **dry mass, mass budget, combined (bipropellant) propulsion system**]

launch pad

    (i) A platform from which a **launch vehicle** is launched. Also called a **launch platform** [see figure L3].

    (ii) The platform and its immediate surroundings (e.g. **service tower, flame trench** and miscellaneous **ground support equipment**).

launch platform

    (i) A structure from which a **launch vehicle** is launched; an integral part of a **launch pad** (sense (ii)).

    (ii) An off-shore platform used for **rocket** launches (e.g. **San Marco platform, Sea Launch**).

[See also **mobile launch platform**]

launch profile

    (i) The shape of a **launch vehicle's trajectory** with reference to the surface of the Earth; a graphical representation of the vehicle's **altitude** against its distance **downrange** from the **launch site** [see figure L4].

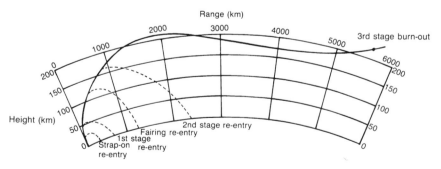

L4: **Launch profile** for an **Ariane** 4 launch vehicle.

    (ii) Any general description of a launch vehicle's **flight** parameters, including **range**, height, velocity, etc.

launch shroud – see **payload fairing**

launch site

    (i) A facility engaged in rocket launches; sometimes called a **spaceport**. See **Alcantara, Baikonur Cosmodrome, Guiana Space Centre, Jiuquan,**

Kapustin Yar, Kagoshima, Kennedy Space Center (KSC), Northern Cosmodrome, Palmachim, Plesetsk, San Marco platform, Sea Launch, Sriharikota, Svobodny, Taiyuan, Tanegashima, Vandenberg Air Force Base (VAFB), Wallops Flight Facility, Woomera, Xichang.

(ii) The general area surrounding a **launch pad**, otherwise known as a **launch complex**.

launch vehicle Any vehicle designed to carry **payload**s into space. A launch vehicle can be categorised in many different ways: it may be expendable or reusable (or a combination of the two); it may use **solid propellant** or **liquid propellant** or both; it may be capable of launching a payload to a **sub-orbital** or interplanetary **trajectory**, **low Earth orbit** or a **transfer orbit**; it may have one **stage** or several; it may or may not be **man-rated**.

[See also **expendable launch vehicle, heavy lift vehicle, aerospace vehicle, orbit, single stage to orbit, Ariane, Atlas, Blue Streak, Delta, Energia, Europa, Jupiter, H, Long March, N, Proton, Redstone, Saturn 1B, Saturn V, Scout, Sea Launch, Space Shuttle, Thor, Titan, Evolved Expendable Launch Vehicle (EELV)**]

launch window A limited period of time during which the **launch** of a particular **spacecraft** may be undertaken. For launches into Earth **orbit** the main constraints are the desired orbit, the point of **injection** into that orbit, and the possibility of a collision with other orbiting spacecraft or **orbital debris**. For launches to interplanetary trajectories, the time of launch is constrained by the relative positions of the planets, at the time of both launch and **encounter** (i.e. the spacecraft must be aimed at a point in space that the planet will have reached by the time of encounter). A spacecraft launched outside the launch window would require more **propellant** and/or more time to reach its target.

launcher

(i) A colloquial term for **launch vehicle**.

(ii) A ground-based mechanical system for launching rockets (i.e. a **rocket launcher**). Generally used for relatively small rockets, such as **sounding rockets** or military armaments.

launcher release gear – see **hold-down arm**

L-band: 1–2 GHz – see **frequency bands**

LEM An acronym for lunar excursion module. See **Apollo**.

lens antenna A type of antenna that focuses RF radiation in a manner analogous to an optical lens: the longer lengths of **waveguide** or **dielectric** medium reduce the velocity of the propagating wave, which alters the form of the wavefront. Variations are known as dielectric and **microwave** lenses.

[See also **antenna**]

LEO – see **low Earth orbit**

Leonids The name given to an annual **meteor** shower, so named because it appears to originate from the constellation of Leo. In fact, the Leonid meteors originate from the tail of the comet Tempel–Tuttle, through which the **Earth** orbits once a year on its path around the **Sun**. About once every 33.25 years, it passes through a particularly dense region of the cloud, resulting in a rise in the number of meteors observed. Although the most recent peak occurred in November 1999, the Leonids received their greatest publicity the previous November, when it was believed that **satellites** orbiting the Earth were at risk; in fact, there were very few reports of suspected Leonid damage.

[See also **space debris**]

LEOP An acronym for **launch** and early **orbit** phase (or launch and early operations), used mainly in connection with the **tracking** of spacecraft before they reach their **operational** orbital position.

Lewis Research Center – see **NASA**

LF (low frequency) – see **frequency bands**

LHCP (left-hand circular polarisation) – see **polarisation**

Liability Convention The colloquial name for the treaty of **space law** whose official title is "Convention on International Liability for Damage Caused by Space Objects". The Liability Convention was adopted by the United Nations General Assembly on 29 November 1971, opened for signature on 29 March 1972 and entered into force on 1 September 1972. It was one of four main space treaties derived from the **Outer Space Treaty** of 1967.

[See also **Rescue Agreement, Registration Convention, Moon Agreement, space object, space insurance**]

libration The act of oscillating (e.g. as performed by a **satellite** in **geostationary orbit**). Solar radiation pressure tends to add to a satellite's orbital velocity as it moves away from the Sun and decelerate it as it moves towards the Sun. This results in an oscillation about the satellite's **nominal orbital position** as viewed from Earth [see **libration period**].

Although the Moon orbits the Earth in 'captive rotation' (nominally with the same face towards the Earth), it too experiences libration, due to the eccentricity of its orbit. When it moves closer to the Earth its orbital velocity increases and its axial rotation falls behind, allowing observers on Earth to see more of its eastern hemisphere; when further away it slows down, its axial rotation moves ahead and more of the western hemisphere is exposed. This makes it possible to see, and conduct line-of-sight **communications** with, about 59% of the lunar surface from Earth.

[See also **solar wind**]

libration period A period of oscillation of a **satellite** about a mean position on the **geostationary orbit**, caused by solar radiation pressure [see **solar wind**]. Assuming the pressure remains constant, the **libration** period is equivalent to the **orbital period**, which for geostationary orbit is approximately 24 hours.

[See also **equilibrium point**]

libration point – see **Lagrange point**

life support The theory and practice of sustaining life in an environment or situation in which the human body is incapable of sustaining its own natural functions. In space this principally involves protection against the vacuum, harmful radiation and temperature extremes, and the provision of a breathable atmosphere at a comfortable atmospheric pressure, a thermal control system and a water supply.

[See also **portable life support system, spacesuit, decompression**]

life support system Any collection of equipment designed to provide **life support** (e.g. that within the Space Shuttle **orbiter** is called an **environmental control and life support system (ECLSS)**). A mechanical or biological system that recycles air, water and food is sometimes called a closed-ecology life support system (CELSS).

[See also **portable life support system, spacesuit**]

lifetime The length of time for which a device, vehicle or system is intended to perform its function. A distinction is made between its 'design lifetime' and its 'operational lifetime'. The design lifetime is the period over which it is designed to be capable of operation, whereas the operational lifetime is the period over which one may actually plan to use it (which may be less) or the period over which it remains in **operational** status (which may be more). For example, a **satellite** with a design lifetime of 15 years may be intentionally employed for only 13 years or it may be allowed to operate for 17 years. Design lifetimes have increased gradually since the 1960s, when satellites became industrial products.

The design lifetime of different types of **spacecraft** is dependent upon different factors. That of satellites in **geostationary orbit** (e.g. **communications satellites**, **meteorological satellites**, etc.) is governed mainly by the amount of **propellant** carried for **station keeping**: once this is depleted, the satellite can no longer hold its **orbital position** and begins to **drift**. Before this occurs, it is generally moved out of orbit to make way for another spacecraft [see **graveyard orbit**]. Commercial communications satellites based in geostationary orbit typically have design lifetimes of around 15 years, while non-commercial (e.g. technology demonstration) satellites and weather satellites usually have shorter lifetimes. The lifetime of satellites in **low Earth orbits** is constrained chiefly by the propellant that can be carried for orbital adjustments, the height of the orbit, which determines the degree of **atmospheric drag**, and the exposure to radiation. The lifetime of **astronomical satellites** which carry infrared detectors is generally constrained by a limited supply of cryogenic coolant. Although the design lifetime of the constituent parts of an irretrievable satellite should at least equal the desired operational lifetime of the whole vehicle, components can be allowed to fail if a **redundancy** provision has been made.

A reusable **launch vehicle**, such as the Space Shuttle **orbiter**, typically has an operational lifetime longer than that of some of its component parts (e.g. **Space Shuttle main engines** are intended to be used for only about ten flights).

[See also **beginning of life (BOL)**, **end of life (EOL)**, **bathtub curve**, **probability distribution**, **cryogen**]

lift The aerodynamic force produced by a **lifting surface**, perpendicular to the air-flow and opposing gravity.

[See also **lifting body**, **aerospace vehicle**]

lifting body An aerodynamic vehicle, usually an unpowered glider, used – among other things – to evaluate proposed designs for future **aerospace vehicle**s.

lifting surface Any surface which provides **lift** (e.g. a wing).

[See also **aerospace vehicle**]

lift-off The moment of **launch**; the initial vertical movement of a **launch vehicle** from the surface of a **launch platform**. More colloquially termed 'blast-off'.

[See also **countdown**, ***T*-minus . . .**, **hold**, **launch window**]

light (velocity of) The velocity at which **electromagnetic radiation** travels in **free space** (i.e. 2.998×10^8 m s^{-1}).

[See also **light-year**]

light baffle – see **baffle**

light-year The distance travelled by light in one mean solar year (i.e. 9.46×10^{15} m), where the velocity of light is 2.998×10^{8} m s^{-1} and 1 year is 3.156×10^{7} s; a unit of distance used in astronomy. Sub-divisions: light-minute and light-second (e.g. 'Earth is 8.3 light-seconds from the Sun').
[See also **parsec**]

line heater A heating device wound onto **propellant** lines to maintain the propellant above its freezing point. For example, the **hydrazine** liquid propellant used in **spacecraft** reaction control systems is particularly prone to low temperatures in the extremities of the system where the volumes are low: in the lines, valves and **thrusters**, the last of which are on the spacecraft exterior.
[See also **heater, thermal insulation, thermal control subsystem**]

line scanner A **scanner** which images a line at a time. See **pushbroom scanner**.

linear aerospike engine – see **aerospike engine**

linear polarisation – see **polarisation**

linearity A characteristic of an **amplifier** which describes how well its **output** power follows its **input** power (depicted graphically as a straight line when linear). A **high power amplifier (HPA)** is inherently non-linear, meaning that the output **power** is not simply proportional to the input power. See **saturated output power** and **back off (from saturation)**.

liner

(i) A material placed between the **motor case** and the **propellant** in a **solid rocket motor** [see figure R7]. See **propellant liner, flame inhibitor**.

(ii) A thin layer of adhesive used to bond **solid propellant** to a motor case or to a layer of **thermal insulation** or thermal blanket.

(iii) An ablative material lining the inner wall of a **combustion chamber** or **nozzle** to reduce heat transfer to the wall or nozzle.

[See also **rocket motor, ablation**]

link A **communications** link or **radio frequency** link (e.g. between a **spacecraft** and an **earth station**). Depending on the application, the link may be known more specifically as a **data** link, **telemetry** link or **command** link, for example.

   Satellite communications systems are designed to link earth stations in three main ways:

• point-to-point: signals transmitted from a single station to another individual station

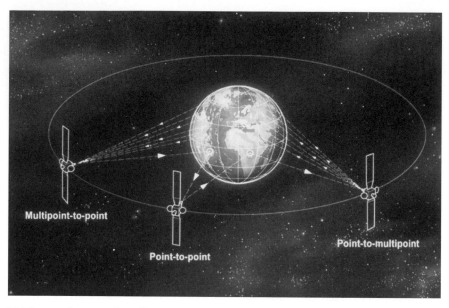

L5: Three main types of satellite communications **link**. [Mark Williamson/Willis Inspace]

- point-to-multipoint: signals transmitted from a single station to an unlimited number of independent receiving stations (e.g. **direct broadcasting by satellite (DBS)**)
- multipoint-to-point: signals transmitted from many individual stations to a single controlling station (e.g. environmental **data**-collection) [see figure L5]. [See also **uplink, downlink, asymmetric link, inter-satellite link (ISL), inter-orbit link (IOL), crosslink, link budget, link margin, unicasting, narrowcasting, multicasting, broadcasting**]

link budget A calculation of signal power in a communications link design which indicates whether or not a signal will be successfully received. A link budget is the 'profit and loss account' of the transmitted signal. 'Profit' is accrued by the amplification of the signal, either in the satellite **transponder** or in the **earth station**'s **amplifier chain**, and by means of the inherent **gain** of the **spacecraft** and earth station **antenna**s. 'Losses' arise from the signal's distance of travel: on its path through space, through Earth's atmosphere and in the communications hardware. The link budget constitutes a 'balance sheet' which combines these profits and losses; the resultant 'balance' is the **margin** in signal power. See **link margin**.

[See also **free space loss, atmospheric attenuation,** I^2R **loss, negative margin**]

link margin The amount by which the **RF power** in a communications **link budget** is above that required to achieve the performance specified for the link under **nominal** conditions. The point where there is no margin will depend on the quality standards set; if the power drops below this point, there is said to be a '**negative margin**'. Usually expressed in dB [see **decibel**].

liquid apogee engine (LAE) A rocket engine used to transfer a satellite from **geostationary transfer orbit (GTO)** to **geostationary orbit (GEO)**, fired when the spacecraft reaches the point in the GTO ellipse furthest from the Earth, the **apogee**. It is usually part of a **combined propulsion system** (combining apogee boost and reaction control functions), and uses **liquid propellants**. [See also **apogee kick motor (AKM)**, **perigee kick motor (PKM)**, **rocket engine**, **rocket motor**, **reaction control system (RCS)**]

liquid cooling and ventilation garment (LCVG) – see **spacesuit**

liquid hydrogen A widely used **cryogenic propellant**; a liquid **fuel** with a boiling point at -252.8 °C and a melting point at -259.2 °C, widely used with **liquid oxygen** as the **oxidiser**. Commonly abbreviated to 'LH$_2$'. [See also **slush hydrogen**, **liquid propellant**, **rocket engine**]

liquid oxygen A widely used **cryogenic propellant**; a liquid **oxidiser** with a boiling point at -182.98 °C and a melting point at -218.76 °C, widely used with **kerosene** or **liquid hydrogen** as the **fuel**. Commonly abbreviated to 'LO$_2$' or 'LOX'. [See also **liquid propellant**, **rocket engine**]

liquid propellant The combustible substance which provides **thrust** in a **rocket engine** (note that rockets which use solid propellant are known as **rocket motors**). There are two main types of liquid propellant: 'storable' and 'non-storable', the former being a storable liquid at terrestrial environmental temperatures and the latter needing various degrees of cooling below these temperatures to render it liquid. The non-storable propellants are otherwise termed 'cryogenic' after the branch of physics concerned with very low temperatures [see **cryogenic propellant**]. Liquid propellants are generally more difficult to store and handle than solids and even some of the storable liquids are, to some extent, hazardous, being highly reactive and toxic. The liquid engine is, however, more flexible in its operation [see **rocket engine**] and liquid propellants typically exhibit a higher **specific impulse** than solids.

A liquid propellant is either a **fuel** or an **oxidiser**. Storable liquid oxidisers widely used for rocket propulsion in the past include red fuming nitric acid

(RFNA) and hydrogen peroxide (H_2O_2). Currently more common is nitrogen tetroxide (N_2O_4). As for storable liquid fuels, the methyl and ethyl alcohols, used in early rockets such as the **V-2** with **liquid oxygen** as the oxidiser, have since been replaced by hydrocarbons, such as kerosene, which provide a more efficient mixture (kerosene was used with liquid oxygen in the first stage of the **Saturn V**). Two fuels which have been used with nitric acid are hydrazine (N_2H_4) and unsymmetrical dimethylhydrazine (UDMH: $(CH_3)_2N.NH_2$). Hydrazine can also be used with hydrogen peroxide (H_2O_2) or N_2O_4, both of which are **hypergolic** combinations; UDMH is widely used with N_2O_4 (e.g. in the first two stages of the **Ariane** 4 launch vehicle).

Early **reaction control system**s used hydrazine as a **monopropellant** (i.e. without an oxidiser), but the hypergolic combination of nitrogen tetroxide and monomethyl hydrazine (MMH: $CH_3NH.NH_2$), used in liquid **bipropellant** systems, is now more common [see **combined propulsion system**].

[See also **propellant, solid propellant, NHMF**]

lithium hydroxide canister – see **environmental control and life support system (ECLSS)**

lithium-ion cell A voltaic cell used in a spacecraft **battery**.

live stage A launch vehicle **stage** containing **propellant** and capable of powered-flight (cf. a 'live round' of ammunition).

[See also **dummy stage**]

LLV – see **Athena**

LMLV – see **Athena**

LNA – see **low noise amplifier**

LNB An abbreviation for low noise block-downconverter. See **low noise converter**.

LNC – see **low noise converter**

LO – see **local oscillator**

load

   (i) Mechanical: a force exerted on a component, mechanism or structure – 'structural load'. Mechanical loads can be either static or dynamic: static loads are exerted on a stationary structure; dynamic loads on a structure in motion.

  (ii) Electrical: the **energy** delivered by, or required by, an item of power subsystem equipment – colloquially termed the 'power load'.

(iii) Electrical: a device which dissipates the power of a **signal**, often used to terminate a **transmission line** to absorb unwanted signal-power on that line. 'A load' is similar to a very high loss **attenuator**.

[See also **load path, thrust structure, payload, power supply, power bus**]

load path The path through which mechanical loads are transmitted in a structure. For example, knowledge of the load path in a spacecraft structure helps to ensure that the structure is capable of surviving the mechanical loads imparted upon it by a **launch vehicle**. The load path between a launcher and a spacecraft is via the launch vehicle **adapter ring** to a rigid **thrust structure** which, along with a number of platforms and panels, constitute the **primary structure** of the spacecraft. The remaining structural elements, such as support brackets for propellant tanks, mountings for **antenna** reflectors, **feed** assemblies, etc., fit into the category of **secondary structure**, thus completing the load path from subsystem equipment, through secondary and primary structures, to the adapter ring.

local oscillator (LO) A device which provides a stable, fixed-frequency source of **radio frequency** energy used in telecommunications equipment (e.g. in a spacecraft **transponder**: see **mixer**, **downconverter**, **upconverter**). Alternatively known as a 'carrier generator'.

When used to supply an **RF beacon**, or similar device where the output is used for transmission in its own right, rather than for frequency-conversion, this type of device tends to be called simply a 'frequency source'.

Long March A Chinese **launch vehicle** developed, in three variants, during the late 1960s and early 1970s. The first two **liquid propellant** stages of the Long March 1 (otherwise known as the CZ-1) were developed from the CSS-2 missile, to which was added a third **solid propellant** stage; it was introduced in 1970. The CZ-2 (originally called the FB-1) was a two-stage vehicle first flown in July 1975 and used as a satellite launcher until 1983. The CZ-3, which made its first orbital flight in January 1984, is composed of the two liquid stages of the FB-1, using **nitrogen tetroxide/UDMH**, and a third stage using **liquid oxygen/liquid hydrogen**.

The CZ-3 was offered to the West, as the commercial Long March 3 launch vehicle, in the late 1980s. From the **Xichang** launch site it has a **payload capability** of 1500 kg to **geostationary transfer orbit (GTO)**, the most common delivery orbit for commercial satellites, while later variants, the LM-3A and LM-3B, have capabilities of 2700 kg and 5000 kg, respectively. The commercial LM-2E [see figure L6] can lift about 3600 kg to GTO with the addition of four liquid propellant (nitrogen tetroxide/UDMH) **strap-on** boosters and a **perigee kick motor (PKM)**.

longeron A main longitudinal structural brace in a **launch vehicle** or **spacecraft** structure.

[See also **stringer, airframe**]

L6: **Long March** LM-2E with four liquid propellant **strap-on** boosters.

LOS An abbreviation for loss of signal. Used typically when a **communications** link with a **spacecraft** has been interrupted (e.g. owing to blockage by a **planetary body**, radio **interference** or during **re-entry** [see **S-band blackout**]). The opposite of 'acquisition of signal' (AOS).
[See also **signal**, **carrier**]

louvre(s) A device used in the thermal control of a spacecraft to shield or expose a **radiator** surface. The angle of the louvres, which resemble a domestic venetian blind, is varied by a bimetallic spring (or, less commonly, by the expansion and contraction of a fluid). Thermal louvres are not generally used on **communications satellites** in geostationary orbit, since other methods of thermal control are more effective. They are, however, sometimes used for spacecraft in **low Earth orbit**, and those on interplanetary trajectories [see figure R2].
[See also **thermal control subsystem, thermal insulation**]

low Earth orbit (LEO) A nominally circular **orbit** of low altitude (typically less than 1000 km), and of short period (approximately 100 minutes) [see figure O1]. Low Earth orbits, commonly abbreviated to LEO, tend to lie below the inner of the two **Van Allen belts** where **radiation** exposure to **spacecraft** can be minimised.

This type of orbit, which includes **polar orbit**, has historically been occupied predominantly by **remote sensing**, military and other satellites which require (i) a close proximity to the Earth's surface for **imaging** and (ii) a short orbital period (typically less than two hours) to allow rapid 'revisits'. Since satellites in low Earth orbits move quickly across the sky and soon disappear below the observer's horizon, they were not considered useful for telecommunications applications and **communications satellites** have tended to occupy the much higher **geostationary orbit**. However, in the 1990s, a number of LEO-based satellite communications constellations were introduced [see **constellation**].

Most manned spaceflight activities have been confined to LEO and it is here that the **International Space Station (ISS)** and most future manned space stations will reside. The Space Shuttle is designed to deliver payloads to LEO.
[See also **orbital decay, drag compensation, re-visit capability**]

low gravity A term used to describe an environment where the acceleration due to gravity has a value less than that on Earth (e.g. the $1/6g$ of the Moon).
[See also **gravity, microgravity, gravitational anomaly, perturbations**]

low noise amplifier (LNA) A type of amplifier designed to deliver a good **signal-to-noise ratio**. In a spacecraft **communications payload** the LNA is usually one of the 'pre-amplifiers' preceding the **channel filter**. As part of a receiving **earth terminal** the amplifier may be combined with a frequency converter: it is then termed a **low noise converter (LNC)**.

low noise converter (LNC) A **low noise amplifier** combined with a frequency converter, usually mounted at the focus of an **earth terminal** antenna [see

L7: **Low noise converter** mounted at the focus of an **earth terminal** antenna. [Mark Williamson]

figure L7]. The unit amplifies the high **frequency** satellite signal and converts it to a lower **intermediate frequency (IF)**. Sometimes called an LNB (or low noise block-downconverter), although LNBs tend to be very **wideband** devices (i.e. they convert the whole of an available band), whereas LNCs can be quite narrow-band devices.

[See also **frequency bands**]

low-pass filter A **filter** with a high-**frequency** cut-off which allows only low-frequency signals to pass.

[See also **band-pass filter**]

low-power DBS – see **direct broadcasting by satellite**

LOX A commonly used abbreviation for **liquid oxygen**.

LRG An abbreviation for laser ring gyro – see **laser gyro**.

LRV – see **lunar roving vehicle**

Luna The name given to a series of Soviet spacecraft designed for lunar research. Luna 2 was the first spacecraft to impact the Moon (in September 1959), Luna 3 was the first to image the lunar far side (in October 1959), Luna 9 made the first 'soft landing' on another **planetary body** (on 3 February 1966), and Luna 10 was the first spacecraft to enter lunar orbit (in April 1966).

[See also **Lunokhod**]

lunar 'lope' A form of locomotion, developed by the **Apollo** lunar **astronaut**s, which proved particularly effective in the 1/6g environment of the Moon; an undulating 'one-legged skip'.

lunar module – see **Apollo**

lunar roving vehicle (LRV) One of three electrically powered motor vehicles used to extend the area of the **Moon** which could be explored during the last three **Apollo** lunar missions (Apollo 15, 16 and 17) [see figure A8]. Variously called a 'lunar rover', 'Moon car' or 'Moon buggy', the LRV was stored in the **descent stage** of the lunar module in a package of volume 0.85 m^3 and unfolded to a length of 3.1 m and a width of 1.82 m. It was capable of carrying two **astronauts**, their equipment and lunar samples at a maximum speed of 14 km h^{-1} (8.7 mph).

luni-solar gravity The collective gravitational attraction of the Moon and Sun; one of the mechanisms that results in the perturbation of a spacecraft's **orbit**. The varying effect of the gravitational fields of the Sun and Moon, as their positions relative to the Earth change, causes a tilting of a satellite's orbit-plane and a north–south oscillation of the spacecraft's position as seen from the Earth. This tendency to wander is corrected by the spacecraft's **orbital control** subsystem.

[See also **perturbations, equilibrium point, triaxiality, solar wind**]

Lunokhod The first **teleoperated rovers** to operate on a **planetary body** other than the **Earth**. The Soviet **Luna** 17 delivered Lunokhod 1 to the **Moon** in November 1970 and Luna 21 deployed Lunokhod 2 in January 1973.

[See also **Sojourner**]

M

M-5 A Japanese **launch vehicle** developed under the auspices of the Institute of Space and Astronautical Science (**ISAS**). The three-stage, **solid propellant** rocket was first launched in February 1997.

[See also **H, N**]

MAC – see **multiplexed analogue component**

Mach number The ratio of the speed of a body in a particular medium to the speed of sound in that medium, such that 'mach 1' corresponds to the speed of sound (named after the Austrian physicist Ernst Mach [1838–1916]). Applicable both to the motion of a body in a fluid and the motion of a fluid in a body (e.g. a rocket **exhaust** in a **rocket engine**). If the Mach number is less than 1 the flow is 'subsonic'; if it is greater than 1 the flow is 'supersonic'; if it is 5 or greater it can also be called 'hypersonic'.

[See also **nozzle**]

magnetometer A device for measuring the intensity and orientation of a magnetic field. Usually carried by planetary **probes** to measure planetary magnetic fields. See **magnetosphere**.

magnetopause – see **magnetosphere**

magnetoplasmadynamic thruster – see **electric propulsion (EP)**

magnetosphere A region surrounding a **planetary body** in which the behaviour of charged particles is dominated by the body's magnetic field. For example, the **Earth**'s magnetosphere takes its general form from the lines of force of the magnetic field which 'loop' from pole to pole like those from an ordinary bar-magnet. The inner region is a 'geomagnetic trap' for charged particles from the **Sun**, which spiral round the lines of force from pole to pole, while the outer is swept into a 'geomagnetic tail' by the **solar wind** [see figure V1]. The edge of the magnetosphere, known as the magnetopause, is compressed to an altitude of about 70,000 km on the sunward side and 'blown' beyond the **Moon**'s orbit on the opposite side. Jupiter also possesses both a magnetic field and a magnetosphere; disruptions caused by its moon Io, which orbits within the magnetosphere, are significant.

[See also **Van Allen belts, ionosphere, atmosphere, magnetometer, geomagnetic storm, space weather**]

magneto-torquer A device which makes use of a planet's magnetic field to stabilise a **satellite**. See **torque coil**.

major axis The long dimension of an ellipse described by a line passing through both foci. The major axis of an elliptical **orbit** about a **planetary body** is the line passing through the body with the **periapsis** at one end and the **apoapsis** at the other. The major axis of an **elliptical antenna** is the longest dimension; the minor axis is the shortest dimension.
[See also **eccentricity, orbital parameters**]

maneuver – American spelling of **manoeuvre**

manipulator arm – see **remote manipulator system**

manned manoeuvring unit (MMU) A self-contained propulsive device carried in the **payload bay** of the Space Shuttle and used by **astronauts** for untethered **extra-vehicular activity (EVA)** [see figure M1]. The MMU is, in effect, a manned space vehicle with its own power supply, **propulsion system**, **inertial guidance** system, controls and displays. Its **propellant** is gaseous nitrogen and it uses 24 fixed **thrusters** to provide position-control with six **degrees of freedom**. When carried, MMUs are stowed in 'flight support stations' aft of the crew compartment. A forerunner to the MMU, called the Astronaut Manoeuvring Unit (AMU), was tested aboard the **Skylab** space station in the mid-1970s.

manned spacecraft Any **spacecraft** designed to carry a **crew**. Any other spacecraft is either an **unmanned spacecraft** or a **man-tended spacecraft**.
[See also **Apollo, Gemini, Mercury, Mir, Salyut, Skylab, Soyuz, Spacelab, Space Shuttle, Space Station, Vostok, Voskhod**]

manoeuvre Any change made to the **attitude**, orbit or **trajectory** of a **spacecraft**, **launch vehicle** or **aerospace vehicle**.
[See also **plane change, mid-course correction, attitude control, orbital control, orbital injection, de-orbit, orbital relocation**]

man-rated An attribute of a **launch vehicle** or **aerospace vehicle** classified as suitable (i.e. sufficiently safe) to carry a **manned spacecraft** or incorporate the facility to carry passengers or **crew**.
[See also **cabin, flight deck**]

man-tended spacecraft A **spacecraft** which usually operates unmanned but is designed to be visited by **astronauts** for short-term activity or the replenishment of **consumable**s; a spacecraft designed for human habitation, but only on a short period basis (cf. a **manned spacecraft**). Otherwise known as a 'man-tended **platform**'.
[See also **unmanned spacecraft**]

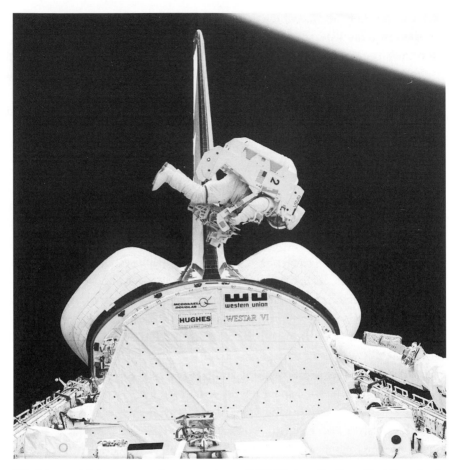

M1: **Astronaut** flying a **manned manoeuvring unit (MMU)** above the **Space Shuttle** payload bay on mission STS 41-B. Note the docking apparatus mounted in front of the MMU (used to manoeuvre satellites into the **payload bay**) and the **sunshield** protecting the Westar VI **communications satellite**. [NASA]

manufacturer – see **prime contractor**

margin An additional amount of a resource, over and above the minimum required, included as a contingency reserve (e.g. a '**propellant** margin', a '**power** margin', etc.) [see **propellant budget**, **power budget**]. Since all spacecraft **hardware** and **consumables** represent mass which must be launched, the magnitudes of the margins have to be very carefully controlled [see **mass budget**]. An amount which is less than the minimum required is referred to as a 'negative margin'.

[See also **link margin**, **beginning of life (BOL)**, **end of life (EOL)**]

Mariner The name given to a series of American planetary exploration **probes** launched towards Mars, Venus and Mercury in the 1960s and 1970s.

Mars Pathfinder A **NASA** spacecraft which landed on Mars on 4 July 1997 carrying the Sojourner **teleoperated rover**, the first to operate on a **planetary body** other than the **Earth** and the **Moon**.

[See also **Lunokhod**]

Marshall Space Flight Center – see **NASA**

Martian meteorite A colloquial term for the meteorite designated **ALH84001**.

mascon An abbreviation for 'mass concentration'; otherwise referred to as a 'gravitational anomaly'. Usually refers to a concentration of mass detected through a change in a spacecraft's **orbital parameters** as it circles a **planetary body**. The term was first used with reference to the Earth's Moon, following studies of the orbital motion of the Lunar Orbiter 5 spacecraft in 1968. High concentrations of material were found to lie beneath many of the lunar maria (seas).

[See also **perturbations**]

maser An acronym for **microwave** amplification by stimulated emission of radiation. See **laser**.

mass budget A method of accounting for the mass of a spacecraft or subsystem with a view to the launch-mass capability of the intended **launch vehicle**. During a spacecraft project, each subsystem engineering discipline aims to design its hardware within its allotted portion of the mass budget. The sum of subsystem masses – the mass budget – allows the overall spacecraft mass to be monitored throughout the project. This sum would usually incorporate a '**balance mass**', typically about 1% of the total, which must be allowed for when balancing the total spacecraft in preparation for launch. As it stands, this budget represents the 'dry mass' of the spacecraft: the addition of propellant for orbital injection and station keeping can more than double the mass.

[See also **power budget, propellant budget, link budget, launch mass, in-orbit mass**]

mass dummy A unit that simulates the mass of **flight hardware**. See **dummy payload**.

mass ratio

(i) The ratio of the mass of a **launch vehicle**'s **payload** to the mass of the total vehicle at **lift-off**. Also called 'payload mass ratio' and 'payload fraction'. Typical figures for a three-stage **expendable launch vehicle** are about 1% to **geostationary orbit** and 2–3% to **low Earth orbit**.

(ii) The ratio of a launch vehicle's initial mass at **lift-off** to the mass of the vehicle without **propellant** (i.e. its **dry mass**). Also called 'propellant mass ratio' and 'propellant mass fraction'.

mass spectrometer An instrument in which **ions**, produced from a sample, are separated by electric or magnetic fields according to their charge-to-mass ratios. This produces a 'mass spectrum', a record of the type and amount of ions present.

mass-limited Jargon: 'limited in terms of the allowable mass'. The mass of anything launched into space is limited by the **performance**, or **payload** capability, of the **launch vehicle**. The mass of all components and **subsystems** is therefore a major **constraint** in any **spacecraft** design [see **mass budget**]. By way of illustration, a **communications satellite** capable of providing sufficient **power** to operate more **transponders** than the mass budget would allow it to carry would be 'mass-limited'.

[See also **power-limited**, **materials**]

mate/de-mate device A framework structure used to mount a Space Shuttle **orbiter** on the **Shuttle carrier aircraft** or remove it from it.

materials One of the most important characteristics for materials used in spacecraft is low density for high structural strength (a high strength-to-weight ratio), owing to the need to limit the launch mass of the spacecraft. However, materials are chosen to match the particular application and other factors are also important (e.g. high stiffness-to-weight ratio, low thermal expansion coefficient, etc.). The following materials are used to varying degrees: **aluminium, beryllium, titanium, carbon composite, ceramics, ceramic–metal composite, metal matrix composite**.

[See also **Mylar, Kapton, Dacron, Nomex, thermal insulation**]

maximum dynamic pressure – see **dynamic pressure**

max-Q

  (i) An abbreviation for maximum dynamic pressure – see **dynamic pressure**.

  (ii) The name given to a band of amateur musicians formed by **NASA** astronauts.

MCC An abbreviation for (i) **mid-course correction** and (ii) **mission control centre**.

MECO An acronym for main engine cut-off, specifically the point at which the **Space Shuttle main engines** cease **combustion** during a **launch** of the Space Shuttle (namely, at about $T + 8$ minutes, 40 seconds).

medium Earth orbit (MEO) A class of orbits with an altitude of approximately 10,000 km; this places them between the two **Van Allen belts** where **radiation** exposure to **spacecraft** can be minimised. Some MEOs are also known as intermediate circular orbits (ICOs); although ICO is often used as a synonym for MEO, the MEO classification is not restricted to circular orbits [see figure O1].

[See also **orbit**]

medium-power DBS – see **direct broadcasting by satellite**

MEMS An acronym for micro-electromechanical system, an extremely small-scale **system** which may form the basis of operation for a **picosat** or **nanosat**.

MEO – see **medium Earth orbit**

Mercury A series of six American **manned spacecraft** launched in the early 1960s, largely to show that an **astronaut** could survive a spaceflight, control a spacecraft, etc. The one-man **capsule** had no room for an ejection seat, but was topped by a **launch escape tower**. **Attitude control** in orbit was accomplished using **hydrogen peroxide** thrusters. A **retro-pack** comprising three **solid propellant** rocket motors was secured to the **heat shield** by metal straps and jettisoned following the retro **burn** and prior to **re-entry**. The shield itself was detached before **splashdown** to expose a rubberised glass-fibre pneumatic cushion, known as a 'landing bag', to absorb the impact.

The first two manned Mercury capsules were launched by **Redstone** launch vehicles onto **sub-orbital** trajectories: the first American in space was Alan Shepard, launched 5 May 1961 on the world's second **manned spaceflight** (of just over 15 minutes duration). The remaining four missions were launched into **low Earth orbit** by **Atlas** boosters: the first American in orbit was John Glenn, on the third Mercury flight, launched 20 February 1962 (duration 4 h 55 m). Only six of the seven pilots chosen for the Mercury flights (known as 'The Mercury Seven') actually flew in a Mercury capsule: Deke Slayton was 'grounded' owing to a suspected heart complaint until 1975 when he flew on the **Apollo–Soyuz Test Project**.

[See also **Gemini, Apollo**]

mesh network A VSAT network in which all terminals are interlinked, and there is no central 'hub' station. The main alternative is a **star network**.

mesopause – see **atmosphere**

mesosphere – see **atmosphere**

MET An abbreviation for:

  (i) **mission elapsed time**;

  (ii) **modularised equipment transporter**.

metal matrix composite (MMC) A material composed of a metal matrix reinforced by threads or fibres of a metallic or ceramic material. Example matrices are aluminium, lead, copper, magnesium and titanium; possible fibres are graphite, boron, silicon carbide, tungsten, molybdenum and alumina. MMCs have high strength and stiffness at high temperatures, good dimensional stability and high thermal and electrical conductivities. They have lower thermal expansion coefficients than the thermosetting polymers and exhibit good resistance to **radiation** damage, little **outgassing** and no low temperature brittleness.

[See also **composites**, **carbon composite**, **ceramic–metal composite**, **materials**]

meteor A meteoroid or **micrometeoroid** made visible by frictional heating in a planet's **atmosphere**.

[See also **Leonids**]

meteorite A **meteor** which has collided with the **Earth**, another **planetary body** or a **spacecraft**. Small versions are called micrometeorites.

[See also **ALH84001**, **micrometeoroid**]

meteoroid A small **celestial body** of natural origin. See **micrometeoroid**.

meteorological satellite A specialised class of **remote sensing** satellite concerned primarily with imaging weather patterns, usually in the visible and infrared parts of the spectrum. The majority of 'metsats' occupy positions in **geostationary orbit**, from where a **constellation** of three satellites can provide full-Earth coverage – except for the polar regions. Weather satellites are also placed in lower altitude, **polar orbit**s which allow regular observations of the polar regions (and less regular observations of the remainder of the Earth).

[See also **Eumetsat**, **NOAA**, **Tiros**, **radiometer**, **sounder**, **space weather**]

Meteosat The name given to a series of **meteorological satellite**s based in **geostationary orbit (GEO)** and operated by **Eumetsat**, the first of which was launched in 1977. The second generation of these satellites was named MSG (for Meteosat Second Generation).

metsat A colloquial abbreviation for **meteorological satellite**.

MF (medium frequency) – see **frequency bands**

MGSE An acronym for mechanical ground support equipment. See **ground support equipment**.

MHz An abbreviation for megahertz – see **hertz**, **frequency bands**.

microgravity A term used to describe an environment in which the acceleration due to gravity is approximately zero (sometimes abbreviated μg); otherwise termed **weightlessness**. The only accelerations are those produced, for example, by **station keeping** manoeuvres, the movements of mechanical devices (e.g. the deployment of equipment on **boom**s), and the movements of any **crew**-members. The microgravity environment of **low Earth orbit**, particularly, is useful for certain types of manufacturing (e.g. that of semiconductor crystals, which tend to grow larger and with fewer defects than they would on Earth).

[See also **gravity**, **low gravity**, **gravitational anomaly**, **perturbations**]

microgravity environment – see **space environment**

micrometeorite A small **meteorite**.

micrometeoroid Any small **celestial body** of natural origin, about the size of a grain of sand or smaller (about 10 μm); a 'small meteoroid' – the distinction is open to discussion! Once the body has collided with the **Earth**, another **planetary body** or a **spacecraft**, it becomes a meteorite or micrometeorite (a meteor is a meteoroid made visible by frictional heating in a planet's atmosphere). Depending on their size and relative velocity, micrometeoroids can degrade spacecraft surfaces (**thermal insulation**, **solar arrays**, etc.) and, at worst, puncture pressurised **cabin**s or **module**s. However, depending on the **orbit**, collision with man-made objects can be more of a problem – see **orbital debris**.

[See also **Leonids**]

microprocessor A single integrated circuit (IC) that performs the basic functions of the central processing unit (CPU) in a computer; the so-called 'computer on a chip'. Microprocessors are invaluable in spacecraft because of their small size and low mass, but can suffer from radiation effects because of their size. See **single event effect (SEE)**.

[See also **radiation hardening**]

microsat A colloquial term for a very small satellite, which may be defined as having a **launch mass** between about 10 kg and 100 kg. The exact definition of a 'microsatellite' is open to discussion, but it is common to define satellite mass ranges as follows:

- smallsat   500–1000 kg
- minisat   100–500 kg
- microsat   10–100 kg
- nanosat   <10 kg.

[See also **smallsat**, **in-orbit mass**, **mass budget**]

microwave The section of the **electromagnetic spectrum** between 0.001 m and 0.3 m (1×10^9–3×10^{11} Hz; 1–300 GHz). See **radio frequency (RF)**.

[See also **waveband**]

microwave lens – see **lens antenna**

microwave link A terrestrial communications system utilising **parabolic antennas**, operating at microwave frequencies (1×10^9–3×10^{11} Hz), typically mounted on tall buildings or towers to increase their range.

microwave monolithic integrated circuit – see **MMIC**

microwave sounder An **active** microwave **payload** designed to measure temperature and/or humidity. The use of **microwave** frequencies allows measurements, known as 'soundings', to be made through cloud. Microwave sounders are usually carried by **meteorological satellites** and **Earth observation** satellites.

[See also **sounder, synthetic aperture radar (SAR), scatterometer**]

mid-course correction (MCC) An adjustment made to the **trajectory** of a **spacecraft** at an intermediate point in its flight, particularly spacecraft travelling between planetary bodies. It is practically impossible to inject a spacecraft into **orbit** around another **planetary body** using only one motor '**burn**'; MCCs allow the initial trajectory to be improved upon as the spacecraft's actual trajectory is determined throughout the flight.

[See also **injection, parasitic station acquisition**]

mid-deck The part of the Space Shuttle **orbiter** directly below the **flight deck** which houses the living quarters, storage lockers and, on some missions, an **airlock** which opens into the **payload bay**.

middle infrared – see **waveband**

mil-spec A colloquial abbreviation for military **specification**, a reference to the high **performance** and high **reliability** characteristics required for military systems.

[See also **hi-rel**]

minisat A colloquial term for a small satellite, which may be defined as having a **launch mass** between about 100 kg and 500 kg. The exact definition of a 'minisatellite' is open to discussion, but it is common to define satellite mass ranges as follows:

- smallsat   500–1000 kg
- minisat   100–500 kg
- microsat   10–100 kg
- nanosat $<$10 kg.

[See also **smallsat, in-orbit mass, mass budget**]

minor axis The shorter axis of an ellipse, **elliptical orbit** or **elliptical antenna**.

[See also **major axis**]

M2: Russian space station **Mir** in **low Earth orbit**. [NASA]

minor planet An alternative term for **asteroid**.

minus-*x* face The face of a **three-axis stabilised** spacecraft which faces west. See **spacecraft axes**.

minus-*y* face The face of a **three-axis stabilised** spacecraft which faces north. See **spacecraft axes**.

minus-*z* face The face of a **three-axis stabilised** spacecraft which faces directly away from the Earth; otherwise known as the anti-Earth face. See **spacecraft axes**.

Mir A former-Soviet **space station**, the name of which is translated as 'peace' [see figure M2]. Its core **module**, launched on 20 February 1986 by a **Proton** launch vehicle, is similar to the **Salyut** 7 space station and has a number of **docking ports** to accept visiting **Soyuz** 'ferries' and **Progress** re-supply 'tankers'. An astrophysics **module** called 'Kvant' ('Quantum') was added in April 1987, and followed by Kvant 2 in 1989, Krystall in 1990, Spektr in 1995 and Priroda in 1996.

[See also **Skylab, gyrodyne**]

mission A specific task assigned to a **spacecraft, launch vehicle** or **crew**. The duration of the mission of an **unmanned spacecraft** is measured in terms of its 'operational **lifetime**'; that of a **manned spacecraft** typically extends from lift-off to touchdown [see **mission elapsed time**]. The mission of a typical launch vehicle is completed with the injection of its **payload** into a designated **orbit** or **trajectory**.

[See also **mission specialist**]

mission control A collection of personnel and computer systems with the task of overseeing and controlling a **launch** or a space **mission**, whether manned or unmanned; the function of the 'mission controllers'; the building or complex in which the function is performed (e.g. **mission control centre**). [See also **launch control centre**]

mission control centre (MCC) A building from which a space **mission** is controlled and monitored [but see **launch control centre (LCC)**].

mission elapsed time (MET) A measure of time since the beginning of a **mission** used by **NASA** in the control and monitoring of **manned spacecraft**. A mission clock is started at the moment of **lift-off** and stopped, in the case of the Space Shuttle, at the instant the **orbiter** comes to rest after landing (on earlier missions at the moment of **splashdown**).

mission specialist A category of person who is part of the crew of the Space Shuttle, but has not been trained as an **astronaut**; typically an engineer or scientist responsible for an aspect peculiar to a particular **mission** (e.g. a scientific **payload** carried by the Shuttle **orbiter**).

[See also **payload specialist**]

mixed oxides of nitrogen – see **MON**

mixer An electrical device in which two or more **input** signals are combined and filtered to produce a single **output** signal. The mixing or 'heterodyning' process is used, among other things, to convert one **radio frequency** to another in a satellite **communications payload**. See **upconverter**, **downconverter**.

[See also **local oscillator, filter**]

mixing products In **telecommunications**, products of the mixing process in an **upconverter** or **downconverter** which may cause **interference** and are therefore undesirable. See **mixer**.

MLI An abbreviation for multi-layer insulation. See **thermal insulation**.

MMC – see **metal matrix composite**

MMH An abbreviation for monomethyl hydrazine ($CH_3NH.NH_2$), a storable liquid **fuel** used in **rocket engines**. See **liquid propellant**.

MMIC An acronym for microwave monolithic integrated circuit, a solid state device combining active **field effect transistors** (FETs) and passive circuit elements that are deposited on a **semiconductor** chip in a single process. MMICs are sometimes used as **gain** stages to drive higher power devices. Typical late-1990s MMICs could, individually, provide output **powers** of up to about 20 W at **C-band** and 5 W at **Ku-band**, although using power combination techniques this could be increased.

[See also **ASIC**]

MMU – see **manned manoeuvring unit**

mm-wave Millimetre-wave: 40–100 GHz. See **frequency bands**.

mobile launch platform In general, any **launch platform** which can be moved to a **launch pad** with a **launch vehicle** mounted upon it. In particular, the mobile launcher platform (MLP) upon which the Space Shuttle is mounted for its transfer from the **vehicle assembly building** to the launch pad [figure C9]. The MLP is 41 m wide, 49 m long, 7.6 m high and weighs 3700 tonnes. It has three openings for the **Space Shuttle main engine** and **solid rocket booster** exhausts, two **tail service mast**s for **propellant** loading and electrical power, and a water-based **sound suppression system**. The MLP was originally used to transport **Saturn** V launch vehicles during the **Apollo** programme, when it also carried a launch tower or '**service structure**'. This is now a permanent fixture at the pad [but see **rotating service structure**].

The main advantage of a mobile platform is that a launch vehicle can be assembled under controlled environmental conditions with access to all ground-servicing equipment and transported to the launch pad only when **integration** and testing is complete. This philosophy, also followed by the **Ariane** vehicles, leaves the launch pad free for actual launches, rather than vehicle integration. The alternative, used almost exclusively for smaller launch vehicles, is to integrate the vehicle inside a mobile environmental housing, which is withdrawn from the launch pad prior to launch.

[See also **Crawler, crawlerway, Sea Launch**]

mobile service structure – see **service structure**

mobile-satellite service (MSS) ITU terminology for a **satellite communications** system providing links between mobile **earth station**s and one or more orbiting spacecraft. The International Telecommunication Union (**ITU**) definition of MSS includes maritime (MMSS), aeronautical (AMSS) and land mobiles (LMSS).

[See also **GMPCS, fixed-satellite service (FSS), broadcasting-satellite service (BSS)**]

modem A contraction of '*mod*ulator–*dem*odulator'; the name usually given to a device for converting digital **data** to audio tones (and vice versa), typically used when connecting two computers via a telephone line. In **communications** terms, a modem is characterised by its ability to transmit data, measured in terms of its **bit rate**.

modularised equipment transporter (MET) A two-wheeled **aluminium** equipment carrier used on the **Apollo** 14 mission and nicknamed 'rickshaw', 'wheel-barrow' and 'caddy cart' by the **astronauts**.

modulation In **telecommunications**, the process whereby a signal is superimposed upon a higher frequency **carrier** wave which 'carries' the signal until it is received and demodulated. Commonly referred to as 'modulating the carrier'. The modulating signal can be analogue or digital, originating from a telephone, TV camera, computer, etc., and is often referred to as the '**baseband**' signal.

There are two primary modulating techniques in general use: **amplitude modulation (AM)** and angle modulation, the latter of which includes **frequency modulation (FM)** and **phase modulation (PM)**. Prior to modulation by one of these techniques, a carrier may undergo a modulation of a different kind, a form of coding, e.g. **pulse code modulation (PCM)**. Other methods less commonly used are differential pulse code modulation (DPCM), **delta modulation (DM)** and adaptive versions of PCM, DPCM and DM (referred to as APCM, ADPCM and ADM, respectively).

An RF signal may also be 'modulated' onto an **electron beam**, for instance in a **travelling wave tube**.

[See also **QPSK (quadrature phase shift keying)**]

modulator A device whose function is to modulate a signal onto a **carrier** wave. See **modulation**.

module

  (i) A self-contained unit or item which can be assembled and tested independently from other items and 'integrated' with the rest of the **spacecraft**, **launch vehicle**, etc., at a later stage [see figures M3, M4]. The term is equally applicable to the smallest 'electronics modules' and the largest **subsystem** components [e.g. **payload module**, **service module**, **propulsion module, antenna module, docking module**].

  (ii) A self-contained, and usually separable, part of a spacecraft or launch vehicle with a specified task [e.g. **command module, lunar module, crew** module, **habitable module, re-entry** module, **extension module**].

[See also **International Space Station (ISS), Mir, integration**]

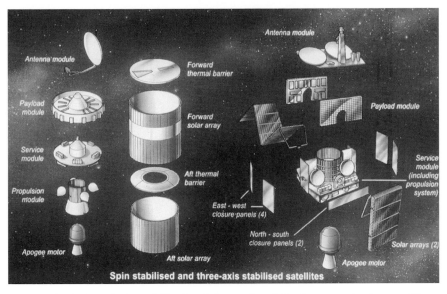

M3: Modular design of **spin-stabilised** and **three-axis stabilised** communications satellites [see **module**]. [Mark Williamson/Willis Inspace]

Molniya

(i) The name given to a series of Soviet/Russian **communications satellites** which operate from **Molniya orbit** [see figure M5].

(ii) A Russian **launch vehicle** developed in the late 1950s and first flown in 1961. It had a **payload** capability of about 1500 kg to Molniya orbit.

[See also **Proton, Energia, Angara, Tsyklon**]

Molniya orbit An **elliptical orbit** with an inclination of about 63°, an **apogee** and **perigee** of approximately 40,000 km and 500 km, respectively, and a period of 12 hours [see figures M5, O1]. The **orbit** was named after the Soviet 'Molniya' **communications satellites** which pioneered its use. The Molniya orbit's high inclination allows satellites to provide coverage of high latitudes, while its 12-hour period produces two **coverage area**s on opposite sides of the Earth, as a result of its two apogees. Full-time coverage of a given area is typically provided by three satellites synchronised in similar orbital paths so that each one contributes about eight hours of continuous service as it approaches and recedes from its apogee. A disadvantage is the orbit's low perigee, which increases **atmospheric drag** on the spacecraft and increases radiation exposure, owing to its passage through the **Van Allen belts**.

[See also **tundra orbit, geostationary orbit**]

M4: Integration of Skynet 4 payload and service **module**. [British Aerospace]

momentum bias A property given to a spacecraft by an **attitude control**
subsystem incorporating a momentum wheel; a system using a momentum
wheel to stabilise a **three-axis stabilised** spacecraft. See **momentum wheel**.
momentum dumping The practice of reducing the angular momentum of a
momentum wheel or **reaction wheel** which has reached its maximum

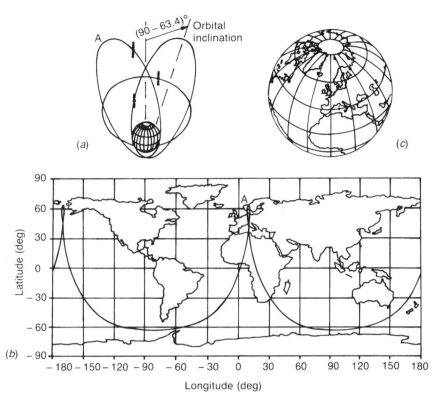

(b)

M5: **Molniya orbit**: (a) three-satellite **Molniya** system; (b) **ground track** of three satellites; (c) view from **apogee** (A).

allowable rotation speed. Since any change in the wheel's rotation speed causes the spacecraft to rotate, a **thruster** is fired as the wheel is decelerated to maintain the spacecraft's **attitude**. The process is alternatively known as 'off-loading momentum' or 'momentum off-loading'.

momentum off-loading – see **momentum dumping**

momentum wheel A wheel mounted in a **three-axis stabilised** spacecraft used to provide gyroscopic stability [see figure M6].

A spinning momentum wheel, aligned with the spacecraft's north–south (**pitch**) axis, acts like a **gyroscope** to provide a resistance to perturbing forces in **roll** and **yaw** while allowing the body to rotate once every 24 hours about the pitch axis [see **spacecraft axes**]. An alternative system uses two wheels set at a slight angle to each other but symmetrically on either side of the N–S axis. Run together the wheels will act as a double-sized single wheel, but

M6: Spinning **momentum wheel** (with top cover removed). [Teldix]

changing the speed of one will alter the balance to provide a degree of **attitude control**. These arrangements are examples of the so-called 'momentum bias' system.

A momentum wheel spins constantly in one direction with typical speeds between 6000 and 12,000 rpm, depending on the type of wheel, with a variation of plus or minus 10% of the nominal value for control adjustments. Other attitude control systems use the **reaction wheel** – which by definition has a nominal speed of zero and can be rotated in either direction – as part of a 'zero momentum' system. Both types of wheel are sometimes called inertia wheels. A momentum wheel mounted in a gimbal is known as a **control moment gyro (CMG)**.

[See also **momentum dumping**, **gyrodyne**]

MON An acronym for 'mixed oxides of nitrogen', a **liquid propellant** comprising **nitrogen tetroxide** (N_2O_4) with an added proportion of nitric oxide (NO). The percentage (by weight) of added NO is indicated by a figure in a particular propellant designation, e.g. 'MON3'.

monocoque A support structure which alone carries all or most of the structural stresses placed upon it (e.g. the **thrust structure** of a simple **spin-stabilised** satellite).

[See also **balloon tank**]

monomethyl hydrazine – see **MMH**

monopropellant A chemical **propellant** which comprises a single component.
Liquid propellants can be either monopropellants or **bipropellants**. **Solid
propellants** are considered by some as bipropellants because they have two
components, but can also be regarded as monopropellants since they are
handled and used as a single-component propellant: the **fuel** and **oxidiser**
are mixed during the manufacturing process (cf. liquid bipropellants which
are stored separately and mixed in a **combustion chamber**). **Hydrazine**
(stored as a liquid and decomposed to form a gas) is an example of a single-
component monopropellant used in **reaction control system**s.
[See also **hybrid rocket, hydrazine thruster**]

monopulse RF sensor – see **RF sensor**

Monte Carlo analysis A mathematical or statistical **simulation** method which
uses sequences of random numbers to simulate a **system** or process. The
physical or mathematical system under consideration is described in terms
of probability density functions (PDFs) and the Monte Carlo simulation
operates by random sampling from these PDFs. Many simulations or 'trials'
may be performed and the desired result is taken as an average of the results.
The name Monte Carlo was coined during the World War II Manhattan
Project (to develop the atom bomb) because of the similarity between
statistical simulation and games of chance, as played in Monte Carlo, the
capital of Monaco.

Moon The natural **satellite** of the **Earth**. Usage: when used as a proper name
(when referring directly to the Earth's moon), Moon should have an initial
capital (e.g. **Moon Agreement**); when referring to any other natural satellite,
it should have a lower case initial (e.g. 'the moons of Mars). A related adjective
is 'lunar' ('of the Moon', from luna, Latin for Moon) – see, for example, **luni-
solar gravity**.
　　Physical data: equatorial diameter 3476 km; mass 7.15×10^{22} kg; **escape
velocity** 2.38 km s^{-1}; surface **gravity** 1.62 m s^{-2}; average distance from Earth
3.844×10^5 km; average light travel time from Earth 1.25 s; axial tilt 1.5°;
orbital inclination 5.145°; **orbital period** 27.32 days.
[See also **light (velocity of)**]

Moon Agreement The colloquial name for the treaty of **space law** whose official
title is "Agreement Governing the Activities of States on the Moon and Other
Celestial Bodies". The Moon Agreement was adopted by the United Nations
General Assembly on 5 December 1979, opened for signature on 18 December

1979 and entered into force on 12 July 1984. It was one of four main space treaties derived from the **Outer Space Treaty** of 1967.

[See also **Liability Convention, Registration Convention, Rescue Agreement, celestial body**]

motor – see **rocket motor**

motor case The outer structure or pressure shell of a solid rocket motor (SRM). The case is, in effect, a receptacle for the **propellant**, which is poured in during manufacture and 'moulds itself' to the contours of the case [see figure R7].

[See also **rocket motor, solid rocket booster, propellant grain, flame inhibitor**]

MOU An abbreviation for memorandum of understanding, industrial terminology for a document signed by two or more parties who wish to undertake a joint programme of work (usually in cases where a commercial contract would not be applicable and sometimes in advance of a contract).

MPD An abbreviation for magnetoplasmadynamic. See **electric propulsion (EP)**.

MPEG An acronym (pronounced "empeg") for moving or motion picture experts group, a group of people that defines technical standards for full-motion video.

MPLM An abbreviation for Multipurpose Pressurised Logistics Module, a **module** of the **International Space Station (ISS)** supplied by the Italian Space Agency (ASI).

MSG An abbreviation for Meteosat Second Generation. See **Meteosat**.

MSS An abbreviation for:
  (i) **mobile-satellite service**
  (ii) **multispectral scanner**
  (iii) mobile **service structure**.

MTBF An abbreviation for mean time between failures, used in **reliability** assessments. The abbreviation is sometimes used for 'mean time *before* failure', in which case it is equivalent to **MTTF**. Failure-rate is the inverse of MTBF ($\lambda = 1/\text{MTBF}$).

[See also **bathtub curve, probability distribution**]

MTTF An abbreviation for mean time to failure, used in **reliability** assessments.

MTTR An abbreviation for mean time to repair.

multicasting A subset of **broadcasting**, in which **data** is transmitted to many recipients at the same time, but not necessarily to all receiving stations in the network (for example, some subscribers may choose to receive only a part of a service). The term is widely used in connection with **Internet** communications; the equivalent in **satellite communications** is 'point-to-multipoint' [see **link**].

[See also **narrowcasting, unicasting**]

multi-layer insulation (MLI) A type of thermal insulation colloquially termed a thermal blanket. See **thermal insulation**.

[See also **Mylar, Kapton, Dacron, Nomex**]

multimedia A **satellite communications** application involving the delivery of information with different content formats: e.g. video, audio, still images, graphics, animation, sound and text. In general usage, the term is typically linked with **broadband**, as in 'broadband multimedia applications', since wide bandwidths are usually required to handle multimedia content.

[See also **digital compression, digital video broadcast (DVB), Internet**]

multipaction An undesirable effect which can occur, in high-power **radio frequency (RF)** equipment operating in a vacuum, if the electric field in a **transmission line** or other component is strong enough to accelerate **electron**s to a velocity at which they are capable of causing the 'secondary emission' of other electrons (e.g. when in collision with an RF electrode or **waveguide** wall). Under the right conditions, the secondary electrons can cause subsequent multiple emissions resulting in an electron avalanche or 'breakdown'.

Multipaction occurs most readily in **microwave** components operating at around 1 GHz, particularly **filter**s. Its effects generate **broadband noise** and may, in some equipment, lead to a gas discharge which can cause physical damage. The severity of the effect depends on parameters including electric field strength, RF power level, **frequency** and the dimensions of the transmission line. Multipaction is sometimes called 'multipactor' (chiefly US).

multipath An undesirable effect caused when a radiated telecommunications **signal** takes more than one path from **transmitter** to **receiver**. The intended path is always the shortest, 'line-of-sight' path, but in built-up areas particularly, signals may be reflected from solid objects causing them to take a longer path to the receiver. Since both signals travel at the speed of **light**, the reflected signal reaches the receiver fractionally later than the main signal.

Considering the signal as a simple waveform, multipath constitutes the addition of two waves which combine 'constructively' or 'destructively', to produce a variation in signal level. In the case of a complex TV signal, multipath is manifested on the screen as a faint 'ghost image' or images offset to the right of the main image, since the **electron beam** which produces the picture scans from left to right across the tube. The secondary signals reach the receiver slightly later than the primary signal, having taken a longer path.

multiple access A coding method for information transmission between a potentially large number of users. There are three principle types: see **frequency division multiple access (FDMA), time division multiple access (TDMA), code division multiple access (CDMA)**.

multiple beam A pattern of several **spot beams** formed by a **satellite** communications **antenna** with multiple **feedhorns**, placed in a cluster at the focus of the antenna **reflector**. The multiple feedhorn configuration has two principal uses. If different feedhorns are fed with different power levels and/or different phases, the beam can be shaped to illuminate an irregularly shaped area more efficiently. This can save wasting power over areas of empty sea, for instance, and obviate possible interference due to **overspill**, advantages which are not possible with a simple circular or elliptical beam. Alternatively, multiple feedhorns can provide sophisticated communications links between individual beams by switching signals from one beam to another and back in quick succession.
[See also **beam-forming network, shaped reflector, satellite switching, beam-hopping**]

multiplexed analogue component (MAC) A **television** transmission standard developed by the UK Independent Broadcasting Authority (IBA) for use with European DBS (**direct broadcasting by satellite**) in the 1980s. The MAC system electronically compressed the luminance and chrominance (brightness and colour) information and transmitted them sequentially rather than simultaneously as with previous terrestrial TV standards. Separating the two information-sets eliminated interference between them, producing a clearer picture, but since MAC was incompatible with standard (terrestrial TV) receivers, an adapter or converter was required. MAC has since been superseded by **digital compression**.

multiplexer A device for combining or separating different signal frequencies; in effect, a multiple **channel filter**. For example, where a number of **channels** share the **bandwidth** available in a satellite **repeater**, an input multiplexer separates the channel-frequencies and routes each **carrier** to its own

amplifier chain. Once amplified, the channels are recombined in an output multiplexer for the return transmission. The individual devices are colloquially known as the input 'mux' and output 'mux', or 'Imux' and 'Omux', respectively. A two-channel multiplexer is called a 'diplexer'.

multiplexing In **telecommunications**, the process of placing more than one **signal** on a **carrier**, **subcarrier** or **channel**. See **frequency division multiplexing (FDM)** and **time division multiplexing (TDM)**.

multipoint-to-point In **satellite communications**, a method of linking **earth stations** such that **signals** are transmitted from many individual stations to a single controlling station. See **link**.

Multipurpose Pressurised Logistics Module (MPLM) A **module** of the **International Space Station (ISS)** supplied by the Italian Space Agency (ASI).

multi-purpose satellite A **satellite** which carries payloads for different applications (e.g. **communications** and meteorology).

multispectral Composed of several parts of the spectrum (typically ten or less); images composed of more than ten spectral bands are known as **hyperspectral**. In **remote sensing**, the term is used to describe the response of a film or **sensor** that produces an **image** derived from distinct parts of the spectrum; usually displayed as a colour image. Multispectral images are typically of lower **resolution** than **panchromatic** (monochromatic) images, but can be used to identify particular characteristics such as moisture content and crop-health.

[See also **waveband, electromagnetic spectrum, spectral resolution**]

multispectral scanner A **payload** on an **Earth observation** satellite which gathers **data** composed of several parts of the spectrum; often abbreviated to MSS.

[See also **pushbroom scanner**]

multistage rocket A **launch vehicle** with several **stage**s.

mux – a colloquial abbreviation for **multiplexer**

Mylar A plastic material used for spacecraft **thermal insulation** (a trademark). Used in **multi-layer insulation (MLI)**, it is typically aluminised on one side only so that it acts as a low conductivity spacer. A more complex, high performance construction uses Mylar film metallised (with aluminium or gold) on both sides separated by low conductance interlayers (e.g. silk, **Nomex** or **Dacron** net).

[See also **Kapton**]

N

N (launch vehicle) A three-stage Japanese **launch vehicle** based on the American **Delta** and built under license in Japan. The propellants used are **liquid oxygen/kerosene** in the first stage, **nitrogen tetroxide/aerozine-50** in the second and a **solid propellant (CTPB)** in the third stage and **strap-on** solid boosters. The N launch vehicle was produced in two variants which are no longer **operational**: the N-I, which used three strap-ons, had a **payload** capability of 130 kg to **geostationary orbit (GEO)**; the N-II, which used nine strap-ons, had a capability to lift 350 kg to GEO (or about 700 kg to **geostationary transfer orbit (GTO)**, the equivalent of a Delta 2910). The N-I made its first flight, from the launch site on **Tanegashima** Island off the south coast of mainland Japan, in September 1975, and its seventh and last in September 1982; the N-II made its first of eight flights in February 1981 and its last in February 1987.
[See also **H (launch vehicle)**]

N-1 A Soviet **launch vehicle** developed in the 1960s as part of an ill-fated manned lunar programme (N stands for 'nositel' or 'carrier'). All four launch attempts, made between February 1969 and November 1972, were failures and the programme was cancelled.

NACA An abbreviation for National Advisory Committee for Aeronautics, the precursor to **NASA**.

nadir The point on the celestial sphere directly below an observer and diametrically opposite the **zenith**; the point directly beneath an orbiting **spacecraft** (used particularly for **Earth observation** satellites).

nanosat A colloquial term for a very small satellite, which may be defined as having a **launch mass** less than about 10 kg. The exact definition of a 'nanosatellite' is open to discussion, but it is common to define satellite mass ranges as follows:

- smallsat   500–1000 kg
- minisat   100–500 kg
- microsat   10–100 kg
- nanosat   <10 kg.

[See also **picosat, smallsat, in-orbit mass, mass budget**]

narrowcasting The dissemination or transmission of a broadcast-type signal to a closely defined audience with a particular interest, as opposed to **broadcasting** which does not imply application to a particular audience. [See also **multicasting, unicasting, link**]

NASA An acronym for National Aeronautics and Space Administration, a body formed in October 1958 to coordinate United States research and development into aeronautics, **space science** and **space technology**. NASA was based on the structure of its predecessor, the National Advisory Committee for Aeronautics (NACA).

NASA's main establishments are as follows:

- Headquarters, Washington, DC (includes six programme offices: Office of Aeronautics and Space Technology (OAST); Office of Space Flight; Office of Space Science and Applications (OSSA); Office of Space Station; Office of Space Tracking and Data Systems; Office of Commercial Programs);
- Ames Research Center*, Moffett Field, California (aeronautics research, computer science, flight simulation, biomedical research, etc.);
- Dryden Flight Research Center*, **Edwards Air Force Base**, California (flight research [see **X-1, X-15, Enterprise**]);
- Glenn Research Center (until 1999, Lewis Research Center), Cleveland, Ohio (space **power** and **propulsion system**s, etc.);
- Goddard Space Flight Center (GSFC), Greenbelt, Maryland (meteorological, communications and scientific satellites);
- Jet Propulsion Laboratory (JPL), Pasadena, California (interplanetary spacecraft);
- Johnson Space Center (JSC), Houston, Texas (manned spaceflight, **astronaut** training and **mission control**); JSC also operates the White Sands Test Facility, New Mexico (**propellant** and **rocket engine** testing, etc.).
- **Kennedy Space Center**, Cape Canaveral, Florida (launch facilities);
- Langley Research Center, Hampton, Virginia (advanced aeronautical and **materials** research);
- Marshall Space Flight Center, Huntsville, Alabama (**launch vehicle** development, space science, etc.);
- Stennis Space Center (formerly National Space Technology Laboratories), Bay St. Louis, Mississippi (static tests of large propulsion systems);

---

* Ames and Dryden were merged in 1981 to form the NASA Ames Dryden Flight Research Facility, referred to as 'Ames–Moffet' and 'Ames–Dryden'.

- Wallops Flight Facility, Wallops Island, Virginia (launch facilities for **sounding rockets**, scientific balloons, etc.).

NASDA An acronym for National Space Development Agency of Japan, a body responsible for the development of Japanese space technology and the implementation of its applications. NASDA was established in October 1969 and its first satellite, ETS-1/Kiku, was launched in 1975. Facilities include the Tsukuba Space Center (research and development, **tracking** and control), Kakuda Propulsion Center (**propulsion system**s) and the **Tanegashima** Space Center (**launch site**).

[See also **ISAS**]

National Space Technology Laboratories – see **NASA**

navigation satellite A satellite, usually part of a system providing global coverage, which gives accurate position information to mobile terrestrial **receivers**. The definition includes military and civilian systems serving maritime, aeronautical and land mobiles.

[See also **Navstar, Global Positioning System (GPS), Glonass, COSPAS-SARSAT, mobile-satellite service (MSS)**]

Navstar An acronym for *Nav*igation *s*ystem with *t*iming *a*nd *r*anging; the name given to a series of American **navigation satellites** which formed the first-generation **Global Positioning System (GPS)**. Later satellites were simply named GPS.

NDT An abbreviation for **non-destructive testing**.

near infrared – see **waveband**

Near-Earth object A small **asteroid**al body, commonly abbreviated to **NEO**.

Near-Earth space The **space environment** close to the **Earth**; in essence, the opposite of **deep space**. The term near-Earth space is inexact, but typically refers to space within the Earth–**Moon** system.

[See also **space**]

negative margin – see **margin**

NEO An acronym for near Earth object, a small asteroidal body, the orbit of which brings it relatively close to the **Earth**. This has brought NEOs into consideration as potential 'Earth impactors' with great destructive force. They are also known as 'Earth-crossing **asteroid**s' because their orbits cross the Earth's **orbit** around the **Sun**.

[See also **space debris**]

NERVA An acronym for nuclear engine for rocket vehicle application, the name given to an American nuclear-powered rocket development programme of the 1960s, which was cancelled largely due to lack of funds. See **nuclear propulsion**.

[See also **nuclear reactor**]

neutraliser A device attached to an **ion engine** which emits a flux of **electron**s to prevent ions being attracted back towards the engine and to avoid a build up of positive charge in the vicinity of the spacecraft.

Newtonian telescope A type of astronomical reflecting telescope in which light is reflected from a parabolic primary mirror onto a plane secondary and through an aperture in the side of the telescope to an eyepiece [see figure N1]. Named after Sir Isaac Newton [1642–1727], the English scientist and mathematician.

[See also **Cassegrain reflector**, **Coudé focus**, **Schmidt telescope**, **Gregorian reflector**, **Hubble Space Telescope**]

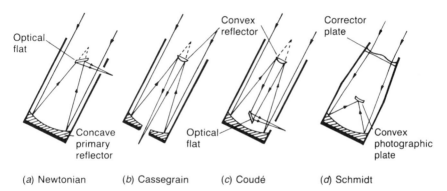

N1: Telescope optics [see **Newtonian telescope**, **Cassegrain reflector**, **Coudé focus**, **Schmidt telescope**].

NGSO An abbreviation for non-geostationary orbit. Usually refers to **low Earth orbit (LEO)** or **medium Earth orbit (MEO)**.

[See also **geostationary orbit**]

NGST An abbreviation for next-generation space telescope, the **NASA** project to develop a follow-on to the **Hubble Space Telescope (HST)**.

NHMF An abbreviation for non-polluting **hypergolic** miscible fuel, a rocket fuel developed by the US Navy in the late 1990s. It is a non-toxic alcohol-based catalyst which ignites spontaneously when mixed with an oxidiser, such as hydrogen peroxide (H_2O_2), and offers an alternative to hydrazine (N_2H_4), which is highly toxic.

nickel–cadmium (NiCd) cell A commonly used voltaic cell in a spacecraft **battery**.

nickel–hydrogen (NiH$_2$) cell A commonly used voltaic cell in a spacecraft **battery**.

nickel-metal-hydride (NMH) cell A voltaic cell used in a spacecraft **battery**.

NIR An abbreviation for near infrared, part of the **electromagnetic spectrum** adjacent to the red end of the visible spectrum; commonly defined in terms of the **wavelength** range 0.7–1.4 μm.

[See also **waveband**]

nitrogen tetroxide (N$_2$O$_4$) A storable liquid **oxidiser** used in **rocket engines**. See **liquid propellant**.

NMH An abbreviation for **nickel–metal-hydride**, a type of **battery**.

NOAA An acronym for National Oceanic and Atmospheric Administration, a US Government body that operates **meteorological satellites**; a series of weather satellites based in **heliosynchronous orbit** and operated by NOAA. The system evolved from a long line of **experimental** and **operational** satellites beginning with Tiros 1, launched on 1 April 1960. The Tiros-N/NOAA series was introduced in 1978 and subsequently evolved into an advanced **generation**, beginning with NOAA 9 which was launched in December 1984. The Administration also operates the GOES (Geostationary Operational Environmental Satellite) series.

[See also **Eumetsat**]

nodes

(i) The two points at which the **orbital plane** of a **spacecraft** or **celestial body** intersects a reference plane, such as the equatorial plane of a **planetary body** or (for planetary bodies) the ecliptic. If the orbiting body moves across the reference plane from south to north, the node is referred to as an ascending node; if it moves from north to south, the node is a descending node.

(ii) Elements of the **International Space Station (ISS)** designed to interconnect the pressurised modules. Each node has six **docking port**s, one at either end and four around the circumference of the cylindrical **module**.

noise An undesirable electrical disturbance which, if sufficiently severe, can mask the **signal** in a communications system (see **signal-to-noise ratio**, **carrier-to-noise ratio**). Noise has a number of origins: the equipment itself (see **thermal noise**, **shot noise**); other electrical devices such as motors and ignition systems; natural terrestrial or atmospheric noise such as that from electrical storms (significant only below about 30 MHz); and extraterrestrial noise, particularly from the Sun. Noise caused by natural electrical discharges in the atmosphere is sometimes called 'static'.

[See also **noise figure, noise power, noise power density, noise temperature, Boltzmann's constant**]

noise factor (*F*) – see **noise figure (NF)**

noise figure (NF) A measurement of the **noise** contribution of a device (e.g. an **amplifier**) expressed relative to a theoretical noise-free amplifier at a reference temperature. The noise figure of a device is its 'noise factor' expressed in dBW [see **decibel**] – noise factor, *F*, can be calculated using the expression below:

$$F = P_N \,/\, GkT_aB$$

where P_N is the **noise power** at the output, *G* is the **gain**, and kT_aB is the noise power of the theoretical source resistance at ambient temperature (where *k* is **Boltzmann's constant**, T_a is the normal ambient temperature of the source resistor, taken to be 290 K, and *B* is the effective noise **bandwidth** of the system in **hertz (Hz)**). If no noise is contributed by the device, the noise factor is 1 and the noise figure is 0 dBW.

noise power A quantitative measure of **noise** in a communications system (units dBW), analogous to **carrier** power; the quantity used to calculate the **carrier-to-noise (power) ratio**.

Noise power, P_N, is given by the expression

$$P_N = kTB$$

where *k* is **Boltzmann's constant**, *T* is the noise temperature of the source (measured in **kelvin (K)**), and *B* is the effective noise **bandwidth** of the system in **hertz (Hz)**. See **noise temperature**.

[See also **noise figure**]

noise power density (NPD; N_0) In **telecommunications**, a quantitative measure of the level of **noise** in every 1 Hz of **bandwidth** (units dBW Hz^{-1}). NPD is given by the expression

$$N_0 = P_N \,/\, B = kT$$

where P_N is the noise power (dBW), *B* is the bandwidth (dBHz), *k* is **Boltzmann's constant** and *T* is the **absolute temperature** (dBK).

[See also **noise temperature, noise figure, decibel, kelvin, hertz**]

noise temperature In **telecommunications**, a concept which relates the **noise power** at the source (or other reference point) of a system to the temperature of a reference resistance from which the same 'thermal power' is available. Put another way, the noise temperature is the temperature at which the reference produces an equivalent noise power to that produced by the system under consideration, over the same **frequency** range (or **bandwidth**). It is given by the expression:

$T = P_N / kB$

where P_N is the noise power (dBW), k is **Boltzmann's constant** and B is the bandwidth (dBHz). The units of noise temperature are dBK (in the same way that **power** is referenced to the watt in dBW, temperature is referenced to the **kelvin** in dBK).

For an orbiting **satellite** 'looking at' the Earth, the noise temperature is the sum of the **thermal noise** of the **receiver** equipment and the noise due to the Earth. The Earth's temperature from space is an average of 290 K, so its noise temperature is 24.6 dBK ($10 \log_{10} 290$). The noise temperature of an **earth station** 'looking at' the satellite is increased in the event of rainfall since it 'sees' warm rain as opposed to cold sky [see **rain attenuation, subreflector**].

The Sun is a strong variable noise source with a noise temperature of about 10^6 K at 30 MHz and at least 10^4 K at 10 GHz under quiet Sun conditions – large increases occur at times of greater solar activity.

[See also **noise figure, hertz**]

Nomex A low conductance/insulating material (a trademark) used, for example, as an interlayer between successive layers of **Mylar** film in spacecraft **thermal insulation**.

[See also **multi-layer insulation (MLI), thermal protection system (TPS)**]

nominal

(i) 'Theoretical' or 'intended' (a 'nominal value': e.g. 'nominal **capacity**', 'nominal **orbit**', etc. [see **nominal orbital position**]). Usage, e.g.: 'its deviation is nominally zero'.

(ii) Jargon: used to describe the performance of a device, subsystem, etc., which is 'within specification' (i.e. operating as intended). Usage, e.g.: 'power subsystem is nominal'.

nominal orbital position The intended, average **orbital position** of a **satellite** in **geostationary orbit**. An allocated orbital position cannot be maintained precisely, owing to the spacecraft's susceptibility to a variety of perturbing forces – chiefly **gravitational perturbation**s and the **solar wind** [see **libration**]. The gravitational attraction of the Sun and Moon, for example, tends to 'pull' the satellite from its nominal position while the solar wind 'pushes'. The mass and surface area of the spacecraft and the relative

positions of the astronomical bodies govern the magnitude of these effects
and vary the amount of **thruster** fuel utilised for **station keeping**.
[See also **equilibrium point, perturbations, orbit, orbital spacing, orbital
control, collocate**]

non-destructive testing (NDT) A type of testing conducted on **hardware** required
to maintain its mechanical and **operational** characteristics after testing (e.g.
safety-critical hardware such as **rocket engines** and **rocket motor**s).
Techniques include X-ray, ultrasound and thermal imaging.

non-operational The opposite of 'operational'. See **operational**.

NORAD An acronym for North American Aerospace Defense Command, a
military agency based in the United States (within Cheyenne Mountain, near
Colorado Springs) and supported by Canada. Until 1986, in addition to
providing warning of attacks by intercontinental ballistic missiles (**ICBMs**)
and other guided missile systems, NORAD tracked and monitored **spacecraft**
and other objects in Earth **orbit**. Since then, the US **Space Command** has
operated the US **Space Surveillance Network (SSN)** to maintain the official
'Satellite Catalog'.
[See also **orbital debris**]

normal distribution – see **probability distribution**

Northern Cosmodrome A Russian **launch site** near the town of Plesetsk (at
approximately 62.8° N, 40.1° E) used predominately for the launch of
satellites into high **inclination** orbits. The Northern Cosmodrome, which
conducted its first launch on 17 March 1966, is largely military in nature, the
Russian equivalent of the US **Vandenberg Air Force Base**.
[See also **Baikonur Cosmodrome, Kapustin Yar, Svobodny**]

north–south station keeping – see **orbital control**

nose cone A colloquial term for the forward end of a **rocket**, the section which
generally contains a **payload**. The pointed tip of a rocket booster is
technically known as the '**ogive** section'. The section containing the payload
is referred to generally as a **fairing** and specifically as a **payload fairing**,
payload shroud or launch shroud.

nose fairing An alternative term for **nose cone**.
[See also **ogive**]

nozzle The part of a **rocket engine, rocket motor** or **thruster** which converts the
thermal energy of the **combustion** gases to the kinetic energy of the exhaust
plume [see figures N2, R7]. Known more fully as an 'exhaust nozzle' or
'expansion nozzle'.

N2: **Space Shuttle main engine** showing **nozzle** with **regenerative cooling**. [Rockwell International]

A typical nozzle comprises two sections: a convergent section and a divergent exit cone. The narrowest part of the nozzle, called the throat, is designed to maintain the required pressure within the **combustion chamber** and regulate the outflow of combustion gases. The gases move naturally from the high pressure of the chamber to the vacuum of space, expanding and accelerating rapidly as they leave the chamber. The throat is the region of transition from subsonic to supersonic flow. The exit cone controls the expansion of the exhaust plume [see **expansion ratio**]. Some nozzles are extendable, to increase their efficiency at higher altitudes.

The convergent–divergent shape of the nozzle is due to the variation in the rates of change of the velocity and specific volume (volume per unit mass) of the combustion gases: since the rate of increase of specific volume is at first less than, but finally greater than, the rate of increase of velocity, the nozzle must first converge and then diverge.

[See also **exhaust velocity, flow separation, regenerative cooling, Mach number, plug nozzle**]

NTO An abbreviation for **nitrogen tetroxide (N_2O_4)**.

NTR An abbreviation for **nuclear thermal rocket**.

nuclear propulsion A form of rocket propulsion which uses the heat from a nuclear reactor to vaporise a **propellant** (typically **liquid hydrogen**), the resultant gas being ejected from a **nozzle** to produce **thrust** in the same manner as a chemical propulsion system. Although there are currently no operational systems, nuclear propulsion remains an attractive prospect for the future since it promises a very high **specific impulse** (around 1000 s compared with an average of about 400 s for **liquid hydrogen/liquid oxygen**). [See also **nuclear thermal rocket (NTR), NERVA, chemical propulsion, electric propulsion**]

nuclear reactor A device in which a nuclear reaction is maintained and controlled to produce **energy**. Reactors have flown on a number of spacecraft, chiefly Russian in origin, but are no longer common. See **radioisotope thermoelectric generator (RTG)**.
[See also **nuclear propulsion, NERVA, nuclear thermal rocket (NTR)**]

nuclear thermal rocket (NTR) A thermodynamic **propulsion system** that uses a **nuclear reactor** to heat **propellant**, such as **liquid hydrogen**, to extremely high temperatures (e.g. 3400 K). Not **operational**.
[See also **nuclear propulsion, NERVA**]

nutation A periodic variation in the **precession** of a spinning body (e.g. a planet, **satellite** or **gyroscope**); a 'nodding' motion superimposed on the precession of the spin axis [see figure N3].
[See also **nutation damper**]

nutation damper A device designed to reduce the **nutation**, or 'nodding' motion, of a spinning spacecraft. Most important for **spin-stabilised** spacecraft, but

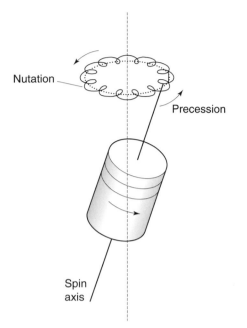

Nutation

Precession

Spin
axis

N3: Spin axis **precession** and **nutation**.

also used for **three-axis stabilised** satellites which are spin-stabilised during their **transfer orbits**. Active nutation damping is realised using **reaction control thrusters**: since the spacecraft is spinning, most thruster firings have to be made in **pulse-mode**, pulsing each time the spacecraft's rotation brings the thruster into alignment with the desired **thrust vector**. Passive damping involves a device containing either a viscous liquid in a sealed circuit or a steel ball in a gas-filled tube: the energy of nutation is converted to heat in the working-fluid, typically mercury in the first instance and neon in the second, and the motion is damped out.

[See also **DANDE**]

Nyquist frequency The minimum sampling frequency required to reconstruct a **signal** (i.e. at least twice the maximum **frequency** contained in the signal); also known as the Nyquist rate. In practice, a factor of two is too low to achieve an accurate representation of the **data** and a more realistic sampling rate would be five or more times the maximum frequency. For **remote sensing** systems, the Nyquist frequency is the highest frequency that can be detected by the imaging system.

O

OBDP – see **on-board data processing**

OBP – see **on-board processing**

observatory Usually an institution or building designed and equipped to perform observations of astronomical or meteorological phenomena. The term is also applied to spacecraft designed to conduct astronomical observations from space (e.g. Orbiting Solar Observatory (OSO), Solar and Heliospheric Observatory (SOHO)).
[See also **Hubble Space Telescope**, **Lagrange point**]

OEM An abbreviation for original equipment manufacturer; industrial terminology for a company which manufactures its products by combining basic parts supplied by other companies. For example, an **earth station** OEM may buy-in **antenna**s and **amplifier**s to build its products.

offset-fed (antenna) An **antenna** system in which the **feedhorn** is placed away from the principal axis of the antenna reflector: it is 'offset' from the centre [see figures A3, D3]. In most designs the horn is also placed outside the **beam** formed by the reflector, to eliminate the **aperture blockage** and diffraction effects caused by a horn which intercepts an antenna beam. An antenna with a feedhorn on the principal axis is known as 'centre-fed'. Both **spacecraft** and **earth station** antennas can be either offset-fed (asymmetric) or centre-fed (axisymmetric).

ogive A section of a **launch vehicle** fairing or tank, shaped like a pointed (Gothic) arch: usually the forward end (e.g. the forward end of the Space Shuttle **external tank** [see figures A6, E6]).
[See also **fairing**]

omni A colloquial abbreviation for **omnidirectional antenna**.

omnidirectional antenna Theoretically, an **antenna** with coverage in all directions (in three-dimensional space), commonly used on spacecraft as a **telemetry**, **tracking and command** antenna. Often abbreviated to 'omni'. In practice the 'antenna' may consist of two or more separate radiating elements, mounted at opposite ends of the spacecraft, to provide two (or more) overlapping hemispheres of coverage. Because of this, and the general difficulties in providing a perfect **antenna radiation pattern**, the omni cannot be described as **isotropic** since the pattern is not equal in all directions. See figure S14 for an example of a deployable omni and figure S2 for one mounted on a satellite's **feedhorn** tower.

[See also **isotropic antenna**]

OMS – see **orbital manoeuvring system**

OMT – see **orthomode transducer**

Omux A colloquial abbreviation for output multiplexer. See **multiplexer**.

OMV – see **orbital manoeuvring vehicle**

on-board data processing (OBDP) Data processing operations performed on board a spacecraft, as opposed to within an **earth station** or elsewhere in the **earth segment**. The term is most often used with reference to scientific or **remote sensing** data, rather than the **data communications** traffic handled by **communications satellites**, although it may refer to **attitude** data and other similar inputs from a satellite's **subsystems**.

[See also **data**]

on-board processing (OBP) Signal processing operations performed within a satellite's **communications payload**, as opposed to within an **earth station** or elsewhere in the **earth segment**. It may take the form of simple **channel** switching, performed at **baseband** instead of **microwave** frequencies, or it may be more complex, including error correction coding, **spread spectrum** decoding or on-board **decryption**.

[See also **regenerative repeater, satellite switching, forward error correction (FEC)**]

one-and-a-half stage – see Atlas

ONET An acronym for **output** network, a combination of output **multiplexer** and switch network.

on-station The condition of a satellite in **geostationary orbit** located at its **nominal orbital position**.

operational Generally: 'capable of, or actually involved in operations'; or 'pertaining to operations' (e.g. operational aspects, operational requirements). For example, a new **launch vehicle** is typically deemed 'operational' following the successful completion of a number of test launches; once the vehicle has been retired, it could be termed non-operational. A **satellite** or **spacecraft** generally becomes operational when it has reached its allocated **orbital position** (or attained its intended **trajectory**) and successfully completed a series of performance tests [see IOT]. In some cases **satellite communications** systems are considered 'operational' when

there are sufficient resources (i.e. satellites and satellite **channel**s) to offer a continuous and guaranteed service. For a system in **geostationary orbit** this usually requires at least two satellites, providing **redundancy** should one of them fail, while for a **constellation** system in a lower orbit, it would typically require the whole system to be in place, including a number of **in-orbit spare**s. Prior to this, the systems would be considered **pre-operational**.
[See also **ground spare**, **lifetime**]

operational lifetime – see **lifetime**

OPF – see **orbiter processing facility**

optical bench An accurately aligned arrangement of optical components used either for test purposes (on the laboratory bench) or within an **operational** optical device (e.g. **COSTAR**).

optical fibre fibres of glass, usually manufactured in continuous lengths of several kilometres and packed in bundles to form a cable, used for the transmission of **signal**s modulated onto a **laser** beam; often referred to as **fibre optics**. The fibres are generally made from silica glass (with the chemical formula SiO_2) with dopants such as germania (GeO_2) or fluorine (F) added to produce small changes in the refractive index. The fibre core is doped to a higher index than the surrounding cladding to confine the light within the core by total internal reflection.

Terrestrial optical fibre offers competition to **communications satellite** systems over long-distance trunk routes, due in part to its high transmission capacity and relatively low cost. Its advantages over **coaxial cable** include higher capacity, wider **repeater** spacings and immunity from **electromagnetic interference (EMI)**. In common with coaxial cable, however, physical connections must be made to all users, in contrast with satellite systems, which provide potential links to anyone in the **coverage area** who wishes to install receiving equipment (see **direct broadcasting by satellite**). Optical fibre is also used in some spacecraft payloads for data transmission, again because of its high capacity and EMI immunity.

optical gyroscope – see **laser gyro**

optical solar reflector (OSR) A material which reflects the Sun's visible spectrum. This term is often used synonymously with **second surface mirror (SSM)**, but is less definitive since it includes other solar reflectors, such as white paint, silver-coated silica sheets, etc. [see figure S2].
[See also **solar spectrum**]

optical tracking – see **kinetheodolite**

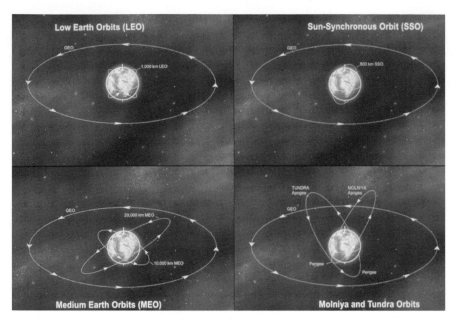

O1: Types of **orbit** shown approximately to scale with the **Earth** and with **geostationary orbit (GEO)**. [Mark Williamson/Willis Inspace]

orbit The path followed by an object (**spacecraft** or **planetary body**) in its motion around a **celestial body**; to follow such a path [see figure O1].

Two periods of revolution can be defined for an orbiting body, dependent on the **frame of reference**: the orbital period with respect to the stars is known as the 'sidereal period'; the period with respect to the celestial body is called the 'synodic period'. Consider, for example, a satellite in an equatorial, **low Earth orbit** and an observer fixed in space; both the Earth and the satellite revolve from west to east. The observer, ignoring the Earth entirely, notes the time the satellite takes to describe a complete circuit against the background of stars – this is its sidereal period of revolution. In this time, however, the Earth has rotated to a new position and the line of longitude that was directly beneath the satellite at the beginning of the sidereal period is now ahead of it. The observer notes the time the satellite takes to catch up with the original **sub-satellite point** and adds this to the sidereal period – this is the synodic period of revolution. For any orbit the synodic period can be defined as the time between two passes over a reference longitude.

In the early days of spaceflight, the practice arose of referring to the synodic periods as 'revolutions' and the sidereal periods as 'orbits' (e.g. Gemini 7 completed 206 revolutions and 220 orbits during its 14-day mission).

[See also **geostationary orbit (GEO)**, **geosynchronous orbit**, **supersynchronous orbit**, **heliosynchronous orbit**, **medium Earth orbit (MEO)**, **high Earth orbit (HEO)**, **Molniya orbit**, **tundra orbit**, **transfer orbit**, **geostationary transfer orbit (GTO)**, **Hohmann transfer orbit**, **drift orbit**, **prograde orbit**, **retrograde orbit**, **graveyard orbit**, **halo orbit**, **orbital position**, **orbital parameters**, **plane change**, **orbital debris**, **perturbations**, **apogee**, **perigee**, **trajectory**]

orbital control The ability to maintain or change a spacecraft's **orbital parameters**; the subsystem or process by which this control is effected. Most spacecraft required to orbit a **celestial body** have an orbital control subsystem, since they need to counter the forces which combine to perturb that orbit [see **perturbations**]. The system usually comprises a set of **reaction control thrusters**.

The satellite in **geostationary orbit (GEO)** is, however, a special case, since it is allocated an **orbital position** measured in degrees of longitude, which must be maintained to limit the possibility of **interference** with other services using GEO, to allow the use of fixed ground **antenna**s and to avoid collisions. This is referred to as 'station keeping': the two main components of orbital control are east–west station keeping and north–south station keeping. The former relates to a drift in longitude along the geostationary arc away from the **nominal orbital position**, the latter to a drift north or south of the equatorial plane [see **inclined orbit** and **free-drift strategy**]. Orbital control is usually combined with **attitude control** in a common subsystem, the 'attitude and orbital control system' (AOCS).

The orbital parameters of a spacecraft can be determined using **star sensors** and an on-board computer, or by the measurement of distance and direction from a ground station [see **tracking**].

[See also **sun sensor**, **earth sensor**, **equilibrium point**]

orbital debris A blanket term for any man-made artefact discarded, or accidentally produced, in **orbit** (chiefly around the Earth). This includes satellites which have reached their **end of life (EOL)**, spent **launch vehicle** stages, hardware accidentally released by **astronaut**s and the remnants of spacecraft and launch vehicles which have exploded [see **passivation**] or been hit by other debris.

Objects in **low Earth orbit (LEO)** at least 10 cm in diameter can, in principle, be tracked on a routine basis by the US Space Surveillance Network (SSN): about 10,000 objects are tracked, some 8,600 of which are officially catalogued. Objects greater than 1 m in diameter can be detected in **geostationary orbit (GEO)**, but the ability to detect objects depends on many factors, including shape, material composition and the eccentricity of the orbit. About 2000 objects in GEO, including active and inactive satellites, rocket bodies, etc., are tracked by the SSN.

According to **NASA** estimates, the population of particles between 1 cm and 10 cm in diameter is greater than 100,000, while the number of particles smaller than 1 cm probably exceeds tens of millions. Individual objects as small as 3 mm can be detected by ground-based radars, providing a basis for a statistical estimate of their numbers, while assessments of the debris population smaller than 1 mm can be made by examining impact features on the surfaces of returned spacecraft (although this has been limited to spacecraft operating at altitudes below 600 km). Most orbital debris resides within 2000 km of the Earth's surface, although the amount of debris varies significantly with altitude, with concentrations at altitudes of around 800 km, 1000 km and 1500 km.

The average impact speed of orbital debris with other space objects in LEO is approximately 10 km s^{-1}, which means that even small debris collisions involve considerable energy. The probability of collision with orbital debris has been estimated, for a Space Shuttle **orbiter** on a typical 7-day mission in LEO, to be about 4×10^{-6}; for a satellite in GEO it lies between 10^{-6} and 10^{-5} in one year (about a one-in-a-million chance). The first validated collision between two catalogued objects occurred in July 1996 when the French Cerise microsatellite was hit by part of an Ariane rocket.

The best way to reduce the probability of collision is to prevent the accumulation of debris, which for geostationary satellites is achieved by removing spacecraft from orbit at the end of their lives [see **graveyard orbit**]; satellites in LEO may be **de-orbit**ed and burn up in the Earth's atmosphere. [See also **space debris, NORAD, NEO, micrometeoroid**]

orbital decay The gradual reduction in the altitude of an **orbit** due to **atmospheric drag**. If it remains uncorrected, the decay of an orbit results eventually in an uncontrolled **re-entry**.
[See also **drag compensation**]

orbital elements – see **orbital parameters**

orbital environment – see space environment

orbital inclination The angle between the plane of an orbit and the equatorial plane of the orbited body.

[See also plane change, orbital control]

orbital injection The process of placing a spacecraft in orbit, also known as orbital insertion.

[See also injection, aerobraking, aerocapture, combined (bipropellant) propulsion system]

orbital manoeuvring system (OMS) A propulsion system on the Space Shuttle used for changing orbits, orbital insertion, rendezvous and de-orbit. It comprises two 26,700 N (6000 lb) engines mounted in two removable 'OMS pods' (commonly pronounced 'ohms pods') on either side of the orbiter's tail [see figure S9].

The pods also house part of the orbiter's reaction control system (RCS): twenty-four 3870 N (870 lb) primary thrusters and four 111.2 N (25 lb) vernier thrusters (an additional 14 primary and 2 vernier thrusters are located in a module near the orbiter's nose). All OMS and RCS thrusters use the hypergolic propellants monomethyl hydrazine (MMH) and nitrogen tetroxide.

orbital manoeuvring vehicle (OMV) The concept of a spacecraft designed to reposition payloads in orbit; more colloquially known as a 'space tug'. In its ultimate form such a vehicle would be reusable, by being refuelled, and would remain in orbit throughout its operational life, rather than be returned to Earth.

[See also orbital transfer vehicle]

orbital parameters The parameters which define the size, shape and orientation of an orbit in space (or the position and motion of a spacecraft in orbit); also known as orbital elements (when applied to a celestial body). The six classical elements (for a celestial body) are eccentricity, semi-major axis, angle of inclination (i), longitude (or right ascension) of the ascending node (Ω), argument of perihelion (ω) and epoch (or time of perihelion passage/true anomaly). The orientation of the orbital plane is defined by three angles: i (the angle between the plane of the orbit and the equatorial plane of the orbited body); Ω (the angle between the vernal equinox and the radius vector passing through the ascending node); and ω (the angle measured in the orbital plane from the line of nodes to the perihelion end of the major axis). Note: for a satellite orbiting the Earth, perihelion is replaced by perigee.

[See also **orbital inclination, apogee, perigee, drift, nominal orbit position, ephemeris**]

orbital period Of a spacecraft, in general taken to be the **sidereal period** (as opposed to the **synodic period**). See **orbit**.

orbital plane The plane in which a **spacecraft** orbits. See **orbit**.

[See also **plane change, orbital inclination, orbital parameters**]

orbital position The position of a **satellite** in **geostationary orbit** referred to the Earth's longitude at the **sub-satellite point** and measured in degrees east or west of the Greenwich Meridian [see figure O2]. The allocation of orbital positions (as well as frequencies) is coordinated through the International Telecommunication Union (**ITU**). The concept of an orbital position is not relevant to non-geostationary spacecraft, since they do not remain over a particular line of longitude.

[See also **nominal orbital position, orbit, orbital spacing, orbital control**]

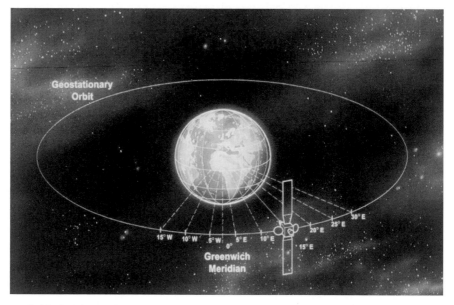

O2: Orbital position of a satellite in **geostationary orbit (GEO)**. [Mark Williamson/Willis Inspace]

orbital relocation The movement of a spacecraft in **geostationary orbit** from one **orbital position** to another. This occurs typically when a **communications satellite** is required to provide coverage to an area different from that which was originally intended. The manoeuvre may be performed using the satellite's orbital control **thruster**s or the spacecraft may be allowed to drift until the required position is reached, dependent upon the relative positions. See **equilibrium point**.

[See also **coverage area**, **service area**]

orbital replacement unit (ORU) Any equipment designed to be transported from Earth to a **space station**, for example, to replace a similar unit already in **orbit**. This type of hardware (e.g. a science experiment, computer, etc.) is designed specifically to be easy to change once the **spacecraft** is **operational**.

[See also **International Space Station (ISS)**]

orbital slot A colloquial term for **orbital position**.

orbital spacing The separation of satellites in **geostationary orbit** measured in degrees of longitude. Although precise separation values vary around the **orbit**, the general rule for satellites operating in the **fixed-satellite service** is two degrees for **Ku-band** and three degrees for **C-band**, while for satellites operating in the **broadcasting-satellite service** (typically in Ku-band) the nominal separation is six degrees.

[See also **orbital position**]

orbital track The projection of a satellite's **orbit** on the surface of a **planetary body**, otherwise known as a 'ground track'; the locus of **sub-satellite point**s drawn out as the satellite orbits the body.

orbital transfer vehicle (OTV) The concept of a spacecraft designed to transfer a **payload** from one **orbit** to another, typically from a **low Earth orbit (LEO)** to a higher altitude orbit (e.g. **geostationary orbit (GEO)**); or from an **equatorial orbit** to a **polar orbit**. In its ultimate form such a **vehicle** would be reusable, by being refuelled, and would remain in orbit throughout its **operational** life, rather than be returned to Earth. The closest contemporary vehicle is ESA's **automated transfer vehicle**.

[See also **plane change**, **orbital manoeuvring vehicle**, **CTV**, **CRV**]

orbiter Another name for the Space Shuttle [see figure O3]; the part of the **space transportation system** which attains **orbit** and operates as a **spacecraft** in orbit, as opposed to being simply part of a **launch vehicle** [see **external tank**, **solid rocket booster**]. The Shuttle orbiter is 37.18 m long, 17.39 m high and has a 23.77 m wingspan. It is designed to operate in **low Earth orbit (LEO)**, i.e. between altitudes of 185 and 1110 km (115–690 miles).

[See also **payload bay**, **flight deck**, **Space Shuttle main engine (SSME)**, **auxiliary power unit (APU)**, **orbital manoeuvring system (OMS)**, **remote manipulator system (RMS)**, **thermal protection system (TPS)**]

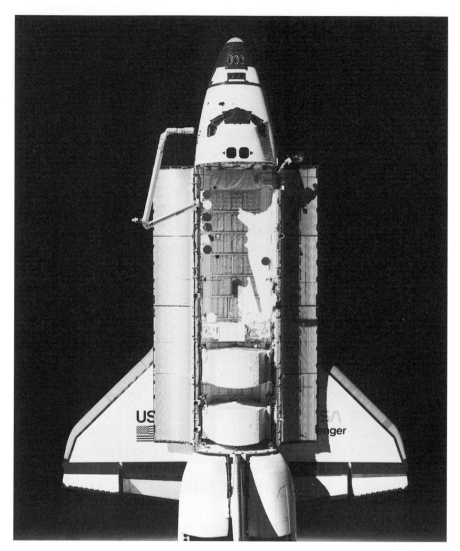

O3: Space Shuttle **orbiter** Challenger. Note **sunshields** protecting two **communications satellites** in **payload bay**, a row of **Getaway Special** canisters on left sill, deployed **remote manipulator system (RMS)**, and **thruster** ports of **reaction control system** on orbiter's nose. [NASA]

orbiter processing facility (OPF) The building at **Kennedy Space Center** in which Space Shuttle **orbiters** are prepared for flight; the Shuttle-equivalent of an aircraft hangar. It is divided into two 'high bays', each capable of accommodating a Shuttle orbiter in a horizontal attitude, and an interconnecting 'low bay' containing ancillary equipment and offices, etc. **Payload**s which require horizontal **integration** are loaded in the OPF; those which can be integrated with the orbiter in a vertical attitude are loaded at the **launch pad** using a **payload canister**. The orbiter is towed from the OPF to the **vehicle assembly building (VAB)** for assembly with the **external tank** (ET) and **solid rocket booster**s (SRBs).

[See also **processing**]

Orlan The name given to a Russian **spacesuit**.

orthogonal axis A perpendicular axis. Commonly used to refer to **spacecraft axes** (pitch, roll and yaw) or the axes of individual pieces of hardware (e.g. the major and minor axes of an **elliptical antenna** are 'orthogonal'). Also used for axes of radiation **polarisation** (e.g. vertical and horizontal polarisations are 'orthogonal').

orthomode transducer (OMT) A component of a **waveguide** circuit, known more fully as an orthogonal-mode transducer, that separates or combines two orthogonally polarised signals. An OMT allows the same **antenna** and **feedhorn**(s) to be used with transmitted and received signals of opposite polarity.

ORU – see **orbital replacement unit**

OSR – see **optical solar reflector**

OTV – see **orbital transfer vehicle**

outdoor unit – see **head-end unit**

outer space A colloquial term for '**space**' or 'the **space environment**'.

Outer Space Treaty The colloquial name for the treaty of **space law** whose official title is "Treaty on Principles Governing the Activities of States in the Exploration and Use of Outer Space, including the Moon and other Celestial Bodies". The Outer Space Treaty was adopted by the United Nations General Assembly on 13 December 1966, opened for signature on 27 January 1967 and entered into force on 10 October 1967. It is considered the basic treaty of space law, since the subsequent four space treaties – the **Rescue Agreement**, **Liability Convention**, **Registration Convention** and **Moon Agreement** – were derived from its provisions.

outgassing The release of a gas from a material when it is exposed to an ambient pressure lower than the vapour pressure of the gas. Many materials (resins, adhesives and even some metals) evaporate in a vacuum. Since there is no force available to carry the particles away from a spacecraft, they remain in a cloud around it or condense onto its colder surfaces. Spacecraft **materials** are chosen to limit outgassing, since the outgassed molecules can degrade the optical surfaces of **sensors**, **solar cells** and **radiators**, and can even cause shorting of electrical circuits and promote **corona** or electrical discharge. [See also **electrostatic discharge (ESD)**]

output

(i) The exit point or path through which information or energy is removed from a device or system (refers to the hardware).

(ii) The information or energy (a **signal**, **carrier**, **RF-wave**, etc.) passed out of an electronic or **microwave** circuit or component.

[See also **output section, output filter, data**]

output filter A **filter** at the output end of a **transmit chain** in a communications **transponder** which confines the signal to its appointed **bandwidth**. The typical output filter may comprise more than one individual filtering device, but its major constituent is known generically as a **band-pass filter**. [See also **input filter**]

output multiplexer – see **multiplexer**

output section The section of a spacecraft **communications payload** from the signal's input to the **travelling wave tube amplifier** to its output to the transmit antenna. The division of payload components into an **input section** and an output section is an alternative to their classification as part of a **receive chain** or a **transmit chain**. [See also **saturated output power**]

over-design In space technology the term usually refers to an item which has been designed to be stronger or generally more capable than necessary (e.g. a structural item designed to withstand a much greater structural **load** than it can be expected to experience is 'over-designed', as is a **sensor** designed to operate at much greater accuracies than will ever be required). In general, the practice of over-design increases the cost, and in some cases the mass, of the item, and is therefore best avoided.

overexpanded – see **expansion ratio**

overspill The unavoidable tendency of a satellite **beam** to cover more than the intended **service area**. Sometimes called spillover, but this term is more often used in connection with antenna radiation patterns – see **spillover**.

Overspill results from the necessity to design satellite **footprints** to provide sufficient **power flux density** at the edge of the service area even under the worst weather conditions (when the **signal** is most highly attenuated). This built-in **margin** means that, under 'clear sky' conditions, the effective

coverage area is increased and the footprint spills over into adjacent service areas, states or countries. If the degree of overspill is too large, **interference** with other services may result.

[See also **attenuation**]

oxidiser The component of a rocket **propellant** which allows the **combustion** of a **fuel** in the absence of atmospheric oxygen [e.g. **liquid oxygen, hydrogen peroxide** (H_2O_2), **nitrogen tetroxide** (N_2O_4), **ammonium perchlorate** (NH_4ClO_4)].

[See also **solid propellant, liquid propellant**]

P

pad Engineering slang for an **attenuator** (e.g. 'a 3 dB pad'). Also used as a verb: 'to pad a **transmission line**', etc.

[See also **launch pad**]

PAHT An abbreviation for power-augmented hydrazine thruster. See **hiphet thruster**.

pallet A support structure for **payloads** (**flight hardware** as opposed to **ground support equipment**). For example, the portable U-shaped support and carrying structures designed to support payloads in the **payload bay** of the Space Shuttle [see figure P1]. The use of pallets decreases the time required for Shuttle **payload integration** and means that complex payloads can be assembled (on a pallet) and tested independent of Shuttle **processing**. They are available as single 2.9-m-long pallets or 1.4 m 'half-pallets', which can be combined to form 'pallet-trains' dependent on the size of the payload.

P1: **Pallet** carrying scientific equipment prepared for flight on the **Space Shuttle**. [ESA]

Although they have many potential applications, the pallets are sometimes referred to as 'Spacelab pallets', since they were originally designed for use with Spacelab habitable modules (as carriers of instruments requiring direct exposure to space). Pallets are intended primarily to remain within the payload bay, but it would be possible to release them as orbiting spacecraft in their own right (so-called 'free-flying pallets').

[See also cradle, airborne support equipment]

Palmachim The location of an Israeli launch site (on an air force base near the town of Yavne) at approximately 31.9° N, 34.7° E. It conducted its first launch, of Ofeq-1 (Horizon-1), on 19 September 1988.

PAM – see payload assist module

panchromatic Sensitive to all colours. In remote sensing, the term is used to describe the response of a film or sensor that produces a monochromatic (black-and-white) image from data acquired in the visible and/or near infrared parts of the spectrum. Panchromatic images are typically of higher resolution than multispectral images.

[See also waveband, electromagnetic spectrum]

paper satellite A colloquial term for a satellite which exists only as a design concept. The term is used pejoratively in connection with requests for frequencies and orbital positions made to the International Telecommunication Union (ITU) by prospective operators who have no immediate intention to procure and launch a satellite; they wish only to reserve these limited resources for later, possible use.

parabola A conic section formed by a plane that cuts a cone parallel to its side.

[See also conic sections]

parabolic antenna An antenna with a parabolic cross-section; an antenna shaped like a paraboloid (i.e. 'paraboloidal').

parabolic trajectory – see ballistic trajectory

paraboloid A geometric, three-dimensional surface which has parabolic cross-sections in two of three orthogonal planes and either elliptical or hyperbolic sections in the third. For example, the main reflector of a Cassegrain reflector.

[See also Gregorian reflector, orthogonal axis, conic sections]

parallel tubes Two or more travelling wave tubes (TWTs) coupled together in parallel to provide a higher power than that available or conveniently obtained from a single TWT. Paralleling two similar tubes does not produce twice the power of an individual tube owing to losses suffered in the combination process, but if one tube should fail the system can still operate, albeit on reduced power.

parametric amplifier A type of **low-noise amplifier** (often abbreviated to 'paramp') which has one oscillator circuit tuned to the receive frequency and another 'pumping' oscillator at a different frequency. The pumping oscillator periodically varies the parameters of the primary circuit (capacitance or inductance), thereby transferring energy to and thus amplifying the input signal.

paramp A colloquial abbreviation for **parametric amplifier**.

parasitic station acquisition A technique used to deliver satellites with liquid **bipropellant propulsion systems** to **geostationary orbit** with a number of separate **apogee engine firings**, as opposed to the single 'burn' produced by a **solid rocket motor**. This philosophy allows the engine to be calibrated on the first burn, thus increasing the accuracy of subsequent burns. By varying the timing and magnitude of the burns to adjust the periods of intermediate **transfer orbits**, the final burn can be performed at the desired **on-station** longitude, injecting the satellite directly into its final orbit at the correct position. This technique can save a significant amount of **station keeping** fuel and increase the satellite's **lifetime**. The first burn may place the satellite in **supersynchronous orbit**, in which case subsequent burns will lower the **apogee** to geostationary altitude.

[See also **mid-course correction**]

parking orbit A term used for a **low Earth orbit** (or similar orbit around any other **planetary body**) when used as a temporary 'holding position' for a spacecraft, before it is injected into another **orbit** or **trajectory**.

[See also **geostationary transfer orbit**, **trans-lunar injection**]

parsec The distance at which an object displays a parallax of 1 second of arc across a baseline of 1 **astronomical unit** (i.e. 3.086×10^{16} m); a unit of distance used in astronomy. Derivation: measurements taken of relative star positions, viewed from either side of the Earth's orbit about the Sun (i.e. six months apart), show a movement of the nearer stars, against the distant background-stars, known as parallax. The observation baseline is two astronomical units; if observations are adjusted for a baseline of 1 AU, a parallax of 1 second indicates a distance of 1 parsec (from '*par*allax *sec*ond'). One parsec is equivalent to about 3.262 **light-years**.

partial loss A term used in **space insurance** to indicate an amount of degradation in the capability of a **spacecraft** lower than that required to constitute a **total loss**. Partial loss of capability may be based on any of the following aspects: a loss of station-keeping propellant which reduces a spacecraft's **lifetime** [see **orbital control**]; a loss of electrical **power** which reduces a spacecraft's ability to operate a specified number of **transponders** for a given lifetime; the failure of part of a spacecraft's **payload** (and thereby revenue-earning capability).

[See also **constructive total loss, TLO, deductible**]

passband The range of frequencies a **filter** allows to pass. Frequencies outside this range are in the 'stopband'.

[See also **bandwidth, low-pass filter, high-pass filter, band-pass filter, channel filter, input filter, output filter**]

passivation An operation performed on a **propulsion system**, typically a **launch vehicle** stage, to avoid an in-orbit explosion; a so-called 'debris mitigation measure'. Propellant remaining in a tank when the propulsion phase has been completed is subject to solar heating, which can eventually lead to the failure of the tank structure and/or an explosion that produces **orbital debris**. The passivation procedure involves venting the tanks of **residuals** to render it inert; the system is then 'passivated'.

passive A term applied to any device or system which is receptive or responsive to external stimuli in a pre-arranged manner, but makes no individual contribution to the stimulus; the opposite of **active**. For example, a passive device in a **communications system** contains no power source to augment the system's output power; all power is derived from the input **signal** (e.g. a **filter, attenuator** or **hybrid**). As another example, **surface coatings** and **thermal insulation** are passive components in a **thermal control subsystem**.

passive communications satellite An early **communications satellite** with no **active** components: e.g. the Echo series of metallised balloons designed to reflect incident radio signals towards a ground-based **receiver**.

passive intermodulation (PIM) An unintentional and undesirable **modulation** between two or more **radio frequency** (RF) signals which can occur in a **receiver** or satellite **transponder**.

payload Any equipment or cargo carried by a **spacecraft** or **launch vehicle** or mounted on an orbiting **space platform** or **space station**. In terms of the **mass budget**, the term payload also includes the crew of a **manned spacecraft**. For **unmanned spacecraft**, the term is used to differentiate between equipment concerned directly with the spacecraft's **application**

(the payload) and that which supports the operation of that payload (the **subsystem**s contained within the **service module**).

[See also **communications payload, payload module**]

payload assist module (PAM) A **solid rocket motor** designed for use with the Space Shuttle, as a **perigee kick motor** (PKM) or 'perigee stage', to inject satellites from **low Earth orbit** into a **geostationary transfer orbit**; also used as the third **stage** of the **Delta** launch vehicle. Alternatively known as a Spinning Solid Upper Stage (SSUS), since it uses **solid propellant** and is capable of spin-stabilising its **payload** during the transfer orbit. The PAM was originally available in two main versions: PAM-A for payloads in the **Atlas-Centaur** class (about 2000 kg to **geostationary transfer orbit**) and PAM-D for smaller payloads suited to the Delta (1000–1250 kg to GTO). A second generation **variant**, the PAM-DII, had a **payload capability** similar to that of the PAM-A. PKMs have been largely superseded by satellites with **bipropellant propulsion systems**.

[See also **upper stage, spin table, spin-stabilised**]

payload bay The 'cargo hold' of the Space Shuttle **orbiter** [see figure O3]. It is 18.28 m long and 4.57 m in diameter and can accommodate either one large or several smaller **payloads**: e.g. up to five Spacelab **pallets**, or the equivalent in pallets and **habitable modules**, can be carried.

[See also **remote manipulator system (RMS), Spacelab, Spacehab**]

payload canister A container about the same size as the **payload bay** of the Space Shuttle used to transfer **payloads** to the **launch pad** under 'clean room' conditions. The canister is transferred to the pad in a vertical orientation, matching the orientation of the payload bay when the **orbiter** is on the pad.

[See also **rotating service structure (RSS)**]

payload capability A measure of a **launch vehicle**'s ability to lift **payload** to a given **orbit**; also known as payload capacity. Launch vehicles can be split into four groups based on launch capability: small, medium, large and heavy-lift. As a general guide, 'small' vehicles are capable of launching up to about 2000 kg to **low Earth orbit** (LEO) and generally less than 1000 kg to **geostationary transfer orbit** (GTO). The 'medium' category contains those which can launch between 1000 kg and 2000 kg to GTO, 'large' vehicles can lift between 2000 kg and 5000 kg to GTO, while heavy-lift vehicles exceed this value.

Figure P2 shows how the payload capability of a given launch vehicle would

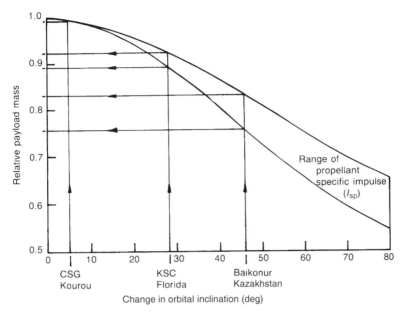

P2: Variation of launch vehicle **payload capability** with latitude of **launch site**.

vary with the latitude of the **launch site** by plotting the percentage of mass in GTO that can be injected into **geostationary orbit (GEO)** against the required change of **orbital inclination**. If the launch site is on the equator, for example, 100% of the mass in GTO can reach GEO because no plane change is required [see **plane change**]. The two curves encompass a family of curves relating to the **specific impulse** (I_{sp}) of the **propellant**, and cover a range from about 300 s to 450 s.

[See also **rocket equation**]

payload changeout room – see **rotating service structure (RSS)**

payload envelope The dimensional constraint on the volume available for a **payload** within a **launch vehicle** fairing [see figure P3]. The envelope constrains the size and shape of a spacecraft and its various appendages, such as **antenna**s and **solar array**s, with the result that most of them have to be folded against the side of the spacecraft for launch and **deploy**ed once in **orbit**.

[See also **payload fairing**]

payload fairing A protective **fairing** which houses the **payload** of a **launch vehicle**; also known as a 'payload shroud', 'launch shroud' or, more

271

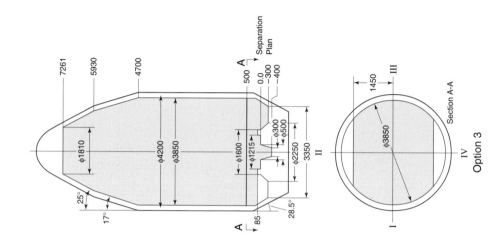

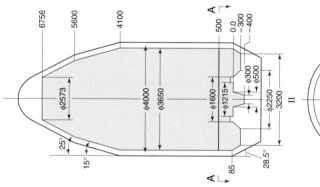

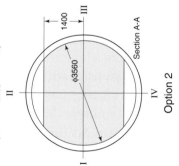

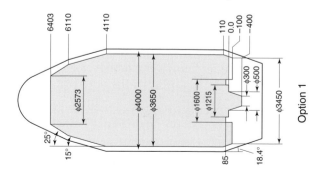

Option 1

Option 2

Option 3

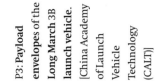

P3: **Payload envelopes** of the **Long March 3B launch vehicle.** [China Academy of Launch Vehicle Technology (CALT)]

P4: Separation of **Ariane 5 payload fairing**. [ESA–CNES–Arianespace/D. Ducros]

colloquially, 'nose cone'. The fairing protects the payload from the environmental characteristics of the atmosphere prior to **launch**, and from frictional heating and other dynamic effects during its passage through the atmosphere. Most fairings are divided pyrotechnically into two equal halves and jettisoned once the vehicle is above the denser part of the atmosphere [see figure P4].

[See also **payload envelope**, **SYLDA**, **SPELDA**, **SPELTRA**, **pyrotechnics**]

payload fraction – see **mass ratio**

payload integration

   (i) The loading of one or more **spacecraft** onto a **launch vehicle**.

  (ii) The assembly of the components of a spacecraft **payload** (e.g. **communications payload**).

 (iii) Sometimes the mating of a spacecraft's **payload module** with its **service module**, but this is usually called **spacecraft integration**.

[See also **assembly, integration & test (AIT)**]

payload mass ratio – see **mass ratio**

payload module A self-contained section of a modular spacecraft containing the payload (e.g. the **communications payload** of a **communications satellite**, otherwise known as the communications module) [see figures M3, M4]. A satellite payload module may include the communications **antennas** or they may be contained within a separate antenna module, depending on the design.

[See also **module, service module, propulsion module**]

payload shroud – see **payload fairing**

payload specialist A category of person who is part of the crew of the Space Shuttle, but has not been trained as an **astronaut**; an engineer or scientist responsible for a particular **payload** carried by the Shuttle **orbiter**.

[See also **mission specialist**]

PCM – see **pulse code modulation**

PDR An abbreviation for preliminary design review. It is followed, at a later stage in the programme, by a critical design review (**CDR**) at which the design is finalised.

P-code Precision-code, the most accurate mode of operation of the **Global Positioning System (GPS)**.

PCS – see **personal communications system**

PCU An abbreviation for **power** control unit.

[See also **power supply**]

peak The beam-centre or RF **boresight** of an **antenna radiation pattern**; the point on an **antenna gain** plot (a graphical representation of antenna gain) at which the gain is a maximum [see figure B4]. A distinction is made between the **radio frequency** boresight referred to here and the geometrical boresight (axis) of the antenna, the angle between the two being known as the **squint** angle. Also used as a verb (e.g. 'to peak', 'to peak up', etc.) particularly when aligning an **earth station** antenna with a satellite in **orbit**.

[See also **beamwidth**]

Pegasus An American air-launched solid propellant **launch vehicle** developed in the late 1980s to deliver small payloads to **low Earth orbit (LEO)**. The standard version (with a payload capability of about 375 kg) was first launched in 1990; the XL **variant** (capable of delivering about 460 kg to LEO) was first launched in 1995. The 19-tonne Pegasus was deployed from a B-52 bomber (the same aircraft that launched the **X-15** rocket-planes in the 1960s), while the 23.6-tonne Pegasus XL is launched from a converted Lockheed L1011 Tristar, named 'Stargazer' [see figure P5]. The aircraft is effectively the 'first **stage**' of the Pegasus launch system.

[See also **Taurus, Athena, Atlas, Delta, Titan, Scout, solid propellant**]

P5: **Pegasus** air-launched **launch vehicle** carried by Stargazer 'first **stage**'. [Orbital]

Peltier effect The phenomenon whereby a current in a circuit made by two different conductors produces a temperature difference between two junctions; the reverse of the **thermoelectric effect**. The effect can be used to cool a sample or a detector, for example, in a process known as 'Peltier cooling'.

percussion primer A mechanical device in an **ignition system**, whereby the impact of a firing pin against an anvil ignites an impact-sensitive charge, which then ignites the **propellant**.

performance In general, a term used to express the quality of functioning of a component or **system** (e.g. '**payload** performance', 'RF performance'). The performance of a **rocket engine** or **rocket motor** defines that of the **stage** or **launch vehicle** to which it is attached: it is typically measured by quantities such as **thrust** and **specific impulse**.

[See also **uprate, partial loss**]

periapsis The point in an **orbit** closest to the centre of gravitational attraction; the opposite of **apoapsis**. From this general term, specific terms relating to a given **planetary body** can be derived: e.g. **perigee** for Earth, **perihelion** for Sun, perilune for Moon, perijove for Jupiter, etc.

perigee The point at which a body orbiting the Earth, in an **elliptical orbit**, is at its closest to the Earth [see figure G1]; the opposite of **apogee**. The word is derived from the Greek 'perigeion': the prefix 'peri' meaning 'near or adjacent'; the suffix 'gee' referring to 'Earth'.
[See also **periapsis, orbit, perigee kick motor**]

perigee kick motor (PKM) A **rocket motor** used to place a satellite in a **geostationary transfer orbit (GTO)**, fired when the spacecraft reaches the point in the GTO ellipse closest to the Earth, the **perigee**. Sometimes called a perigee stage. Geostationary satellites launched by NASA's **Space Shuttle** require a PKM, since the Shuttle delivers spacecraft to a **low Earth orbit** (most spacecraft launched by the Shuttle require some form of **upper stage**). Some **expendable launch vehicle**s (ELVs) also require the use of a PKM (e.g. **Delta**); others (e.g. **Ariane**) inject their **payload**s directly into GTO using their uppermost **stage**, thus removing the need for a PKM.
[See also **payload assist module (PAM), inertial upper stage (IUS), Centaur, apogee kick motor (AKM), rocket engine**]

perigee stage A **launch vehicle** stage which injects a spacecraft into a **transfer orbit** by firing its **engine**(s) at the **perigee** of that **orbit**. Sometimes used as an alternative term for **perigee kick motor**.
[See also **stage**]

perihelion The closest point to the Sun in an elliptical solar orbit (from the Greek for 'near to the sun'); the opposite of **aphelion**.
[See also **periapsis**]

period In physics and maths, the time taken to complete one cycle of a repetitive action or motion; the reciprocal of **frequency**.
[See also **orbital period, sidereal period, synodic period, libration period**]

personal communications system (PCS) A **communications system** designed to provide a wide range of **telecommunications** services to individual mobile telephone handsets. The services may be delivered by **satellite** systems, conventional terrestrial means, or a combination of both.
[See also **GMPCS**]

personal rescue sphere (PRS) A self-contained, one-man **life support system** (0.86-m-diameter inflatable sphere) designed to allow crewmembers not protected by pressure suits to be transferred from a disabled Space Shuttle **orbiter** to a rescue vehicle.

perturbations Deviations in the **orbit** or **trajectory** of a **spacecraft** or **planetary body** caused by disturbing forces such as gravitational attraction [see **gravity**]. Three main perturbation mechanisms are important for satellites in **geostationary orbit** [see **triaxiality**, **luni-solar gravity** and **solar wind**]; more important for satellites in **low Earth orbit** is **atmospheric drag**.
[See also **gravitational anomaly**, **conic sections**]

PFD – see **power flux density**

PGSE An abbreviation for payload ground support equipment. See **ground support equipment**.

phase

(i) The position in the cycle of a periodic quantity expressed as an angle (e.g. 'the two waveforms are 90 degrees out of phase'). See, for example, **phase modulation**.

(ii) A stage in the development process for a **spacecraft**, **system** or **mission** (whereby 'Phase A' generally involves a feasibility study; 'Phase B' covers detailed design and hardware prototyping; 'Phase C' involves hardware manufacturing, implementation and validation; and 'Phase D' is the **operational** phase).

(iii) In astronomy, the state of illumination (e.g. new, crescent, half, gibbous, full) of the Moon or one of the 'inferior planets' (Mercury and Venus), as viewed from Earth. For space travel, the concept is extended to other planetary bodies (e.g. the phases of the Earth as viewed from the Moon or Mars).

(iv) In chemistry, a distinct state of matter characterised by homogeneous composition and properties and the possession of a clearly defined boundary.

phase modulation (PM) A transmission method using a modulated **carrier** wave, whereby the phase of the carrier is varied in accordance with the amplitude of the input signal. The PM technique relies on the fact that two or more samples of the same periodic waveform (or carrier) can be arranged to have different time origins (i.e. they are 'out of phase'). In the **demodulation** process the phase changes can be detected and the original signal reconstructed.

For example, carrier waveforms 180 degrees out of phase can be used to represent the digits '1' and '0', a technique commonly referred to as PSK modulation, or binary phase shift keying (BPSK). Thus, in two-phase **modulation** (BPSK), one **bit** of data is encoded on the carrier for each phase change.

By employing more than two phases it is possible to encode more than one bit per phase, thereby increasing the **bit rate** without altering the modulation rate. Thus four-phase modulation – more accurately termed quadrature phase shift keying (QPSK) – codes two bits per phase change, allowing the representation of '00', '01', '11' and '10' for phase-changes of 0, 90, 180 and 270 degrees respectively. This allows **modem**s employing QPSK to generate and transmit bit streams at burst-rates of 66 Mbits s^{-1} or greater. QPSK is usually preferred over BPSK since, for a given bit rate and **bit error rate**, it requires the same **power** yet only half the **bandwidth**. Higher order modulation schemes, such as 8PSK, are also used: 8PSK requires two-thirds the bandwidth of QPSK, but since the phase states are 45 degrees apart they are more difficult for the **receiver** to distinguish (and for a given BER, a higher power is required). For even higher bit rates, 16QAM (16-phase quadrature amplitude modulation) is available.

[See also **modulation, amplitude modulation (AM), frequency modulation (FM), digital compression**]

phase shift keying (PSK) – see **phase modulation (PM)**

phased array antenna An **antenna** consisting of a number of radiating elements designed to form multiple beams or to allow electronic steering of a given **beam** [see figure D5].

photon A quantum of **electromagnetic radiation**; an indivisible unit of electromagnetic energy. The energy of a photon is the product of the **frequency** of the radiation (measured in **hertz (Hz)**) and the Planck constant, h (6.626×10^{-34} J s). Thus, for example, the energy of an X-ray photon is greater than that of an infrared photon.

[See also **electromagnetic spectrum**]

photon engine A hypothetical **rocket engine** in which **thrust** is created by a flow of **photon**s. Such a device has the highest **specific impulse** imaginable since, by definition, the photon travels at the velocity of **light**.

photovoltaic cell – see **solar cell**

picosat A colloquial term for an extremely small satellite, which may be defined as having a **launch mass** between 1/10 kg and 1 kg. The exact definition of a 'picosatellite' is open to discussion, since satellites with masses less than about 10 kg are generally known as **nanosat**s.

[See also **smallsat, in-orbit mass, mass budget**]

PIM An acronym for **passive intermodulation**.

pin-wheel louvre(s) A device used in the thermal control of a spacecraft to shield or expose a **radiator** surface; an early example of a **louvre** system comprising a circular, segmented wheel controlled by a bimetallic spring which rotates to enhance or inhibit the radiation of thermal energy.

Pioneer A series of American lunar, **deep space** and planetary exploration **probe**s. Pioneer 1, launched in October 1958, failed to reach the Moon but set a 'distance record' of 113,854 km (less than a third of the way to the Moon). Pioneer 4, launched in March 1959, passed within 60,000 km of the Moon. Pioneer 10, launched in March 1972, was the first spacecraft sent to Jupiter (it made a close approach of 131,400 km in December 1973) and the first to escape the solar system (by passing the orbit of Pluto in June 1983). Pioneer 11, launched in April 1973) was the second spacecraft to Jupiter and the first to Saturn (**encounter**s December 1974, September 1979; escaped solar system February 1990). Pioneer 12 and 13 were also named **Pioneer Venus** 1 and 2, respectively.

[See also **gravitational boost**, **Voyager**, **Van Allen belts**]

Pioneer Venus A pair of American spacecraft, otherwise known as **Pioneer** 12 and 13, launched to Venus in May 1978 and August 1978, respectively. Both arrived in orbit in December 1978 and Pioneer Venus 2 launched a series of five 'multi**probe**' entry vehicles which impacted the planet's surface.

[See also **entry-probe**]

pitch A rotation about the pitch axis – see **spacecraft axes**.

[See also **pitch-over**]

pitch axis – see **spacecraft axes**

pitch-over The moment at which a **launch vehicle** leaves the vertical part of its **launch profile** and begins a slow rotation about the **pitch axis**, which decreases the angle between its longitudinal axis and the ground and allows it to follow a **trajectory** that will culminate in an orbital **injection**.

pixel An abbreviation for 'picture element', usually in reference to an electronically generated **image**, where a pixel is the smallest definable element of picture information. For instance, if an image scanning system has a **resolution** of 400 horizontal lines and each line contains 500 individual **bit**s of picture data, the total number of pixels will be 200,000. Thus the number of pixels gives an indication of the definition, or clarity, of the image. Pixel-size is analogous to the grain-size of photographic emulsion, except that the latter is neither constant across the image nor regularly organised in the manner of a grid.

[See also **charge-coupled device (CCD)**]

PKM – see **perigee kick motor**

plane change A change in the plane of an **orbit**; a change in **orbital inclination**. Plane change **manoeuvre**s are kept to a minimum, because they involve a relatively high **propellant** consumption. Since the plane into which a spacecraft is initially launched is related to the latitude of the launch site, it is desirable for the two to be as closely matched (co-planar) as possible [see **payload capability** and figure P2].

 Many satellites are required to operate from **geostationary orbit (GEO)**: for injection into this and other **equatorial orbit**s, a launch site on the equator is ideal, since no plane change is necessary. For this reason, the majority of Western launch sites are located as close to the equator as possible: e.g. **Kennedy Space Center** at 28.3° N in Florida, the USA's most southerly mainland state, and the **Guiana Space Centre** in French Guiana at 5.23° N. A launch from the latter offers a more efficient route to GEO owing to the smaller plane change. It also makes greater use of the Earth's inherent (west-to-east) rotational velocity, which is greatest at the equator (about 1670 km s^{-1}); the corresponding velocity at latitude L is 1670 cos L.

 Since most of the former USSR's territories were far from the equator, early Russian **communications satellite**s were launched into the highly elliptical **Molniya orbit** rather than GEO.

[See also **rocket equation, restartable upper stage, engine re-start (capability), dispenser**]

planetary albedo – see **albedo**

planetary body Any astronomical body in **orbit** around a star, including solid and gaseous planets, moons and asteroids; a subset of **celestial body**.

planetary environment – see **space environment**

planetoid An alternative term for **asteroid**.

plasma An ionised gas, i.e. a mixture of positively charged ions and negatively charged **electrons**. See **ionisation**.

plasma rocket – see **electric propulsion**

plasma sheath A layer of ionised air surrounding a spacecraft re-entering the Earth's atmosphere, or the atmosphere of another **planetary body**. Generated by frictional heating, the plasma sheath effectively blocks all radio communication with the spacecraft.

[See also **S-band blackout, ionisation, ionosphere**]

plasma thruster A type of Hall effect thruster – see **ion engine**.

platform
    (i) A general term for a spacecraft, or part of a spacecraft, usually with the implication that a stable reference or support structure is being provided (e.g. '**space platform**', 'orbital platform', 'observation platform').
    (ii) More specifically, the **service module** of a spacecraft, colloquially termed a '**bus**'.
    (iii) The **antenna platform** or **antenna module** of, for example, a **communications satellite**.
    [See also **inertial platform, launch platform, San Marco platform**]

Plesetsk The location of a Russian **launch site** known as the **Northern Cosmodrome**, which was established in 1957 near the town of Plesetsk (at approximately 62.8° N, 40.1° E). The site was developed to provide **ICBM** launch facilities and the first space-related launch from the cosmodrome was that of **Cosmos** 112 on 17 March 1966.
    [See also **Baikonur Cosmodrome, Kapustin Yar, Svobodny**]

PLSS An acronym (pronounced "pliss") for **portable life support system** and primary life support system (or **subsystem**).

plug nozzle An alternative to the bell-shaped **exit cone** used in the **nozzles** of contemporary **rocket engines** and **rocket motors**, whereby a ring-shaped **combustion chamber** discharges combustion gases against the outer surface of a central truncated cone (giving the appearance of a conventional nozzle with a plug in it). The ring shape of the combustion chamber and its nozzle-outlets gives rise to the alternative term 'annular nozzle'.

    The concept was developed in the 1960s for possible applications on future **single stage to orbit (SSTO)** launch vehicles, largely because it offered an automatic adjustment to the variation in atmospheric pressure between ground level and the upper atmosphere. For example, if a bell nozzle is designed for optimum efficiency at high altitude (i.e. in vacuum or near-vacuum conditions), ambient air pressure at lower altitudes will compress the exhaust **plume** and reduce the efficiency of the device [see **flow separation**]. In contrast, the flow from the plug nozzle adapts automatically to the changing external pressure and always operates at or near optimum efficiency. In most designs the central cone was truncated to optimise the vehicle for the heating effects of **re-entry**. The central stream of exhaust gases followed the line of the cone forming a system of shock-waves below the nozzle, known as an 'aerospike' (from 'aerodynamic spike'). An alternative, simplified design, in which the nozzle-outlets are arranged in a line, is the linear aerospike engine – see **aerospike engine**.

[See also **expansion ratio**]

plugs-out test Jargon: a test of a **launch vehicle** conducted at the **launch pad** with the **umbilicals** disconnected as they would be just prior to a real **launch**.

plumbing Engineering slang for any assembly of **waveguide**-based components. The term is derived from the resemblance of lengths of waveguide, bolted together at their flanges, to industrial pipework. The term is also used for the piping in a liquid **propulsion system**.

plume A stream of **combustion** products ejected from a rocket exhaust **nozzle** [see figure P6].

[See also **plume impingement, degradation, shock diamond**]

P6: **Apollo** 11 launch showing exhaust **plume** of the **Saturn V** first **stage**. Note the beginning of a **shock diamond** effect at lower right. [NASA]

plume impingement An undesired contact between a **thruster** plume and a spacecraft component or surface. The main effect is the production of unwanted forces which disturb the **attitude control** of the spacecraft: for example, the impingement of a thruster plume on the **solar array** panels of a **three-axis stabilised** spacecraft produces a so-called 'windmill torque'. The effect can be reduced by careful siting of the thrusters on the spacecraft body, but this is difficult since the exhaust gases are not emitted in a narrow, pencil beam (about 90% of the gas flow is released in a 60° cone). As a solution, thruster firings can be timed for favourable array positions, or a system for automatic correction of the disturbances can be used.

In addition to unwanted torques, thruster plumes can produce heating effects, erosion due to high speed droplets of unburnt **propellant**, corrosion due to the propellant and its combustion products, and the contamination of thermal control surfaces [see **surface coatings**] and optical surfaces.
[See also **plume shield, degradation**]

plume shield A device mounted adjacent to a spacecraft's **rocket engine, rocket motor** or **thruster**, designed to protect the spacecraft surfaces from the heating effects of a rocket's **plume** and the **combustion** products, which would otherwise cause **degradation** of its **surface coatings** [see figure B5].
[See also **plume impingement**]

plus-x face The face of a **three-axis stabilised** spacecraft which faces east. See **spacecraft axes**.

plus-y face The face of a **three-axis stabilised** spacecraft which faces south. See **spacecraft axes**.

plus-z face The face of a **three-axis stabilised** spacecraft which faces directly towards the Earth; otherwise known as the Earth-pointing face. See **spacecraft axes**.

PM – see **phase modulation**

PMD – see **propellant management device**

pogo A rhythmic longitudinal (vertical) oscillatory motion (or vibration) experienced by a **launch vehicle**. It can be generated by an interaction between the **propellant** flow and **combustion** process, and the structure of the vehicle. The term is derived from a similarity with the motion of a pogo stick.
[See also **pogo suppressor, sloshing**]

pogo suppressor A device which absorbs the longitudinal vibrations that can occur during the **flight** of a **launch vehicle** [see **pogo**]; typically a sealed tube containing a liquid which damps out the oscillation through friction with its container (in a similar way to the **nutation damper** used on some spacecraft).

pointing
   (i) The process of aligning a communications **antenna**, an astronomical **telescope** or a **spacecraft** (usage e.g. 'an antenna requires pointing . . .').

(ii) An attribute of an antenna (usage e.g. 'the antenna pointing is good, so optimum performance is assured ...').

[See also **pointing loss**]

pointing loss A loss in signal power in a communications link due to the mispointing of an **antenna**. The concept is applicable to all links: terrestrial, Earth-to-space and space-to-space. The example below concerns the **uplink** and **downlink** paths of a **satellite communications** system.

The **earth station** antenna cannot always remain pointed precisely at the **satellite** because the spacecraft does not remain stationary with respect to the earth station (see **perturbations** and **box**). This means that the satellite is not located on the **boresight** of the ground-based antenna and does not receive the peak power transmitted from the earth station. The degree of loss depends on the deviation from boresight. In addition the satellite antenna is not always maintained accurately pointing towards the earth station because it requires the expenditure of **reaction control thruster** fuel (but see **RF sensor**). This reduces the received power even further. By the same token, since the earth station is not pointed directly at the satellite and is not on the boresight of the spacecraft antenna, it too receives a lower signal power.

[See also **beamwidth**]

point-to-multipoint In **satellite communications**, a method of linking **earth stations** such that **signals** are transmitted from a single station to an unlimited number of independent receiving stations. See **link**.

point-to-point In **satellite communications**, a method of linking **earth stations** such that **signals** are transmitted from a single station to another individual station. See **link**.

polar mount A structure for the support and guidance of an **antenna** which has its axis aligned with the Earth's axis of rotation, with the result that steering along the **geostationary arc** is possible by rotation about this single axis (as opposed to the two-axis steering required with the alternative **altitude-azimuth mount**.) The polar mount is based on the astronomical 'equatorial mount', whose axis of rotation is known as the **hour-angle** axis; the polar mount is therefore also known as an hour angle-**declination** or 'HA-DEC' mount.

polar orbit An **orbit** at a high **inclination** to the equatorial plane of a **planetary body**, such that a **satellite** in that orbit passes over the body's polar regions [see figures H2, O1]. The **orbital track** of the satellite does not have to cross the poles exactly for the orbit to be defined as 'polar': orbital tracks which pass within 20 or even 30 degrees of the pole are still classed as polar orbits.

polarisation The physical phenomenon whereby electromagnetic waves are restricted to certain directions of vibration. There are two main types of polarisation: linear, in which the electric vector of the radio wave is confined either to the vertical or the horizontal plane, and circular, where the electric vector is rotated either clockwise or anticlockwise to give right- or left-hand circular polarisation (RHCP or LHCP).

The orientation of the vertical and horizontal polarisation vectors depends on the particular communications system: for instance, 'horizontal' may be defined as parallel to or at 45 degrees to the equatorial plane. The precise definition of circular polarisation depends on its source: the former **CCIR** (now **ITU**) recommendation (used in Europe) defines polarisation 'looking away from the transmitter' whereas the IEEE (Institute of Electrical and Electronic Engineers) definition, used in the United States, defines it 'looking towards the transmitter'.

To produce circular polarisation, the **antenna feed** from which the RF radiation originates is fitted with a 'polariser' designed to cause the electric field vector to rotate clockwise or anticlockwise. Only a small amount of the radiation of one polarisation is receivable in a feed designed for the opposite hand, so polarisation represents an effective method of **interference** reduction. (Although opposite polarisations can give as much as 30 dB of **isolation** between channels on the same frequency, this can be degraded, away from the axis of the beam centre, by conversion to elliptical polarisation). See **polarisation conversion**. In satellite communications, signals can be transmitted on opposite polarisations in order to double the use of a **frequency band**. This is known as **frequency re-use**.

polarisation conversion A reversal of **polarity** suffered by an **RF wave** on its passage through the atmosphere under adverse weather conditions or within a **feed** or **antenna** system.
[See also **cross-polar(ised)**]

polarisation frequency re-use – see **frequency re-use**

polariser A device attached to the **feedhorn** of an **antenna** which polarises the RF radiation. See **polarisation**.

polarity In **radio frequency** communications, an attribute of an **RF wave** whereby it may be assigned one of two opposite polarisations to provide **discrimination** between similar signals which might otherwise interfere. Used in the sense of 'RF waves of opposite polarity ...'. See **polarisation**.

polybutadiene A type of solid **fuel** used in **rocket motors** which also acts as a **binder**. It has a basic chemical formula of the type: $CH_2-CH=CH-CH_2$. Two variants used in **solid rocket boosters** are carboxyl-terminated polybutadiene (CTPB) and hydroxyl-terminated polybutadiene (HTPB) (also written as carboxy-terminated and hydroxy-terminated polybutadiene). CTPB has the monovalent group $-$COOH added to the basic formula and HTPB contains the $-$OH group.

[See also **solid propellant**]

portable life support system (PLSS) A **spacesuit** 'back-pack', or other portable device, which contains everything necessary to maintain a comfortable working environment within the suit (i.e. an oxygen supply for breathing and suit pressurisation, and a water supply for thermal control and drinking). The suits used on the **Space Shuttle**, for example, contain supplies for up to seven hours in the suit, which includes a nominal six-hour EVA, 30-minute reserve and a secondary 30-minute supply for use as a back-up in an emergency. Astronauts can also 'plug themselves into' the **orbiter** supplies in the **airlock**.

PLSS is commonly pronounced 'pliss' and is also an acronym for primary life support system or subsystem: the main supply in the portable life support system is known as the primary life support subsystem.

[See also **pre-breathe**]

power

(i) A measure of the 'rate of doing work', measured in watts (W); or the energy per unit time ($J\,s^{-1}$).

(ii) More colloquially, in both engineering and everyday use, a 'commodity' that can be generated, stored and used – more properly termed **'energy'** (e.g. as in 'nuclear power', 'wave power', 'power supply', etc.).

[See also **AC power, DC power, RF power**]

power budget A method of accounting for the **power** requirements of a spacecraft or subsystem with a view to the capability of the spacecraft **power supply**; a sum of subsystem power requirements. During a spacecraft project, each subsystem engineering discipline aims to design its hardware within its allotted portion of the power budget. The power budget allows changes in the overall spacecraft power requirement to be monitored throughout the project.

[See also **mass budget, propellant budget, link budget**]

power bus The main electrical power distribution circuit in a spacecraft. Usually abbreviated to 'bus', a short-form of 'bus-bar'.

[See also **power**]

power flux density (PFD) The level of **radiated power** (measured in watts) received per square metre [units: dBW m^{-2}]. Although relevant for any radiated-telecommunications system (between any two points), PFD is most often quoted from the viewpoint of a receiving **ground station**, which has a fixed effective **antenna** aperture (measured in m^2). If the PFD provided by a spacecraft is too low, only a larger antenna will be able to collect sufficient power to constitute a 'good signal'.

PFD may be calculated using the expression

$$\text{PFD} = 10 \log_{10} \frac{PG}{4\pi R^2}$$

where P is the **transmitter** power (W), G is the **antenna gain** (expressed as a pure number) and R is the distance (m) between the spacecraft and the Earth. If P and G are available in **decibels**, the PFD would be $(P+G-162.1)$ for a spacecraft in **geostationary orbit** where R is about 3.6×10^7 m.

Removing the gain, G, from the expression would give the PFD for an **isotropic antenna**. The product PG is the **equivalent isotropic radiated power (EIRP)**.

power subsystem A spacecraft **subsystem** which provides **power** to the spacecraft. It typically comprises a generating device, such as a **solar array**, **radioisotope thermoelectric generator (RTG)** or **fuel cell**, a storage device such as a **battery**, a **power bus** and associated control and distribution electronics.

[See also **primary power supply, secondary power supply, regulated power supply, unregulated power supply, bus voltage**]

power supply Any device which provides electrical power, usually matched to the specific requirements of the device or system in receipt of that power. See **power**.

[See also **primary power supply, secondary power supply, regulated power supply, unregulated power supply, bus voltage**]

power-augmented hydrazine thruster (PAHT) – see **hiphet thruster**

power-limited Jargon: 'limited in terms of available **power**'. The power consumption of all components and **subsystems** in any **spacecraft** design is constrained by the capability of its **power supply** and must be tailored to suit [see **power budget**]. By way of illustration, a **communications satellite** designed to carry a given number of **transponders**, but with insufficient power to operate them simultaneously, would be 'power-limited'.

[See also **mass-limited**]

pre-breathe The practice of breathing pure oxygen in preparation for an EVA in a **spacesuit** with a low-pressure, pure oxygen supply. For example, the suits used on the **Space Shuttle** are pressurised to only 27 kPa (4.3 psi), compared with 101.4 kPa (14.7 psi) for Earth's surface pressure and nominal **orbiter** pressure. This would normally require pre-breathing for 3.5 hours prior to EVA to eliminate **decompression sickness** ('the bends') when exposed to the lower pressure. This time has, however, been reduced in most cases to about 40 minutes by reducing the orbiter's **cabin pressure** to about 70 kPa (10 psi) 12 hours before an EVA. A proposed 'zero pre-breathe suit' would use oxygen at a pressure closer to cabin pressure and would have to be stronger as a result.
[See also **extra-vehicular activity (EVA)**]

precession The motion of a spinning body (e.g. a planet, **satellite** or **gyroscope**) whereby its axis of rotation sweeps out a cone (like a wobbling spinning-top); also called 'spin axis precession'.
[See also **nutation**]

pre-emphasis A method of improving the **signal-to-noise ratio** in an **frequency modulation** system by increasing the deviation at higher **baseband** frequencies.

pre-launch A term used in **space insurance** to define a period of insurance coverage prior to launch, as in 'pre-launch policy'. See **launch**.

pre-operational Generally: 'prior to becoming capable of or actually involved in operations'. For example, with reference to satellite communications systems this means that there are insufficient resources (i.e. satellites and satellite **channels**) to offer a continuous and guaranteed service. See **operational**.
[See also **in-orbit spare**, **ground spare**]

pre-preg An abbreviation for 'pre-impregnated': generally refers to a woven fibre material, used in the manufacture of low-density spacecraft structures, which has been pre-impregnated with a matrix material [see **carbon composite**].

In a process known as 'laying up', the pre-preg is applied in layers to a former, which governs the final shape. The 'lay-up' is then cured in an autoclave (a type of oven), under conditions of high temperature and pressure. This process, whereby a material hardens permanently after an application of heat and pressure, is known as thermosetting.
[See also **tape-wrapping**]

pressurant A pressurising gas; a gas (typically helium) which provides the pressure (in a pressure-fed **propulsion system**) to force **propellant** from its storage tank to a **rocket engine**. In some **launch vehicle** stages, where the outer **skin** of the **stage** itself forms the **propellant tank**, the pressurant also helps to keep the tank rigid.

[See also **liquid propellant, ullage pressure, blowdown system, balloon tank**]

pressure expansion ratio – see **expansion ratio**

pressure suit – see **spacesuit**

pressurisation

(i) In pressurised **propulsion system**s, the act of increasing the pressure of the vapour above a **liquid propellant**, or of introducing another **pressurant**, chiefly to provide a force to deliver the **propellant** to an outlet valve (particularly the sequence of operations which pressurises the **propellant tank**s of a **launch vehicle** prior to **lift-off**).

[See also **ullage vapour**]

(ii) The pressure to which a **spacesuit** or the cabin of a **manned spacecraft** is 'pressurised'.

[See also **decompression, pre-breathe**]

primary life support system – see **portable life support system (PLSS)**

primary power supply The main source of **DC power** available aboard a **spacecraft** or **launch vehicle**. For most Earth-orbiting satellites it is the **solar array**, used whenever the spacecraft is in sunlight. Some manned space vehicles use the **fuel cell** as the primary source, while interplanetary spacecraft venturing beyond the orbit of Mars tend to use the **radioisotope thermoelectric generator (RTG)**.

[See also **secondary power supply, battery, bus voltage**]

primary structure The main **load**-carrying structure of a **spacecraft** or **launch vehicle**, to which secondary structural components are attached. See **load path**.

[See also **thrust structure**]

prime contractor Industrial terminology for the leading contractor in a project or programme; a contractor which accepts prime responsibility for delivering the product to the customer and is responsible for guiding and monitoring the work of **subcontractors**. The prime contractor is colloquially known as 'the prime' and can be said to 'prime' a project or programme. Where manufacturing is involved, the prime contractor may be referred to less formally as 'the manufacturer'.

[See also **OEM**]

prime focus The point of focus of the primary **reflector** of a communications **antenna** or astronomical telescope; the location of the **feedhorn**, recording device (film or detector) or secondary reflector (if the system includes a **subreflector**). See **Cassegrain reflector, Coudé focus, Gregorian reflector, Newtonian telescope**.

prime spacecraft The **spacecraft** intended for use on a particular **mission** (as opposed to a **back-up** spacecraft).

[See also **ground spare, in-orbit spare**]

prime system Any **system** designated as the main provider of a particular function or operation (as opposed to a **back-up** or **redundant** system).

Priroda A module of the **Mir** space station.

probability distribution In statistics, a continuous distribution of a random variable illustrated by a symmetrical bell-shaped curve [see figure P7] and known as a normal distribution or Gaussian distribution (after German mathematician Karl Friedrich Gauss [1777–1855]). The curve, which has relative frequency as its ordinate and standard deviation (from the mean) as its abscissa, represents the probability density function of a normal distribution. The figures within the curve relate to the percentage of the total area under the curve (i.e. percentage of the total population).

In **space technology**, probability distribution is a useful concept for estimating component or system **reliability**. For example, wear-out mechanisms are governed by a Gaussian distribution, where the probability of failure, $P(t)$, increases to a maximum at a specified time, t_0, and decreases

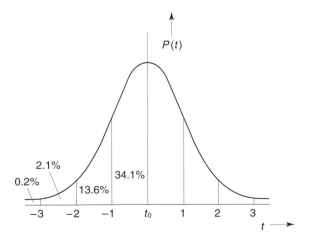

P7: Normal or Gaussian **probability distribution**.

again thereafter (as shown in the figure). A **system** is normally designed so that the mean wear-out times are long compared with the design **lifetime**.

Sigma (σ) notation, a mathematical notation for 'standard deviation from the mean', is used as a measure of the probability of realising a stated value or **specification**. With a Gaussian distribution, 68.27% of all samples of a quantity are within $\pm1\sigma$ of the mean value, 95.45% within $\pm2\sigma$ and 99.73% within $\pm3\sigma$, so if a quantity is quoted 'to three sigma' there is a greater probability of its being achieved than if it is quoted 'to two sigma', for example.

[See also **bathtub curve, MTBF**]

probe
  (i) A colloquial term for an **unmanned spacecraft**, usually designed for scientific investigation of a **planetary body** (e.g. **Viking**, **Voyager**), **interplanetary** space or other celestial object (e.g. **Giotto**). Scientific, and other, spacecraft in Earth orbit tend to be termed **satellite**s rather than **space probe**s.
  (ii) To examine with a probe; any device designed to 'probe' or test a medium under investigation (e.g. the surface of a planetary body, a planetary atmosphere, an electromagnetic field, etc.).
  (iii) The male part of a docking mechanism. See **docking probe**.

processing A blanket term for the preparation of a **spacecraft** or **launch vehicle** for **spaceflight**, used particularly with regard to the American **space transportation system** (e.g. 'Shuttle processing', '**orbiter** processing'). See **orbiter processing facility (OPF)**.

[See also **on-board processing**]

prograde orbit An **orbit** in which the orbiting body has the same sense of rotation as the orbited body (also known as a 'direct orbit'); the opposite of **retrograde orbit**.

Progress A Russian **spacecraft** designed to deliver **payload**s to an orbiting **space station**, particularly the **Mir** station; essentially a **Soyuz** with the descent **module** replaced by a cylindrical compartment containing tanks for fluids (dry cargo being stored in racks within the orbital module).

PROM An acronym for programmable read only memory. See **ROM**.

propagation delay – see **signal delay**

propellant Any substance or combination of substances which constitute a mass to be expelled at high velocity to produce a propulsive reaction force or **thrust**. Propellants comprising a single component are called

monopropellants. Propellants with two components (a **fuel** and an **oxidiser**) are called **bipropellant**s.

Propellants can also be classified according to their physical state, either solid, liquid or gaseous. **Solid propellants** generally consist of a mixture of fuel and oxidiser (some consider them to be bipropellants because they have two components, but to most they are monopropellants since they are handled and used as a single component propellant). **Liquid propellant**s may be monopropellants or bipropellants (the latter are stored separately and combined in a **combustion chamber**). Although solid and liquid propellants are usually employed separately, they are used together in the **hybrid rocket**, the most common combination being 'solid fuel + liquid oxidiser'. The monopropellants used in the **cold-gas thruster** are unusual in that they are generally stored as a gas.

Propellants are also classed as either 'storable' or 'non-storable', the distinction being one of temperature. If a propellant retains its physical properties at normal terrestrial environmental temperatures, it is 'storable'; if it needs to be cooled significantly below these temperatures to render it usable in a propulsion system, it is 'non-storable'. All solids and some liquids are storable; **cryogenic propellant**s are non-storable.

[See also **hypergolic, electric propulsion, nuclear propulsion, rocket engine, rocket motor**]

propellant budget A method of accounting for the amount of **propellant** in a spacecraft's **propulsion subsystem** available for the various phases of its operational **lifetime**. More colloquially termed the 'fuel budget'. There are three principal methods for estimating the amount of fuel remaining in the tanks:

(i) the 'gas law' method: the pressure and temperature of inert gas in the tanks is measured by transducers, and the volume computed from knowledge of the pressure and temperature at launch. The volume of the remaining propellant can then be deduced and its mass determined from a knowledge of its density as a function of temperature. Corrections are applied for the expansion of the tanks and propellant vapour pressure.

(ii) the 'book keeping' method: the **thrust** duration for each **manoeuvre** is measured and, using an empirical model, propellant consumption is calculated from the mass flow rate expressed in terms of pressure.

(iii) the 'dynamical' method: this more sophisticated method is based on measuring the dynamics of the spacecraft following station keeping manoeuvres, which determines its total mass and thereby propellant mass depletion.

Propellant is required, by various different types of **spacecraft**, for **orbital injection**, **plane change**s, **mid-course correction**s, **apogee motor firing**, **attitude control**, **orbital control**, **orbital relocation** and **de-orbit**ing. About half of the launch-mass of a typical **satellite** destined for **geostationary orbit** consists of propellant, the majority of which is used for the **apogee** burn. [See also **fuel, oxidiser, liquid propellant, solid propellant, apogee kick motor, reaction control thruster, retro-rocket**]

propellant charge – see **propellant grain**

propellant dispersal system An arrangement of **pyrotechnic** charges mounted on a **launch vehicle** stage which can be used in an emergency to open the **propellant** tank(s), disperse the propellant and destroy the vehicle. Charges are typically arranged in a continuous line and tend to 'unzip' the tank when ignited. Also known as a **flight termination system (FTS)**.

propellant estimation – see **propellant budget**

propellant grain A quantity of **solid propellant** shaped to give the required combustion characteristics (namely, thrust against time) for a particular **rocket motor**. Also called a propellant 'charge'.

The **thrust** available from a block of propellant is proportional to the area of the combustion surface. One method of burning the propellant is by 'cigarette-combustion', where the propellant block is ignited at one end and burns to the other like a cigarette (also known as 'end-burning'). In this case the cross-section of the block gives the surface area of active combustion and limits the thrust. The active surface area can be increased by a cylindrical hole through the centre of the block so that, as the propellant burns, the hole enlarges radially and the thrust increases. Constant thrust, which is more often required, can be provided by a central hole which is star-shaped in cross-section [see figures P8, R7]. A variation in thrust throughout the **burn** can be arranged by varying the cross-section along the length of the propellant grain. This is generally a requirement for a **launch vehicle** solid rocket motor (SRM) which needs to reduce thrust during the period of **maximum dynamic pressure**.

[See also **propellant, ignition system**]

propellant liner A material placed between the **motor case** and the **propellant** in a solid rocket motor which protects exposed parts of the case from the

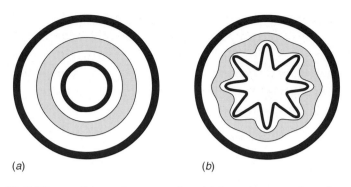

(a)   (b)

P8: Solid **propellant grain** cross-sections: (a) thrust increases throughout burn; (b) constant thrust.

heat of combustion and supports the **propellant grain** during manufacture, as it is poured into the case [see figure R7]. Also known as a **'flame inhibitor'**. [See also **rocket motor**]

propellant management device (PMD) A device or structure inside a **propellant tank** which uses surface tension (capillary) forces to position and hold the **propellant** over the tank outlet, thus allowing propulsion in weightless conditions (i.e. in the absence of inertial or gravitational forces). PMDs vary in design from simple vanes and perforated metal sheets to more complex tubular structures with circular openings, known as collectors or galleries. [See also **ullage rocket**]

propellant mass fraction – see **mass ratio**

propellant tank Any container or reservoir in a **spacecraft** or **launch vehicle** (or on the ground as part of the **ground support equipment**) used to hold **liquid propellant** (**fuel** or **oxidiser**) [see figure P9]. The equivalent item for **solid propellant** is termed a **motor case**.
[See also **external tank**, **vent and relief valve**, **pressurant**, **ullage**, **pogo**, **sloshing**, **propellant dispersal system**]

propulsion bay The part of a **launch vehicle** or **spacecraft** which houses the **rocket engine**s or **rocket motor**s, typically incorporating a **thrust frame** and bordered by engine **cowling**s and other **fairing** components. Also called an **engine bay**.

propulsion module A self-contained section of a modular spacecraft containing the **propulsion subsystem**.
[See also **service module**, **payload module**, **combined (bipropellant) propulsion system**, **propellant**]

P9: Interior of a **propellant tank** for the **Evolved Expendable Launch Vehicle (EELV)**.
[Lockheed Martin]

propulsion subsystem A **subsystem** which provides a spacecraft's propulsion
requirements throughout its **lifetime**. It may use a combination of motors
and engines (using **solid propellant** and **liquid propellant**, respectively), or
thrusters using liquid or gaseous propellant. The term 'propulsion

subsystem' could also be used to differentiate between the various subsystems of a **launch vehicle**, but since the *raison d'être* of a launch vehicle is propulsion this is not common practice.

[See also **propulsion system, rocket motor, rocket engine, apogee kick motor, liquid apogee engine, ignition system, combustion chamber, nozzle, reaction control thruster, cold-gas thruster, combined (bipropellant) propulsion system, electric propulsion**]

propulsion system Any **spacecraft** or **launch vehicle** system that provides propulsion. Compared with **propulsion subsystem**, 'propulsion system' is the more general term: a launch vehicle 'en masse' can be considered a 'propulsion system'; a 'propulsion subsystem' is a part of a vehicle (as is its **guidance** subsystem, **thermal control subsystem**, etc.). See **subsystem**.
[See also **chemical propulsion, electric propulsion, nuclear propulsion, rocket engine, rocket motor, propellant, combined (bipropellant) propulsion system, gravity propulsion**]

protection ratio In **satellite communications** systems, the ratio of the power of the wanted signal to that of the unwanted (interfering) signal (abbreviated C/I, for **carrier**-to-**interference** ratio). It is measured in **decibels** (dB); e.g. if the value of the protection ratio is 3 dB, the wanted signal has twice the power of the unwanted signal. This is for 'single-entry' interference (only one interferer); for more than one interfering signal, the ratio is simply the power of the wanted signal divided by the 'power-sum' of the interfering powers.

Proton A Russian **launch vehicle** developed in the 1960s and first flown in 1968; also known as the SL-12. Different variants have two, three or four **stages**: the three-stage Proton K, offered to the West as a commercial launcher in the late 1980s, had a **payload** capability of 4700 kg to **geostationary transfer orbit (GTO)** and 2000 kg to **geostationary orbit (GEO)**. A later commercial variant, the Proton M (equipped with a restartable Breeze M **upper stage**), was designed to deliver up to 6220 kg to GTO.
[See also **Energia, Angara, Molniya, Tsyklon**]

proving stand A ground-based facility for testing (and 'proving the design' of) **rocket engines** and **rocket motors**. Also called a test stand.

PRS – see **personal rescue sphere**

PSK An abbreviation for phase shift keying. See **phase modulation**.

PSLV An abbreviation for Polar Satellite Launch Vehicle, an Indian **launch vehicle**, designed mainly to launch satellites into **heliosynchronous orbit**.

First launched in 1992, the PSLV's **payload capability** is about 2900 kg to **low Earth orbit (LEO)** and 450 kg to **geostationary transfer orbit (GTO)**.
[See also **ASLV, GSLV, Sriharikota**]

PTT An abbreviation for Posts, Telegraph and Telecommunications organisation; a country's provider of **telecommunications** and other associated services.

pulse code modulation (PCM) A transmission method using a modulated **carrier** wave, whereby the amplitude of the original analogue input signal is sampled at discrete time-intervals to create a representative digital translation of the signal. PCM was devised specifically to enable (analogue) speech to be transmitted in a digital form, using the principle that if an analogue signal is sampled at a rate at least twice that of the highest frequency present, the original speech signal can be reconstructed with acceptable quality from the discrete sample values.

Most PCM systems conform to **CCITT** recommendations and sample at a rate of 8 kHz (8000 times a second). Each sample provides a voltage level which is encoded into a 7-**bit** code, allowing 32 possible values, an eighth bit being added to represent the sign. The **bit rate** for a PCM channel is therefore $8 \times 8000 = 64$ kbits s^{-1}.

pulsed thrust A **thrust** produced in the operation of a **rocket engine** (specifically a **reaction control thruster**), which consists of a sequence of short bursts; an operational alternative to a continuous thrust. Pulses as short as about 12.5 ms are possible, but the minimum duration is limited by the time taken for the thruster valves to open and close (about 5 ms in each case), since this limits the repeatability of the process.

pumped fluid loop – see **cooling system**

purge The removal of contaminants (e.g. 'to purge a **propellant tank**'). Dry nitrogen and helium are gases commonly used for purging a **launch vehicle**'s tanks prior to **propellant**-loading and **lift-off**. The **payload bay** and parts of the fuselage of the Space Shuttle **orbiter** are purged both 'pre-flight' and 'post-flight' to remove contaminants and toxic gases, and to maintain specified temperatures and humidities. The purge gas is monitored for contaminants by a **mass spectrometer**.

pushbroom scanner A **payload** on an **Earth observation** satellite which gathers **data**, using a **charge-coupled device (CCD)**, a line at a time; named by analogy with a broom, which sweeps an area of a given width [see **swath, swath-width**]. A line of data is obtained by sampling the response of the detector elements along the CCD, with successive lines being gathered as the **satellite**

moves along its **orbit**. The pushbroom scanner has advantages over the optical-mechanical line scanner in that it has no moving parts, lower mass, lower **power** requirements, longer operational **lifetime** and higher geometric **resolution**. For example, the scanners designed for the **SPOT** Earth observation satellites deliver resolutions of 10 m in **panchromatic** mode and 20 m in **multispectral** mode (which involve sampling 6000 and 3000 detector elements, respectively).

pyrophoric fuel A **fuel** that ignites spontaneously in air [compare with **hypergolic**].

pyrotechnic Appertaining to **pyrotechnics**, the theory and practice of devices which use an explosive force in their operation.

pyrotechnic cable-cutter A type of pyrotechnic **actuator**; a device which uses an explosive charge to sever a cable connecting two items of **hardware** (e.g. for releasing **solar array** panels folded against the side of a **satellite**). A pyrotechnic bolt-cutter performs the same function for different applications (e.g. separating the **stages** of a **launch vehicle**).

[See also **thermal knife, split-spool device**]

pyrotechnic valve A type of pyrotechnic **actuator**; a device which uses an explosive charge to either close or open a liquid or gas supply-line (e.g. to isolate a **liquid apogee engine** from the rest of a **combined (bipropellant) propulsion system** after the **apogee engine firing**(s) have been made).

pyrotechnics

(i) The theory and practice of devices which use an explosive force in their operation.

(ii) A class of devices which use an explosive charge to separate one item of **hardware** from another [see **pyrotechnic cable-cutter**], isolate part of a **propulsion system** [see **pyrotechnic valve**], or ignite the **propellant** in a **rocket motor** or **rocket engine** [see **ignition system**].

[See also **flight termination system (FTS)**]

Q

QM An abbreviation for qualification model (often abbreviated to 'qual model'); a version of a spacecraft constructed to verify its **performance** to a given set of specifications, to which it must comply to be deemed **space-qualified**.

QPSK (quadrature phase shift keying) A method for modulating the radio frequency **carrier** in digital **satellite communications** links; a type of **phase modulation**. Also called quaternary phase shift keying.

[See also **burst, modulation, digital compression**]

qualification – see **space-qualified**

quarantine facility A terrestrial facility designed to receive and handle extraterrestrial materials (e.g. 'Moon rocks' or Martian soil samples) without contaminating the Earth's environment. Also known as a 'receiving facility'.

[See also **sample return, forward contamination, back-contamination**]

quasi-DBS – see **direct broadcasting by satellite (DBS)**

quaternary phase shift keying – see **QPSK**

R

radar An acronym for *radio detection and ranging*; a method for the
determination of an object's distance and direction using **radio frequency**
radiation. Distance is derived by measuring the time the reflected energy
takes to return to the transmitter. Radar can be used from space for Earth
remote sensing (e.g. measuring ocean wave-heights) and planetary surface
mapping (e.g. determination of Venus' topography despite the visual opacity
of its atmosphere) [see figure R1]. See **synthetic aperture radar (SAR)**.
[See also **side-looking radar, orbital debris**]

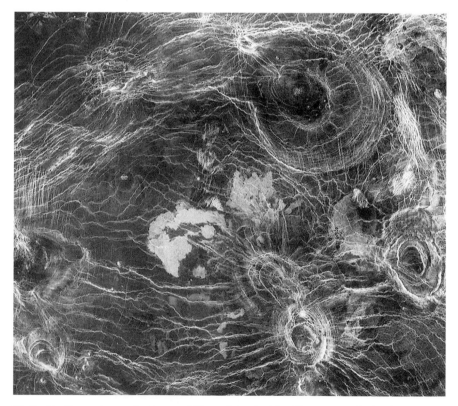

R1: Synthetic aperture **radar** image of Venus from the Magellan spacecraft showing
volcanic features known as arachnoids. [NASA]

radar altimeter An instrument included on some **remote sensing** and planetary exploration spacecraft which uses radar techniques to ascertain the form of the planet's surface features (topography) by measuring their relative height. For example, radar altimeters on **Earth observation** satellites have been used to measure wave-heights, ice surface topography and ocean currents.

[See also **synthetic aperture radar (SAR)**]

rad-hard A colloquial abbreviation for '**radiation**-hardened', a reference to the protection of components, **subsystems**, or **spacecraft** as a whole, from radiation damage. See **radiation hardening**.

[See also **mil-spec, COTS**]

radial thruster A **reaction control thruster** on a **spin-stabilised** spacecraft mounted in alignment with a radius of its cylindrical body (colloquially and collectively termed 'radials'). It produces **thrust** in a direction normal to the spin axis and is typically used for east–west station keeping and orbital repositioning [see **orbital control**].

[See also **axial thruster**]

radiated power **Radio frequency** power emitted from a radiating **antenna**; **electromagnetic radiation** propagating through **free space**. See **equivalent isotropic radiated power (EIRP), power flux density (PFD)**.

radiation

(i) The natural emission or transfer of radiant **energy** as particles or electromagnetic waves – see **electromagnetic radiation**.

[See also **electromagnetic spectrum, space environment, space radiation, radiation hardening**]

(ii) The organised emission or transfer of radiant energy as an electromagnetic wave – see **radio propagation**.

[See also **radio frequency (RF), frequency bands**]

(iii) The emission or transfer of radiant energy as heat. Since convection and conduction are not possible in a vacuum, radiation is the only **heat rejection** mechanism available to spacecraft – see **radiator**.

[See also **thermal control subsystem, thermal insulation**]

radiation effects – see **single event effect (SEE)**

radiation environment – see **space environment**

radiation environment monitor (REM) A device designed to monitor the radiation to which a **spacecraft** is exposed in the **space environment**, as a result of a **solar flare** for example.

radiation hardening The practice of making the components, and thereby the operation, of a **spacecraft** more resistant to radiation damage, particularly from thermonuclear weapons in a military context, but also from natural radiation [e.g. **cosmic radiation**, **solar wind**, **solar flares**, **Van Allen belts**]. Solutions vary from simple shielding to improved design of the affected components, which are predominantly **semiconductors**. Radiation-hardened components or systems are colloquially termed 'rad-hard'; those with a lower tolerance to radiation are termed 'radiation tolerant'.

[See also **single event effect (SEE)**]

radiation pattern – see **antenna radiation pattern**

radiator A device for the removal of heat from a **spacecraft** or components of a spacecraft, usually in the form of a radiator panel, possibly with a number of fins to increase the local surface area.

Radiator panels are of solid or honeycomb construction and are usually covered with **second surface mirrors** (SSMs) on the outside [see figure S2]. The heat source (e.g. **TWTA**, output **multiplexer**, **battery**, etc.) is attached or coupled to the interior surface, which is generally painted black to encourage heat transfer. On a **three-axis stabilised** spacecraft the radiator panels are usually mounted on the north and south faces since they receive solar radiation obliquely; **spin-stabilised** spacecraft generally radiate from their north and south ends, but if additional radiator area is required extra panels are located about the midriff of the cylindrical body. Three-axis stabilised spacecraft may also carry **deployable radiator** panels to further increase the surface area available for radiation.

[See also **heat rejection**, **thermal control subsystem**, **thermal doubler**]

radio (band) – see **waveband**, **radio frequency (RF)**

radio (communications) blackout – see **S-band blackout**

radio frequency (RF) The **frequency** of a radio **carrier** wave: any frequency which lies in the range 3 kHz to 300 GHz (3×10^3–3×10^{11} Hz); frequencies used for radio transmissions, both terrestrially and extraterrestrially. The upper limit is flexible, depending on whether currently available or theoretically possible radio equipment is considered.

[See also **frequency bands**]

radio frequency ionisation thruster – see **ion engine**

radio link – see **link**

radio propagation Transmission of **electromagnetic radiation**, particularly at **radio frequency (RF)**, through **free space** or along a **transmission line**, etc.

radio signal – see **signal**

radio spectrum The radio-frequency portion of the **electromagnetic spectrum**. Also called the radio-frequency spectrum.

[See also **radio frequency (RF)**, **frequency bands**]

radioisotope thermoelectric generator (RTG) A spacecraft power source which derives energy from an array of thermocouples heated by the decay of a radioactive isotope (typically plutonium 238, in its ceramic form of plutonium dioxide, a non-weapons-grade isotope; Pu-239 is weapons-grade). The operation of the RTG is based on the **thermoelectric effect**, typically using two different semiconductor materials, one n-type and one p-type [see **solar cell**]. The 'hot junction' is heated by the radioactive source and the cold junction is attached to a **heat sink**.

One of the first RTGs used in space was the prototype **SNAP**-3 carried by **Transit** 1 in April 1960; it produced a **power** of just 3 W. The RTG is now generally used on spacecraft designed to operate too far from the Sun to use the **solar cell** as a power generator (e.g. planetary exploration **probes** destined for the outer planets) [see figure R2]. It is rarely used in the inner solar system, owing to the mass of shielding required and the potential hazard of an uncontrolled **re-entry**, although the Apollo Lunar Surface

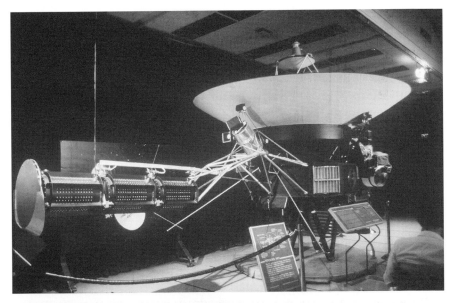

R2: **Engineering model** of **Voyager** spacecraft showing **radioisotope thermoelectric generator (RTG)** at left, imaging system at right and communications **antenna** at top. Note also thermal **louvres** below antenna. [Mark Williamson]

Experiments Package (**ALSEP**) deployed by lunar astronauts was powered by a
SNAP-27 RTG. Spacecraft using RTGs are sometimes referred to as 'nuclear-
powered', but RTGs are not **nuclear reactors**, and use neither fission nor
fusion processes to produce energy.

radiometer Any instrument for the detection and/or measurement of radiant
energy, typically in the visible and infrared **waveband**s. Radiometers are
usually carried by **meteorological satellite**s and **Earth observation** satellites.
The first radiometer, and thus the first meteorological **payload**, to be flown in
space was carried by the scientific spacecraft Explorer 7, launched on 13
October 1959. It consisted of two hemispheres, one painted black to absorb all
wavelengths and the other white to reflect solar energy and absorb only
terrestrial infrared radiation. As a result, it was able to measure the
imbalance between incoming solar energy and outgoing terrestrial
radiation.

[See also **thermal energy balance, surface coatings**]

radiometric resolution A measurement of the ability of an instrument or
recording medium to 'resolve' or distinguish between different levels of
radiant energy. For example, the radiometric resolution of a photographic
film is the smallest detectable difference in exposure.

[See also **radiation, radiometer, resolution**]

rain attenuation The weakening due to rainfall of a radio **signal** propagating
through the Earth's atmosphere. Rain degrades a **satellite communications**
link by attenuating the signal and by increasing the system **noise**
temperature of the **earth terminal** (which 'sees' warm rain as opposed to
cold sky). Attenuation is a result of scattering and absorption of
electromagnetic radiation by the rain drops, an effect which becomes more
pronounced as the **wavelength** approaches the size of the rain drops (about
1.5 mm), as it does at higher transmission frequencies [see **atmospheric**
attenuation]. At 6 GHz (**C-band**), for example, the wavelength is 50 mm,
whereas at 14 GHz (**Ku-band**) it is 21 mm, and the attenuation is
approximately ten times greater.

RAKA An abbreviation for Rosaviakosmos, the Russian Aviation and Space
Agency, formerly known as the Russian Space Agency (RKA).

[See also **RASA**]

RAM An acronym for random access memory, a computer memory to which the
user can add information (as opposed to a read only memory, **ROM**).

[See also **DRAM**]

ramjet A type of jet engine in which **propellant** is burnt in a duct using air compressed by the forward motion of the aircraft. A version known as a supersonic combustion ramjet, or scramjet, has been proposed as the basis of a **combined cycle engine** or air-breathing rocket.

ramp The wedge-shaped surface of a linear **aerospike engine**, against which combustion gases are discharged; also known as a spike.

range

   (i) The distance between a **launch vehicle** or a **spacecraft** and the ground station tracking it.

  (ii) The act or process of **ranging** (i.e. 'to range').

 (iii) The maximum effective distance of a projectile (e.g. on a **ballistic trajectory**), a vehicle or a radio **signal**.

 (iv) A geographical area set aside for **rocket** launches or testing (e.g. the USA's **Eastern Test Range**).

   [See also **slant range, ICBM, IRBM**]

Ranger The name given to a series of American spacecraft designed for lunar research. Ranger 7 (launched in July 1964) was the programme's first success: it produced 4308 images before crashing, as intended, onto the lunar surface.

range safety system A collection of **pyrotechnic** devices on a **launch vehicle** used to destroy it in the event of a malfunction which could threaten the safety of the firing **range** and the surrounding area.

   [See also **propellant dispersal system**]

ranging The determination of the distance between a **launch vehicle** or a **spacecraft** in **orbit** and the ground station tracking it.

   [See also **range, tracking**]

rankine (°R) A temperature scale devised by the Scottish engineer William John Macquorn Rankine [1820–1872] which begins at **absolute zero** but is divided into degrees of equal size to those on the Fahrenheit scale, rather than degrees Celsius as is the **kelvin** scale. The Rankine scale (°R or °Rank) is virtually unheard of in the scientific world, but is sometimes used by engineers.

RARC An acronym for Regional Administrative Radio Conference, a conference convened under the auspices of the International Telecommunication Union (**ITU**) to plan **communications** services (particularly with regard to the allocation of radio frequencies). See **WARC**.

RASA An anglicised abbreviation for the Russian Aviation and Space Agency, Rosaviakosmos (RAKA), which was formed in 1999 when President Yeltsin ordered the transfer of control of the aviation industry from the Economics Ministry to RKA, the former Russian Space Agency.

rat's nest Engineering slang for a tangle of electrical wiring.

raw data Unprocessed data; information or experimental results in their most basic form (e.g. the **telemetry** signals received from a spacecraft before their conversion into measurements of **attitude**, temperature, etc., which can be readily analysed and understood).

reaction control system (RCS) A spacecraft **subsystem** which provides **attitude control** by means of a set of **reaction control thrusters** (typically grouped to provide adjustments in **roll**, **pitch** and **yaw**).

[See also **orbital manoeuvring system (OMS)**, **Apollo**, **Gemini**, **Mercury**, **X-15**]

reaction control thruster A small **rocket engine** used to make fine adjustments to the **orbit** or **trajectory** and the **attitude** of a **spacecraft** by the expulsion of gas molecules at high velocity (hence the more colloquial term 'gas jet'). The term 'reaction control' is derived from the principle of 'action and reaction' (Newton's third law of motion): the momentum (mass × velocity) of the gas molecules causes an equal and opposite momentum of the spacecraft.

The shape and construction of the thrusters is similar to that of the larger engines, in that they have a **combustion chamber** with a narrow **throat** leading to a divergent exit cone or '**nozzle**' [see figure N2]. Thruster dimensions are, however, measured in millimetres rather than metres, and the **thrust** produced may be as low as 0.5 N (as opposed to hundreds of kN). There are three main varieties of thruster commonly used in spacecraft control: the **hydrazine thruster**, the **cold-gas thruster** and the **bipropellant thruster** [see figure B5].

[See also **hiphet thruster, attitude control, orbital control, momentum dumping, plume, exhaust velocity, specific impulse, pulsed thrust, orbital manoeuvring system (OMS)**]

reaction wheel A wheel used to provide **attitude control** in a **three-axis stabilised** spacecraft. Reaction wheels are typically fitted in sets of three – one for each of the orthogonal **spacecraft axes** – although it is quite common to install a fourth wheel, mounted at an angle to all three axes, as a spare. If one of the first three wheels fails, the fourth can be used in conjunction with the remaining wheels to provide full three-axis attitude control.

A reaction wheel is powered by an electric motor and operates under command from an attitude control subsystem. If the wheel is commanded to spin faster, its angular momentum will increase, which, alone, would infringe the law of conservation of momentum, since the angular momentum of the spacecraft as a whole would have increased. However, the law is satisfied by a transfer of angular momentum from the wheel to the spacecraft, which rotates about the wheel-axis in the opposite direction, thereby conserving the total angular momentum of the system. Naturally the spacecraft rotates much more slowly than the wheel, in proportion to their relative moments of inertia. If it was required to rotate the spacecraft in the opposite sense, the wheel would be decelerated: it would lose angular momentum and the spacecraft would gain it by rotating in the opposite direction. In both cases, in the same way that attitude control **thruster** firings require a cancellation force, a further change of wheel momentum would be required to halt the rotation of the spacecraft. It is this 'action and reaction' philosophy that gives rise to the term 'reaction wheel'.

In contrast to a **momentum wheel**, which is normally spinning at a high rate to give stability to the spacecraft, the nominal speed of a reaction wheel is zero and it can be rotated in either direction (typically at $+2500$ rpm). A system using reaction wheels is known as a 'zero momentum' system, as opposed to one using momentum wheels which is called a 'momentum bias' system; both types of wheel are sometimes called inertia wheels.

[See also **momentum dumping**]

reaction wheel assembly (RWA) A number of **reaction wheel**s grouped together.

[See also **assembly**]

reactor – see **nuclear reactor**

real time Jargon: originally the term referred to a data-processing system in which a computer processed **data** as it was generated (i.e. 'in real time'), rather than recording or storing it for later use. The term is now also applied to more general activities.

receive chain A collection of electronic and microwave equipment which receives, filters and amplifies a signal fed from a receive-antenna. In a spacecraft **communications payload**, for example, the receive chain generally comprises an **input filter** or **multiplexer**, **low noise amplifier**, **downconverter**, preamplifier, **channel filter** and associated **RF switch**es and **hybrid**s. Whether or not to include the channel filter or the receive-antenna in the definition of the receive chain depends on the particular design and, to an extent, personal choice.

[See also **receiver, transmit chain, amplifier chain, input section, output section**]

receiver Generally, a device used to detect and amplify incoming electrical
signals or modulated radio waves and convert them into signals to drive an
output device (e.g. audio or video equipment), or to pass them to further
devices prior to retransmission.

In the typical satellite **transponder**, the receiver comprises an **input filter**,
which confines the frequency **bandwidth** of the **signal**, several **pre-amplifier**
stages, and a **mixer** or **downconverter** (to convert to a lower frequency). The
receiver is also known as the 'front end'.

[See also **receive chain, modulation**]

receiving facility Another name for **quarantine facility**.

reciprocity (principle of) The ability of an **antenna** system to both receive and
transmit. It is useful in analysing antenna designs and measuring
performance since receive characteristics can be inferred from transmit
performance and vice versa.

reconnaissance satellite A military satellite, usually in a **low Earth orbit**,
concerned primarily with high **resolution** imaging of 'enemy territory',
although the definition could be extended to included the detection of
ballistic missile launches. Colloquially termed a 'spy satellite' and similar
to a **surveillance satellite**, except that reconnaissance implies a periodic
collection of **data**, whereas surveillance implies continual data
collection.

[See also **remote sensing**]

recovery The retrieval of a space **capsule**, recoverable **solid rocket booster** or any
other item, typically from a body of water (following a **splashdown**) or from a
large area of uninhabited land (following a **touchdown**).

recycle countdown The procedure whereby a countdown clock is re-set to an
earlier point to continue the **countdown**, usually following the solution of a
problem which led to a **hold** in that countdown. The procedure is only
followed when the hold was unforeseen; in the case of a **built-in hold** the
countdown continues from the point at which it was halted.

Redstone An American single-**stage** liquid **propellant** intermediate range
ballistic missile (IRBM) developed by the US Army in the 1950s (**oxidiser:**
liquid oxygen; fuel: 75% ethyl alcohol, 25% water). Used as a **launch vehicle**
for the 15-minute **sub-orbital** flights of the **Mercury** programme (the **Atlas**
booster was used for the orbital flights). No longer **operational**.

[See also **Jupiter, Thor**]

redundancy An attribute of a **system** in which a 'spare' or **back-up** device is available in case of failure of the **prime system**. If a system is to be 'fully redundant', all parts of the prime system must be duplicated in the back-up system: this is known as 'two-for-one redundancy'. It is unusual to duplicate all the devices in, for example, a spacecraft **communications payload** owing to the restriction on mass, but redundancy is typically provided for critical components (thus two **amplifiers** connected in parallel provide '2-for-1 redundancy'; a 'redundancy scheme' where one spare is provided for each pair of amplifiers is '3-for-2 redundancy', and so on).

[See also **ring redundancy, single point failure**]

redundant Spare: a redundant component, **system**, etc., is one designed to replace the **prime system** in case of failure; otherwise termed a **back-up**. See **redundancy**.

re-entry The return of a **spacecraft** or other man-made body into the Earth's atmosphere. For example, the Space Shuttle re-entry is considered to occur at an altitude of 122 km (400,000 feet).

[See also **entry interface, atmospheric drag, plasma sheath, thermal protection system, ablation**]

re-entry corridor A volume of space through the Earth's atmosphere within which a returning **spacecraft** must pass to ensure a safe **re-entry** [see figure R3]. If the angle of the re-entry **trajectory** relative to the Earth's surface is too shallow, the spacecraft may 'skip off' the atmosphere (like a stone skipping off a water surface); if the angle is too steep, it will burn up because of excessive frictional heating (e.g. the nominal angle for the **Apollo** command module was $6° \pm 1°$, and the corridor was 42 km wide).

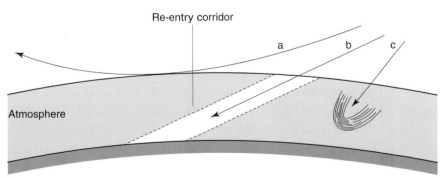

R3: Spacecraft **re-entry corridor**: (a) angle too shallow – vehicle fails to re-enter; (b) correct angle; (c) angle too steep – vehicle burns up.

reflector antenna – see **antenna**

refractory metal A metal able to withstand high temperatures without fusion or decomposition, typically with a melting point above 2200 °C (e.g. columbium, molybdenum, tantalum and tungsten).
[See also **ceramics**]

regenerative cooling The practice of pumping **liquid propellant** through the walls of exhaust **nozzles**, and sometimes **combustion chambers**, before it enters the chamber itself [see figure N2]. Cooling the nozzle enables it to withstand higher exhaust-gas temperatures that it otherwise would without recourse to exotic heat-resistant materials (which can lead to an increase in **specific impulse**). **Cryogenic propellant**s are particularly useful in this regard; regenerative cooling is clearly impossible with **solid propellant** motors.
[See also **rocket engine**]

regenerative repeater A **repeater** that regenerates an input **signal** by stripping the **baseband** signal from its **carrier**, performing various operations upon it, and replacing it on the carrier for onward transmission; sometimes called a regenerative **transponder**. The 'stripping' and 'replacing' functions are called, respectively, demodulation and remodulation (or simply **modulation**) and require the insertion of demodulators and modulators into the transponder chain. The process allows the characteristics of the signal to be 'regenerated' on board a satellite, thus correcting errors caused by **thermal noise** in the preceding equipment, for example, and avoiding the transmission of the errors into the **downlink** beam. Regeneration therefore isolates the **uplink** performance from the downlink performance and allows receive and transmit chains to be designed more independently and with improved performance characteristics.
[See also **transparent repeater, on-board processing**]

Regional Administrative Radio Conference – see **RARC**

Registration Convention The colloquial name for the treaty of **space law** whose official title is "Convention on Registration of Objects Launched into Outer Space". The Registration Convention was adopted by the United Nations General Assembly on 12 November 1974, opened for signature on 14 January 1975 and entered into force on 15 September 1976. It was one of four main space treaties derived from the **Outer Space Treaty** of 1967.
[See also **Rescue Agreement, Liability Convention, Moon Agreement, space object**]

regulated power supply A spacecraft **power supply** which provides a constant **bus voltage**, typically to within plus or minus 1% of the nominal value, whether the spacecraft is in sunlight or **eclipse**, whether relying on its primary or **secondary power supply**. Satellite power subsystems, for instance, can be fully regulated; totally unregulated; or regulated in sunlight, when power can be derived from the **solar array**, and unregulated in eclipse, when the output voltage falls to that provided by the **battery**.
[See also **unregulated power supply**]

reliability A branch of spacecraft engineering which quantifies the likelihood of a failure at component, device, **subsystem**, **system** and **spacecraft**-level and provides feedback to the design teams to reduce the probability of failure. Apart from improved standards of quality control, material and component selection, handling and testing, reliability is improved by incorporating the concept of **redundancy**, thereby minimising the possibility of **single point failure (SPF)**.
[See also **COTS, bathtub curve, probability distribution**]

relief valve – see **vent and relief valve**

re-light – see **engine re-start (capability)**

REM – see **radiation environment monitor**

remote agent – see **artificial intelligence (AI)**

remote manipulator system (RMS) In general, a jointed, mechanical arm (and its associated control systems) attached to a spacecraft and used, by remote control, to handle equipment, **deploy** and retrieve **payload**s, etc. In particular, the mechanical arm mounted on the side of the **payload bay** of the Space Shuttle [see figure R4]. The Shuttle manipulator arm is approximately 16 m long and 0.38 m in diameter, and is capable of handling payloads of nearly 30 tonnes. It comprises two **booms** and an **end effector** (hand) with shoulder, elbow and wrist joints, which give it six **degrees of freedom**. It is operated from the aft **flight deck** of the Shuttle, which overlooks the payload bay.
[See also **grapple fixture, telepresence, International Space Station (ISS)**]

remote sensing Detection and **data** collection from a distance. Remote sensing satellites (otherwise known as **Earth observation** satellites) are usually placed in high-inclination, near-**polar orbit**s, which give full Earth coverage over a period of days with a regular 're-visit capability' to enable data to be updated. Typical **payload**s allow observations in a variety of **wavebands** from the ultraviolet to the infrared, with a multitude of applications including the

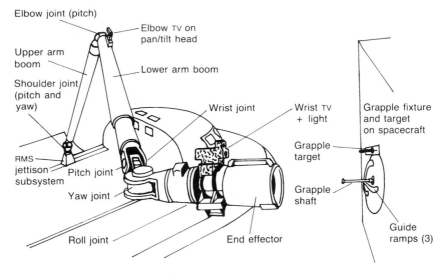

R4: **Remote manipulator system (RMS)** used on **Space Shuttle.**

determination of crop health, pollution detection, map-making and 'intelligence-gathering'. Some satellites carry **radar** payloads, which enable images of land and sea surfaces to be acquired in darkness or through cloud [see **synthetic aperture radar (SAR)**].

In general, the main practical difference between a commercial remote sensing satellite and a military 'spy satellite' is the **resolution** of the **sensor**s, which is generally greater in the latter case. In the late 1990s, however, a number of commercial remote sensing spacecraft with ground resolutions of approximately 1 m, matching that of some military systems, was introduced [see **shutter control**].

[See also **panchromatic, multispectral, hyperspectral, meteorological satellite, reconnaissance satellite, Landsat, link, Nyquist frequency**]

rendezvous A meeting between two or more **spacecraft** engineered by a matching of **orbital parameter**s, usually as a preamble to **docking**. The process is not as straightforward as it might at first appear. The radius of a spacecraft's **orbit** is dependent upon its orbital velocity, such that a spacecraft in a higher orbit has a longer **orbital period**. This means that, for two spacecraft travelling one behind the other in the same orbit, the one behind must decrease its orbital height to reduce its period and 'overtake' the other. Simply increasing its orbital velocity would add energy to the orbit and increase its height; its period would increase and it would lose ground on the first spacecraft. Therefore its orbital velocity/energy must be decreased: height decreases, period decreases and it overtakes. To complete the rendezvous, it must regain its orbital height from a position in front of and below the first spacecraft, which it does by increasing its orbital velocity/energy in a carefully controlled way.

[See also **orbital control**, **equilibrium point**]

repeater In general, a device which receives, amplifies and retransmits electrical signals (e.g. **satellite** repeater, **spacecraft** repeater, **communications** repeater). A spacecraft repeater incorporates all the devices of the **communications payload**, with the exception of the receive and transmit **antenna**s, and usually comprises a number of **transponder** chains. On a multi-band satellite there may be a C-band repeater, a Ku-band repeater, etc., each with several transponders; each repeater generally has a dedicated antenna subsystem because of the differences in **frequency**, **beamwidth**, etc.

The first satellite to carry an **active** radio repeater was the US Army/Department of Defense satellite Courier 1B. It was launched on 4 October 1960, exactly three years after **Sputnik** 1, and operated for 17 days (four less than Sputnik itself). Courier could store up to 360,000 teletype words transmitted from one **earth station** and retransmit them as it passed over another, a mode of operation which identified Courier as a 'delayed repeater' satellite (now termed '**store-and-forward**'). The first instantaneous satellite communications link was provided by **Telstar** 1 in 1962.

Rescue Agreement The colloquial name for the treaty of **space law** whose official title is "Agreement on the Rescue of Astronauts, the Return of Astronauts and the Return of Objects Launched into Outer Space". The Rescue Agreement was adopted by the United Nations General Assembly on 19 December 1967, opened for signature on 22 April 1968 and entered into force on 3 December 1968. It was the first specific agreement derived from the **Outer Space Treaty** of 1967.

[See also **Liability Convention**, **Registration Convention**, **Moon Agreement**]

residual oxygen Oxygen atoms in **low Earth orbit** which can degrade certain spacecraft **surface coatings** by impact [see **degradation**].

residuals Jargon: the **propellant**s or **pressurant**s left in a tank at the end of a **mission**. It is usually not possible to dispense the entire contents of a **propellant tank** and a certain volume of propellant will remain. Since the **lifetime** of many **unmanned spacecraft** is governed by the amount of **station**

keeping propellant carried, it is important to make allowances for 'residuals': an entry for such should therefore appear in a spacecraft's **propellant budget**. [See also **passivation**]

resolution A measurement of the ability of an instrument, a recording medium or an observer to 'resolve' or distinguish between separate parts of an **image**; also known as 'spatial resolution' or 'geometric resolution' and sometimes called 'resolving power' [see figure R5]. The resolution of commercial **remote sensing** satellites of the 1980s and early 1990s was typically around 10 m (**panchromatic**); in the late 1990s, however, this was increased to approximately 1 m, approaching that of some military systems [see **shutter control**].

[See also **pixel, radiometric resolution, spectral resolution, temporal resolution**]

| Quickbird 0.82 m panchromatic | IRS-1D 5m panchromatic | SPOT 10 m panchromatic |

R5: Comparison of simulated sensor **resolution** for three different **Earth observation** satellites. [EarthWatch]

restartable upper stage An **upper stage** with a **liquid propellant** engine which can be started and stopped on **command**, typically to deliver **payload**s to different **orbit**s (e.g. the delivery of **constellation** satellites to differing **orbital planes**).

[See also **dispenser, engine re-start (capability)**]

restraint Any device on or in a **spacecraft** which restricts an **astronaut**'s movement, especially important in the **microgravity** environment. For example, 'foot restraints' used by astronauts on **extra-vehicular activity (EVA)** [see figure E7]; and 'sleep restraints' (sleeping bags used on the Space Shuttle).

retrofit To add components or **systems** after manufacture, often necessary when updating equipment or adding safety systems, etc.

retrograde orbit An **orbit** in which the orbiting body has the opposite sense of rotation to the orbited body; the opposite of **prograde orbit**.

retro-pack In general, a 'package' of **retro-rockets** used to decelerate a spacecraft prior to **re-entry**; in particular, the package of **solid propellant** rocket motors mounted over the **heat shield** of the manned **Mercury** capsule, jettisoned after use.

retroreflector – see **laser ranging retroreflector**

retro-rocket A small, auxiliary rocket engine on a **launch vehicle** or **spacecraft** that produces **thrust** in the opposite direction to the direction of **flight** in order to decelerate the vehicle. Often abbreviated to 'retro'.

As examples, most launch vehicle **stages** have retro-rockets which fire after **stage separation** to prevent the jettisoned stage colliding with the upper stage before its engines are ignited; spacecraft returning to Earth or landing on another **planetary body** sometimes use retro-rockets to make a 'soft landing'. Most retro-rockets use **liquid propellants**, but **solid propellants** have also been used [see **retro-pack**].

[See also **rocket engine, rocket motor**]

reusable launch vehicle (RLV) A **launch vehicle** which can be used more than once; the alternative to an **expendable launch vehicle (ELV)**. In the 1990s, several RLVs were proposed to replace either commercial ELVs or the Space Shuttle, which is only partly reusable. The Shuttle has a reusable **orbiter**, an expendable **external tank (ET)**, and **solid rocket boosters** (SRBs) which can be refurbished for later use.

[See also **X-33, aerospike engine, space tourism**]

re-visit capability – see **remote sensing**

RF – see **radio frequency**

RF chamber – see **anechoic chamber**

RF link – see **link**

RF power Power supplied by devices operating at **radio frequency (RF)**. See **power**.
[See also **radiated power, power flux density, saturated output power, power supply, power bus, power budget**]

RF sensor Radio frequency sensor: a spacecraft **sensor** used to maintain accurate **pointing** of a spacecraft's **antennas**. The sensor detects an RF signal transmitted from a known point in the **service area** and an **antenna pointing mechanism** drives the antenna to maintain the maximum signal level. The most common type is the monopulse RF sensor.
[See also **earth sensor**]

RF signal A radio frequency signal – see **signal**.

RF switch A switch designed to operate at **radio frequency (RF)**. There are two main types: the **coaxial switch** and the **waveguide switch**.

RF wave An abbreviation for **radio frequency** wave: any waveform in the **electromagnetic spectrum** which can be classed as 'RF', i.e. between about 3 kHz and 300 GHz. See **radio frequency (RF)**.

RFNA An abbreviation for red fuming nitric acid, a substance commonly used in the past as a rocket **propellant**.
[See also **liquid propellant**]

RFP An abbreviation for request for proposals, industrial terminology for a document which invites bids for a contract. Similar to a request for quotations (RFQ) or invitation to tender (ITT).

RFQ An abbreviation for request for quotations, industrial terminology for a document which invites bids for a contract. Similar to a request for proposals (RFP) or invitation to tender (ITT).

RHCP (right-hand circular polarisation) – see **polarisation**

ring laser gyro (RLG) – see **laser gyro**

ring redundancy A **redundancy** scheme used for a satellite **communications payload** whereby **travelling wave tubes**, **travelling wave tube amplifiers** or other devices are interconnected in a manner which allows mutual substitution. Instead of installing a dedicated spare for each active device (so-called 2-for-1 redundancy), a group of devices provide mutual back-up so that, if one fails, another in the 'ring' can be activated to replace it.

RKA An abbreviation for the former Russian Space Agency, which was established in February 1992 and later became the Russian Aviation and Space Agency (RAKA).
[See also **RASA**]

RLG An abbreviation for ring laser gyro – see **laser gyro**.

RMS

    (i) an abbreviation for root mean square, the square root of the average of the squares of a set of numbers or quantities, a quantity which provides a measure of the 'spread' of the **data** (i.e. a large RMS implies the data are widely scattered, a small RMS means they are relatively confined).

    (ii) an abbreviation for **remote manipulator system**.

robotics A branch of engineering concerned with automated machines designed and programmed to perform specific mechanical functions with or without human intervention (e.g. a planetary **probe**, such as the **Viking** lander, designed to sample the surface of Mars). When a mechanical manipulator is controlled entirely by human intervention from a distance, the technique is known as **telepresence**.

rocket

(i) A propulsive device using a mixture of **propellants** (**fuel** and **oxidiser**) which, upon **ignition**, provide a reaction force to propel a space vehicle, missile, rocket-assisted aircraft, etc.

(ii) A colloquial term for **launch vehicle**.

[See also **rocket engine**, **rocket motor**, **retro-rocket**, **booster**, **reaction control thruster**, **nuclear thermal rocket (NTR)**, **combined cycle engine**]

rocket booster – see **booster**

rocket engine A propulsive device in which **liquid propellants** are burnt in a combustion chamber to provide a reaction force to propel a vehicle; the part of a chemical **propulsion system** in which propellants are burned and combustion products are used to produce **thrust** [see figure R6]. Although the terms engine and motor are interchangeable in colloquial English, it is

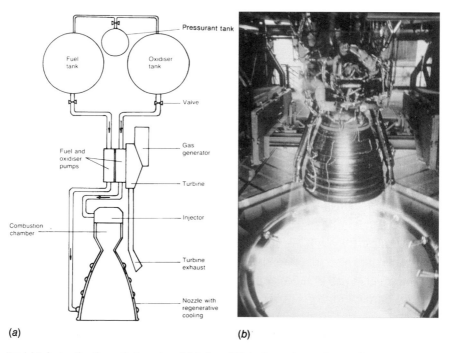

(a)   *(b)*

R6: (a) Schematic of a **rocket engine**; (b) **Ariane** 5 Vulcain engine under test (note two **turbopump** exhausts). [SEP]

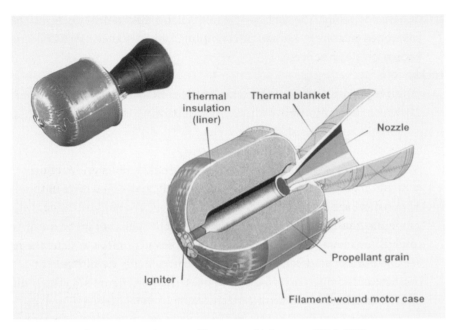

Thermal
insulation
(liner)

Thermal blanket

Nozzle

Propellant grain

Igniter

Filament-wound motor case

R7: MAGE 2 **rocket motor** used as a satellite **apogee kick motor (AKM)**. [SEP]

customary for propulsion devices using liquid propellant to be called 'engines' and for those using **solid propellant** to be called 'motors'.

A typical liquid rocket engine comprises a propellant delivery and **injection** system, an **ignition system**, a **combustion chamber** and an exhaust **nozzle**. Liquid propellant engines are either pump-fed or pressure-fed: although the former are more complex, the latter require thicker walled propellant tanks to withstand the pressure. In general, the liquid engine is more complex than the solid motor: it requires a network of pipes and pumps (and/or **pressurant** tanks) to deliver the propellant to the combustion chamber, whereas a 'charge' of solid propellant makes its own 'chamber'. In operation, however, the liquid engine offers a far greater degree of control since it can be stopped and re-started in-flight, and the flow of propellant can be varied at will. A solid cannot be re-started and its **thrust profile** is set at the manufacturing stage. In addition, whereas rocket engine nozzles and combustion chamber walls can be cooled by piping unburnt propellant through them [see **regenerative cooling**], solids do not have this advantage and their **burn**-times are limited by the thermal properties of the **materials** used in the nozzle.

[See also **rocket motor**, **turbopump**, **gimbal**, **liquid apogee engine**, **ascent engine**, **descent engine**, **combined (bipropellant) propulsion system**, **propellant**]

rocket equation An expression used to calculate the **propellant** mass (Δm) required to perform a spacecraft **manoeuvre** with a given velocity increment (Δv):

$$\Delta m = m_0 \left[1 - \exp \left(-\Delta v / I_{sp} g \right) \right]$$

where m_0 is the initial spacecraft mass, I_{sp} is the **specific impulse** and g is the acceleration due to **gravity**. The equation shows that a reduction in the required velocity increment or an increase in specific impulse reduces the propellant mass required. For example, a velocity increment is required to transfer a satellite from **geostationary transfer orbit (GTO)** to **geostationary orbit (GEO)**, simultaneously removing the inclination and circularising the orbit. Since the transfer orbit has an inclination related to the latitude of the launch site, the velocity increment is lower for launch sites nearer the equator (e.g. 1478 m s^{-1} for an equatorial site, 1502 m s^{-1} for the **Guiana Space Centre** and 1831 m s^{-1} for **Kennedy Space Center**). Thus a given **launch vehicle** will have a greater **payload capability** when launched from a site closer to the equator.

rocket motor A propulsive device in which **solid propellant**s are burnt to provide a reaction force, or **thrust** to propel a vehicle. Also called a 'solid rocket motor' (SRM). Although the terms motor and engine are interchangeable in colloquial English, it is customary for propulsion devices using solid propellant to be called 'motors' and for those using **liquid propellant** to be called 'engines'.

A typical solid rocket motor comprises only a few major components: a **motor case** which contains the **propellant grain**, a surrounding insulating blanket or **propellant liner**, an exhaust **nozzle** and an **ignition system** [see figure R7]. The motor case is typically made from a **carbon composite** or **Kevlar composite** by a process called **filament winding**, which produces a strong yet 'light-weight' structure. The propellant liner acts as both a **thermal blanket** and a **flame inhibitor**, and also supports the propellant grain during manufacture, as it is poured into the open-ended motor case. In the terminology of liquid propulsion systems, the motor case is both the **propellant tank** and the **combustion chamber**.

In operation, the solid propellant motor has no need of the complex piping, pumps and **pressurisation** systems of the liquid **rocket engine**: it is like comparing a safety match to a gas-lighter. It does, however, have to be designed to give precisely the required amount of thrust and total **impulse**, since it cannot be 'throttled', easily stopped or re-started. The only way to control the magnitude of the thrust is to design the propellant charge to offer varying surface areas for combustion throughout the '**burn**' [see **propellant grain**]. Thrust vectoring is generally achieved by gimballing the nozzle, but systems using vanes or injecting inert gas into the nozzle to deflect the **plume** have been used.

[See also **solid rocket booster, reaction control thruster, thrust vector, vector steering, gimbal, throttling**]

rocket nozzle – see **nozzle**

rocket stage – see **stage**

Rockot A commercial **launch vehicle** converted from the former-Soviet SS-19 'Stilleto' medium-class intercontinental ballistic missile (**ICBM**) and operated under a German–Russian joint venture, Eurockot Launch Services. The first and second stages are adapted from the first two **liquid propellant** stages of the SS-19 and the third stage is a space-qualified **restartable upper stage** called Breeze. The **payload capability** of the Rockot in its basic version is about 1500 kg to **low Earth orbit (LEO)** and about 1800 kg with a Breeze KM **upper stage**. It is launched from Russia's Northern Cosmodrome at Plesetsk.

[See also **Cosmos, Dnepr, Start**]

roll A rotation about the roll axis – see **spacecraft axes**.

[See also **roll programme, barbecue roll, Dutch roll**]

roll axis – see **spacecraft axes**

roll programme The sequence of commands to a **launch vehicle** which cause it to rotate about its vertical axis (i.e. to **roll**) to provide gyroscopic stability. The roll programme typically begins a short time after **lift-off** and continues until **payload** separation, although the rate of roll may be reduced (or eliminated) prior to separation. The roll programme of the Space Shuttle differs in that it is a single roll **manoeuvre** (clockwise rotation about the vertical axis) of a number of degrees (e.g. 120°, 140°) dependent on the required **orbital inclination**.

[See also **spin-stabilised**]

ROM An acronym for read only memory, a computer memory device to which the user cannot add information (as opposed to a **RAM**). Variants are PROM (programmable read only memory) and EPROM (erasable programmable read only memory).

Rosaviakosmos The Russian Aviation and Space Agency. See **RASA**.

Rosen scale – see **electrostatic discharge (ESD)**

rotating service structure (RSS) A **service structure**, attached to the 'fixed service structure' at the **launch pad**s used by the Space Shuttle, which can be rotated to provide access to the Shuttle's **payload bay**, particularly for loading [see figure S3]. In a typical loading operation, a **payload canister** is delivered to the pad, and lifted into the 'payload changeout room' on the RSS. The RSS is then rotated towards the **orbiter** until the changeout room doors form a seal with the open payload bay, when the payloads can be transferred to the orbiter.

rotational velocity (of Earth) – see **plane change**

rover A manned or unmanned vehicle designed to provide mobility on the surface of a **planetary body**.

[See also **lunar roving vehicle, Sojourner**]

RP1 – see **kerosene**

RSA An anglicised abbreviation for the former Russian Space Agency (RKA), which later became the Russian Aviation and Space Agency (RAKA).

[See also **RASA**]

RTG – see **radioisotope thermoelectric generator**

runway (for Space Shuttle) – see **Shuttle landing facility (SLF)**

RWA An abbreviation for reaction wheel assembly, a number of **reaction wheels** grouped together.

[See also **assembly**]

S

sacrificial shield – see **Whipple shield**

SADA An acronym for solar array drive assembly. See **solar array**.

SADE An acronym for solar array drive electronics. See **solar array**.

SADM (solar array drive mechanism) – see **BAPTA**

safing The process whereby a space vehicle or system is 'made safe'. The term is typically applied to a spacecraft, such as the **Space Shuttle**, following its return to Earth. It includes the removal of hazardous residual **propellant**s, the draining and purging of propellant feed-lines, and the removal of explosive devices such as **pyrotechnic** actuators.

[See also **purge**]

Salyut A series of Soviet **space station**s launched and operated throughout the 1970s and 1980s. Salyut 1, which was launched in April 1971, was followed by two failures (Salyut 2 and a Salyut-type vehicle Cosmos 557) and the successful flights of Salyuts 3–7: 3 and 5 for military applications; 4 and 6 mainly civil. Salyut 7 combined both applications and was significantly redesigned. It had a total habitable volume of about 100 m³ and comprised three **module**s: a forward transfer compartment (basically an access tunnel used for **EVA**s); an operations compartment used as the main living and working area; and a rear transfer compartment to allow connection with visiting **Soyuz** 'ferries' and **Progress** re-supply 'tankers'. After three **crew**s had occupied Salyut 7 for periods of 211, 150 and 237 days, with short-stays by visiting crews, the station was boosted into a higher **orbit** in 1986 and left for possible future experimental use.

[See also **Mir**, **Skylab**]

sample return A concept involving the collection of a sample of another **planetary body** and returning it to Earth for analysis.

[See also **quarantine facility**, **forward contamination**, **back-contamination**]

San Marco platform A **launch platform** in the Indian Ocean, 5 km from the Kenyan coast (at approximately 3° S, 40° E) developed in the mid-1960s to launch the Italian San Marco satellite using **NASA**'s **Scout** launch vehicle. The facility, which conducted its first launch on 26 April 1967, was operated by the Aerospace Research Centre of the University of Rome and launches were controlled from the adjacent Santa Rita control platform.

[See also **Sea Launch**]

Sanger A two stage **aerospace vehicle** proposed in the 1980s by Germany as a next-generation satellite-launcher. It was named after Eugen Sanger [1905–1964] who developed the original concept in 1942. The proposed second stage was to be either a manned shuttle vehicle (called Horus), or an unmanned cargo transporter (Cargus).

[See also **HOTOL, reusable launch vehicle (RLV)**]

SAR An acronym for **synthetic aperture radar**.

satellite

(i) A **celestial body** in **orbit** around another celestial body.

(ii) A **spacecraft** in orbit around a celestial body, such as the Earth, the Sun or any other of its planets and moons. Although the terms 'spacecraft' and 'satellite' are often used synonymously, spacecraft is the more general term: typically only **unmanned spacecraft** intended to orbit a **planetary body** are referred to as satellites; interplanetary **probe**s are not referred to as satellites, unless and until they enter orbit around a planet.

[See also **artificial satellite, applications satellite, communications satellite, Earth resources satellite, meteorological satellite, navigation satellite, astronomical satellite, reconnaissance satellite, tethered satellite, smallsat**]

satellite communications The collective term for information transmitted via a **communications satellite**. Also called satellite telecommunications.

[See also **communications, telecommunications, digital compression, link**]

satellite switching The routing or distribution of messages, **data**, etc. performed on board a **satellite** rather than on the ground. This is a relatively advanced concept in which the satellite is described as 'a switchboard in the sky'. Most satellite **transponder**s operate simply in that they receive an **uplink** signal, amplify and filter it, change its frequency and then transmit it into a single **downlink** beam [see **transparent repeater**]. For major trunk telecommunications links, the downlinked signal is received by a single **gateway station** and routed from there into the terrestrial network. With satellite switching, the uplinked signal is labelled with its destination and the satellite routes it into one of several **beam**s aimed at several **earth**

stations within the **coverage area**. Although more complex, this type of system uses the limited **bandwidth** more efficiently.

[See also **multiple beam, satellite-switched time division multiple access (SSTDMA)**]

satellite-switched time division multiple access (SSTDMA) A version of TDMA whereby **signal** switching is performed on-board the **satellite** rather than at the **earth station**. See **satellite switching** and **time division multiple access (TDMA)**.

saturated output power The output power of an **amplifier** driven at **saturation**; the maximum output power available from a **high power amplifier**. Beyond saturation, any increase in the input power results in a decrease in output power.

[See also **back off (from saturation)**]

saturation The condition of a **high power amplifier** when driven to its maximum output: it is said to be 'at saturation'.

[See also **saturated output power, back off (from saturation)**]

Saturn 1B An American **launch vehicle** developed in the early 1960s to support the **Apollo** lunar programme, allowing spacecraft tests to be conducted in Earth orbit; no longer **operational**. It comprised two **stages**: the first stage, designated S-1B, used the propellants **liquid oxygen** and **kerosene**; the second stage (S-IVB), which was developed as the third stage for the **Saturn V**, used liquid oxygen and **liquid hydrogen**.

The Saturn 1 booster, formerly known as the Saturn C-1, had a first stage developed from early American single-stage rockets: it was effectively eight **propellant tanks** from the **Redstone** clustered round a central tank based on the **Jupiter**. The Saturn 1B was an **uprate**d Saturn 1, comprising an S-1B stage and an S-IVB stage, which allowed it to orbit the Apollo spacecraft. The Saturn 1B was first launched in February 1966 with an unmanned Apollo. It was also used in October 1968 to launch the first manned Apollo (Apollo 7), in 1973 to launch three crews to **Skylab**, and in 1975 to launch the crew for the **Apollo-Soyuz Test Project (ASTP)**.

Saturn V An American **launch vehicle** developed in the 1960s for the **Apollo** lunar programme [see figure P6]; no longer **operational**. Formerly known as the Saturn C-5, it comprised three **stages**: the first stage, designated S-1C, used the propellants **liquid oxygen** and **kerosene**; the second and third stages (S-II and SIV-B) used liquid oxygen/**liquid hydrogen**. The Saturn V was first launched on 9 November 1967 for the unmanned Apollo 4 test-launch, and last used on 14 May 1973 to launch the American **Skylab** space station. [See also **Saturn 1B, specific impulse**]

S-band: 2–4 GHz – see **frequency bands**

S-band blackout A period during the **re-entry** of a **spacecraft** during which radio communication is impossible owing to the **ionisation** of the atmosphere surrounding the vehicle (e.g. for the Space Shuttle its duration is about 15 minutes). 'S-band' refers to the **frequency** band in common use for radio communication between the ground and a **manned spacecraft** [see **frequency bands**].

scanner A general term for a **payload** on an **Earth observation** satellite which gathers **data** by scanning, or sweeping across, an area of interest. Scanning may be performed either mechanically or electronically [see **pushbroom scanner**].

[See also **multispectral scanner**]

scatterometer An **active** microwave **payload** – also known as a wind scatterometer – designed to measure surface wind-speed and direction by detecting reflectivity perturbations of the sea surface; a type of **radar** carried by some **Earth observation** satellites. A typical instrument has a surface **resolution** measured in tens of metres and a **swath-width** of several hundred kilometres.

[See also **synthetic aperture radar (SAR)**]

scatterometry The measurement of wind-speed and direction using a **scatterometer**.

Schmidt camera – see **Schmidt telescope**

Schmidt telescope A specialised telescope for astronomical photography, devised by the Estonian optician Bernhard Voldemar Schmidt [1879–1955]; also called a Schmidt camera. Most telescope optics are unsuitable for photography because of the distortions they cause to objects far from the centre of the field-of-view. This means that high magnification and a large field-of-view are contradictory requirements. To enable undistorted photography of large star-fields, Schmidt devised, in 1930, an optical 'corrector plate' for use in a telescope with a spherical primary mirror. The light passes through the plate, which corrects for the spherical aberration of the primary, and is reflected from the primary onto a photographic plate mounted at its focus [see figure N1].

[See also **Hubble Space Telescope, Newtonian telescope, Cassegrain reflector, Gregorian reflector, Coudé focus**]

Scout An American four-stage solid propellant **launch vehicle** developed in the late 1950s to launch relatively small, typically scientific, **payloads** into **low Earth orbit (LEO)**. It became **operational** in 1960 and its derivatives were still in use in the 1990s. It can be launched from the Western Test Range at **Vandenberg Air Force Base (VAFB)**, Wallops Island, Virginia or the **San Marco platform**. An uprated third **stage** provides a payload capability of about 220 kg to LEO (550 km).

[See also **Atlas, Delta, Titan, solid propellant**]

SCPC – see **single channel per carrier**

scrambler An electronic device used to make radio transmissions unintelligible to unauthorised receivers. See **scrambling**.

scrambling A process by which radio transmissions are 'mixed up' to make them unintelligible to unauthorised receivers, used particularly for **cable TV** and TV broadcast direct from **satellite**s (DBS); also used in telephone systems, etc. Typically, **baseband** signals are scrambled by dividing them into a number of frequency sub-bands and transposing them in a predetermined manner. The authorised receiver has the 'key' to the transposition and can unscramble the signal.

scramjet An acronym for supersonic combustion **ramjet**.

[See also **combined cycle engine**]

scrub American slang for a postponement of a **launch** (usage e.g.: 'the launch has been scrubbed').

SDI – see **strategic defense initiative**

Sea Launch A commercial **launch vehicle** – otherwise known as the Zenit-3SL – developed from the Ukrainian **Zenit** and launched from a sea-going **launch platform** in the Pacific Ocean [see figure S1]. The first commercial launch of the Zenit-3SL, which has a **payload capability** of 5400 kg to **geostationary transfer orbit (GTO)**, occurred in October 1999.

The Sea Launch joint venture was formed between Boeing (USA), KB-Yuzhnoye/PO Yuzhmash (Ukraine), RSC-Energia (Russia) and Kvaerner (Norway). Boeing supplied the payload **shroud** and interface, and was responsible for **spacecraft integration** and development of the home port in Long Beach, California; Yuzhnoye/Yuzhmash supplied the Zenit; Energia supplied its block DM **upper stage**, previously used on the **Proton**; and Kvaerner was responsible for the sea-going hardware (an Assembly and

S1: **Sea Launch** platform and Assembly and Command Ship. [Kvaerner-Sea Launch]

Command Ship which acts as an integration facility in port and a **launch control centre** at sea, and a converted, self-propelled offshore oil-drilling platform which transports the vehicle to the **launch site** and serves as the launch platform).

[See also **San Marco platform**]

second space velocity – see **escape velocity**

second surface mirror (SSM) A thin sheet of glass or quartz, silvered or aluminised on one side, bonded to the exterior surface of a **spacecraft** as part of its **thermal control subsystem** [see figure S2].

The term 'second surface' is derived from its dual function as a thermal emitter and **solar reflector**: glass is an excellent emitter over the infrared spectrum, so thermal energy from the spacecraft can be conducted to the SSMs and radiated into space (from the outer or 'first surface' of the mirror); moreover, since glass is transparent over most of the **solar spectrum**, the majority of the incoming solar radiation reaches, and is reflected from, the coated rear surface (the 'second surface').

A typical SSM has a substrate based on a borosilicate glass doped with cerium dioxide, similar to solar cell **coverslips**. SSMs are not easy to fit to

S2: Artemis advanced relay and technology mission satellite in **cleanroom**. Note reflective **second surface mirrors**, black **thermal insulation**, **omnidirectional antenna** at top and **ion engine** above technician. [ESA]

anything but flat surfaces since they are essentially rigid, but their high resistance to electron and UV irradiation has led to their widespread use.
[See also **optical solar reflector**]

secondary power supply The secondary source of **DC power** available aboard a **spacecraft**. For most spacecraft it is the **battery**, but some manned space vehicles, which use the **fuel cell** as the primary source, may use the **solar array** as their secondary source. There are no rules which place a power source in one category or the other.
[See also **primary power supply**]

secondary structure The secondary **load**-carrying structure of a **spacecraft** or **launch vehicle**, which transfers structural loads to the **primary structure**. See **load path**.
[See also **thrust structure**]

SEE – see **single event effect**

Seebeck effect – see **thermoelectric effect**

seismometer A device for measuring the strength of an earthquake, moonquake or similar ground tremors on any other **planetary body**; also called a 'seismograph' (specifically when it registers *and* records the 'quake).

selective availability The ability to degrade the accuracy of navigation signals transmitted by the satellites of the **Global Positioning System (GPS)**, so that the **performance** of civilian receivers is inferior to that of military receivers.

semiconductor

   (i) A crystalline material, such as **silicon** or germanium, with an electrical conductivity between that of metals and insulators. Conductivity increases with temperature or with the application of light or voltage. Used, for example, to construct **solar cell**s.

   (ii) A device, such as a transistor or integrated circuit (IC), that depends on the properties of such a substance. Also used as a modifier to describe an item of **hardware**, e.g. semiconductor diode, semiconductor detector.

semi-major axis In astronomy, one of the halves of the **major axis** of an **orbit**; the distance from the centre of an ellipse to the edge through one of the foci (for a circle this would be the radius).

   [See also **orbital parameters**]

semi-minor axis In astronomy, the distance from the centre of an orbital ellipse to the edge, measured along a line perpendicular to the line connecting the foci.

sensor Any device that receives a **signal** or stimulus and responds to it with a proportional signal which can be used in a **spacecraft** subsystem. It may be part of a **subsystem** which controls **pointing** [see **sun sensor**, **earth sensor**, **infrared sensor**, **star sensor**, **RF sensor**], or a **detector** in the **payload** of a **remote sensing** or **astronomical satellite** [see **bolometer**]. In general usage, 'sensors' tend to be used for engineering **applications** and 'detectors' for scientific applications.

   [See also **attitude control**, **space astronomy**, **charge-coupled device (CCD)**]

sequencer – see **countdown sequencer**

service area An area on the surface of the Earth, defined for the purposes of **satellite communications**, within which the administration originating a service can demand protection against **interference** from other transmissions, such that a **receiver** in the service area should receive an interference-free signal from that country's satellite. The concept is designed to create a regime where, ideally, no telecommunications service interferes with any other. It is assumed that the service area lies inside the political boundary of the originating country; however, it is also recognised that some political boundaries are subject to dispute.

[See also **coverage area, beam area, protection ratio, ITU**]

service module

   (i) In general, a self-contained section of a modular spacecraft containing the subsystems which support the **payload**, whether 'man or machine' [see figures M3, M4]. The service module of a **communications satellite**, for example, contains the subsystems that support the **communications payload** (e.g. **power, thermal** and **attitude and orbital control system**s). The term may also include the spacecraft's propulsion system or there may be a separate **propulsion module**, depending on the design. The other main modules are the **payload module** and **antenna module**. The chief advantage of the modular design concept is that modules can be assembled and tested separately (by different contractors in different countries if required) without reference to the constraints of the other modules. Final **integration** and testing follows at a later stage.

   (ii) Specifically, a constituent part of the **Apollo** lunar spacecraft [see figure A7] and of the **International Space Station (ISS)** [see figure Z1].

[See also **assembly, integration & test (AIT), bus**]

service structure A tower-framework which allows access to a **launch vehicle** on its launch pad for maintenance, fuelling, **payload** loading, etc. [see figure S3]. Also called an access tower, service tower, umbilical tower or gantry.

   A service structure can be fixed to a **launch pad** or **launch platform** (a 'fixed service structure'), or mounted on rails to enable it to be moved away from the vehicle for **launch** (a 'mobile service structure').

[See also **rotating service structure, swing-arm, access arm**]

service tower – see **service structure**

SEU An abbreviation for single event upset. See **single event effect**.

shake table – see **vibration table**

shaped reflector An **antenna** reflector which has been specially shaped to produce a **beam** to illuminate a given **coverage area**. An alternative to a **parabolic antenna** reflector fed by a **beam-forming network**.

SHAR An acronym for **Sriharikota** High Altitude Range.

shear A form of deformation or fracture in which parallel planes of a body slide over one another; in physics, the deformation of a body expressed as the

S3: **Space Shuttle** Atlantis at **launch pad** 39B, **Kennedy Space Center**. Note fixed **service structure** with **swing-arms** carrying **beanie cap** (near top) and **white room** (centre); rotating **service structure** (at left) with 'payload changeout room'; and **mobile launch platform** with **tail service masts** (below **orbiter**'s wings). [NASA]

lateral displacement between two points in parallel planes divided by the distance between the planes.

[See also **shear web**, **wind shear**]

shear web A structural element which transfers or shares structural (**shear**) loads between panels or other components.

[See also **load path**, **doubler**]

SHF (super high frequency) – see **frequency bands**

ship earth station An **earth station** installed on a ship; a term sometimes used to differentiate a ship-based station from a **land earth station**. The term is not widely used, but is recognised by the **ITU** as an earth station in the maritime **mobile-satellite service (MSS)** located on board a vessel.

[See also **coast earth station**]

shirtsleeve environment An environment which is comfortable for someone wearing typical indoor clothing (literally, 'in their shirtsleeves'); the environment intended for the **cabin** of a **manned spacecraft**, within which protective garments are usually unnecessary. In case of cabin **decompression**, however, a **spacesuit** known as an 'intra-vehicular pressure garment' may be available.

[See also **environmental control and life support system (ECLSS)**]

shock diamonds A pattern of shock waves, in the form of a number of linked diamonds, sometimes visible in a rocket **exhaust** [see figure P6]. The shock waves are formed when the exhaust **plume** continues to expand beyond the **nozzle**, typically when the exhaust pressure is greater than the external atmospheric pressure. See **expansion ratio**.

shock suppression system – see **sound suppression system**

shot noise A source of **noise** in a communications system produced by discrete charges in an electric current. It is therefore impulsive and random, as opposed to the continuum of noise represented by **thermal noise**. For example, shot noise may be produced by the electrons in a thermionic diode, a **klystron** or a **travelling wave tube (TWT)**.

shroud A protective **fairing** which houses the **payload** of a **launch vehicle**. See **payload fairing**.

shunt dump regulator A device which takes excess spacecraft **power** from its **solar arrays**, converts it to heat by dissipation in a bank of resistors and then 'dumps' or radiates it into space. Alternative names for units with the same or similar capabilities are 'shunt voltage limiter' and 'shunt voltage regulator assembly'.

An alternative to the shunt dump regulator is the array shunt regulator, which connects or disconnects sections of the array in accordance with demand. Since power is only drawn into the spacecraft when it is needed, thermal dissipation is kept to a minimum.

[See also **heat rejection**]

shutter control The restriction, typically by governments, of the collection and/or distribution of imagery of specified geographical areas. The term, derived from the opaque shield in a camera which, when opened, allows light to reach a film or **sensor**, has become increasingly prevalent as the **resolution** of commercial satellite imaging devices has improved to the point where it approaches military specifications.

Shuttle – see **Space Shuttle**

Shuttle carrier aircraft (SCA) The specially adapted Boeing 747 used to ferry

Space Shuttle **orbiters** from place to place, typically from the landing site at **Edwards Air Force Base** to the **launch site** at **Kennedy Space Center**. It would also be used in the eventuality of a Shuttle landing at one of the 'emergency runways' in other parts of the world. The SCA was formerly used to carry the orbiter **Enterprise** in the **approach and landing test (ALT)** and to deliver the **flight model** orbiters from the prime contractor's plant in California to KSC. [See also **mate/de-mate device**]

Shuttle landing facility (SLF) The formal name for the runway at **Kennedy Space Center** on which Space Shuttle **orbiters** land. It is 4575 m (15,000 ft) in length with a 305 m (1000 ft) over-run at each end, 91 m (300 ft) wide and oriented northwest–southeast. A similar facility exists at **Edwards Air Force Base** in California, except that it has an 8 km over-run which extends into a dry lake-bed. Orbiters are guided automatically to a landing by a microwave landing system (MLS) which measures the vehicle's position relative to the runway centre-line and its elevation angle, and instructs it to make alterations to its **glide path** as necessary.

Si The chemical designation for silicon, a **semiconductor** material used for the **solar cell**, microprocessor and **charge-coupled device (CCD)**, etc.

side lobe – see **antenna radiation pattern**

side-looking radar A **radar** whose **antenna** is set at an angle to the local vertical (e.g. a spacecraft-borne **synthetic aperture radar (SAR)** whose **boresight** is offset from the **nadir**).

sidereal period An orbital period defined in the **frame of reference** of the 'fixed stars'. See **orbit**.

sigma notation – see **probability distribution**

signal A variable parameter, such as a current or electromagnetic wave, used to convey information through an electronic or microwave circuit, or through **free space**. When the signal is the **output** of an item of equipment such as a telephone, telex, TV camera, computer, etc., it may also be referred to as a **baseband** signal. For transmission via a communications system, the signal is modulated (or superimposed) onto a **radio frequency** carrier wave; it is then known as an RF signal (or, more colloquially, a radio signal). [See also **signal-to-noise ratio, carrier, modulation**]

signal delay The inherent delay in the transmission of a **signal** through a **transmission line** or in **free space**. In **satellite communications**, the main source of delay is the distance between the **satellite** and the **earth station** (because of the finite propagation velocity of **electromagnetic radiation**).

Although this is greater for satellites in **geostationary orbit (GEO)** than in **low Earth orbit (LEO)**, significant delay can also occur in terrestrial switching networks and (if used) **inter-satellite links**. Even greater transmission delays are experienced by **spacecraft** on interplanetary trajectories which take them far from Earth (e.g. **Voyager**).
[See also **light (velocity of)**]

signal-to-noise ratio The ratio of the power-level of a **signal** to the power-level of the **noise** at the same time and place in a circuit or any signal-carrying medium. Abbreviated to S/N or, less frequently, SNR, this quantity refers to a signal at **baseband** as opposed to RF or IF. S/N may also be measured using other parameters, such as the amplitude of a wave displayed on an oscilloscope tube, but these other parameters are simply a convenient illustration of signal and noise power. In practical terms, the higher the S/N of the wanted signal, the less chance there is of **interference** from unwanted signals.
[See also **noise power, carrier-to-noise ratio, radio frequency (RF), intermediate frequency (IF)**]

silicon A **semiconductor** material used for the **solar cell**, microprocessor and **charge-coupled device (CCD)**, etc.; chemical designation Si.

silicon cell – see **solar cell**

silver–zinc cell A commonly used voltaic cell in a spacecraft or launch vehicle **battery**.

sim A colloquial abbreviation for **simulation** or **simulator**.

simulation A reproduction of specific conditions for the purposes of testing or training. See **simulator**.

simulation heater A **heater** designed to simulate a deactivated device which produces heat when it is operating; also called a substitution heater. For example, if a **TWTA** is switched off, perhaps during an **eclipse**, its temperature drops and the **radiator** to which it is attached cools. This creates an undesirable temperature gradient which upsets the **thermal balance** of the spacecraft. To stop this happening, TWTA simulation heaters, bonded to the radiator or adjacent structure, are activated to replace at least some of the heat produced by the amplifier.
[See also **line heater, thermal control subsystem**]

simulator Any device or facility which reproduces specific conditions for the purposes of testing or training. The term can be applied to devices which simulate conditions aboard a **manned spacecraft**, especially with regard to its control in flight; to devices that simulate the environment that will be experienced by **unmanned spacecraft**; and to the operation of ground-based control and monitoring facilities.

[See also **ground support equipment, centrifuge, thermal-vacuum chamber, acoustic test chamber, vibration facility, anechoic chamber**]

single channel per carrier (SCPC) A derivative of the FDMA multiplexing systems in which the transponder **bandwidth** is subdivided so that each **baseband** channel is allocated a separate **transponder** subdivision and an individual **carrier**. SCPC has become less common as satellites have become more sophisticated; it is far more likely that each transponder will carry dozens or even hundreds of channels multiplexed together using a coding scheme, such as TDMA, FDMA or CDMA.

[See also **time division multiple access (TDMA), frequency division multiple access (FDMA), code division multiple access (CDMA), digital compression**]

single conversion transponder A **transponder** which converts an **uplink** frequency directly to a **downlink** frequency, that is without an **intermediate frequency (IF)**. See **downconverter** and figure C5.

[See also **upconverter**]

single-entry interference A term used in connection with **communications systems** to indicate the presence of only one interfering signal. See **protection ratio**.

single event effect (SEE) A term applied to damage caused by the random effects of natural radiation sources on **spacecraft** electronic components, particularly **semiconductor** storage devices (colloquially known as 'memory chips'). The effects are primarily due to particles generated in **solar flare**s and **cosmic radiation** (e.g. protons, neutrons, pions, muons and heavy ions) which cause a 'change of state' or 'polarity reversal' in the device (colloquially termed a 'latch-up'). The particle deposits a charge and causes a transient pulse which can be decoded as an instruction (e.g. to turn it on or off).

The term single event effect (SEE) is often used interchangeably with single event upset (SEU), but SEEs can be split into three main groups:

(i) single event upset (SEU), caused when an incident particle has sufficient charge to change the logic state of a device, but the circuit is such that it can be reset by **telecommand**;

(ii) single event latch-up (SEL): as SEU, but the circuit cannot be reset;

(iii) single event burnout (SEB), caused when an incident particle impinges on the gate region of a power **FET** resulting in its failure and the shorting of the **power supply**.

Other effects include single event **dielectric** rupture (SEDR) and single event functional interrupt (SEFI).

The increasing incidence of SEE is partly due to the ever-decreasing size of chip-features (the critical feature size is about 3 micrometres), but also due to the decrease in the electrical charge required to activate electronic devices, which is becoming closer to that deposited by the particles themselves. This development requires a parallel evolution in the techniques and definition of **radiation hardening**. However, it is difficult to harden devices against all SEE, which can affect satellites in any orbit as well as spacecraft on interplanetary trajectories, so error detection and correction software is becoming increasingly important.

single event upset (SEU) – see **single event effect (SEE)**

single hop A signal route in a **communications satellite** system which includes a single pass through a **satellite**: i.e. there is one **uplink** path and one **downlink**, the remainder of the route being terrestrial. A route which includes two satellites is known as a **double hop**.
[See also **terrestrial tail**]

single point failure (SPF) A concept in **reliability** engineering whereby a single component can cause the failure of a complete **system** because there is no **redundancy** provision (i.e. no spare component or '**back-up**'). For example, although a satellite **communications payload** typically includes **redundant** amplifiers, it outputs to a single **antenna**. If any part of the antenna fails or the reflector itself fails to **deploy**, the section of the **payload** to which it is attached will be unable to function. The antenna, or the antenna deployment mechanism where it exists, represents a potential 'single point failure'. Other SPF examples include a satellite **apogee kick motor (AKM)**, the main mirror of the **Hubble Space Telescope** and the nose-wheel of a Space Shuttle **orbiter**. Space-based systems are designed to keep SPFs to an absolute minimum.

single stage to orbit (SSTO) The concept of a **launch vehicle** capable of delivering a **payload** to **orbit** (usually **low Earth orbit**) using only one rocket **stage**. Although many designs have been postulated, those generally favoured fit into the category of the **aerospace vehicle**, with both rocket **engines** and **lifting surfaces**, as opposed to the conventional vertical take-off launch vehicle.
[See also **reusable launch vehicle (RLV)**]

skin The outermost layer of a **spacecraft** or **launch vehicle** structure; a light-

weight covering on a rocket **stage** supported by **longerons** and **stringers**. Typically a thin metal covering designed to help protect a spacecraft from the high temperatures of **re-entry** or frictional heating experienced during the **launch** phase (usage example: high 'skin temperature')
[See also **face-skin**, **airframe**, **fairing**, **thermal protection system**, **heat shield**, **aerodynamic heating**]

skirt

(i) A structural extension joining two sections of a **spacecraft** or **launch vehicle**, usually flared to join two sections of different diameter (e.g. adapter skirt, skirt assembly, etc.).

(ii) A **fairing** surrounding the **engine bay** at the base of a launch vehicle's first **stage**. Also called a 'cowling', particularly when it is a fairing round an individual engine.

[See also **interstage**]

Skylab An American **space station** designed for terrestrial and astronomical observation, and research into human capabilities in and reaction to the space environment [see figure S4]. Skylab was launched on 14 May 1973 by a **Saturn V** into a 440×427 km, 50° inclination **orbit** and visited by three **crews** for periods of 28, 59 and 84 days, until it was vacated in February 1974. The crews were transported by **Saturn 1B**-launched command and service modules remaining from the **Apollo** lunar programme.

The station comprised an orbital workshop (OWS) and attached instrument unit, adapted from a Saturn V (S-IVB) third stage; an **airlock** module (AM); a multiple docking adapter (MDA), which allowed up to two Apollo spacecraft to **dock**; and the Apollo telescope mount (ATM), designed particularly for solar observations. It had an oxygen/nitrogen atmosphere at a pressure of about 34 kPa (5 psi) and a total habitable volume of about 360 m³. Two **solar panels** with a combined area of 730 m² and capable of producing nearly 12 kW were attached to the OWS and four more of 111.5 m² to the ATM. Launch vibrations caused the dislocation of a combined **meteoroid**/thermal shield which tore away one of the OWS arrays and restricted the deployment of the other. The first Skylab crew were, however, able to restore the station to an **operational** condition. Despite attempts to raise its orbit, Skylab re-entered the Earth's atmosphere on 11 July 1979 and spread its debris over a thinly populated area of Western Australia.
[See also **Salyut**, **Mir**]

slant range The distance between a **launch vehicle** or a **spacecraft** in **orbit** and

S4: **Skylab** space station showing sunshield deployed by the first crew and a single large **solar array** panel; the other panel was torn off during launch. [NASA]

the ground station tracking it. Used particularly for launch vehicles, the term recognises that the **range** is being measured in a direct line to the vehicle with regard to both its horizontal range from the **launch pad** and its altitude (e.g. it may be 3 km east of the pad at an **altitude** of 4 km, which would make its range 5 km along the 'slanted' hypotenuse of the triangle thus formed). By contrast, its horizontal range is defined as its distance **downrange**.

sleep restraint – see **restraint**

slide-wire An escape device attached to a launch vehicle **service structure**, comprising a 'cage' or 'basket' that can slide down an inclined wire, to take **astronaut**s away from the **launch pad** in an emergency.

sling-shot (gravitational) – see **gravity propulsion**

sloshing A rhythmic lateral motion of a **liquid propellant**, particularly in the tanks of a **launch vehicle**, which can be caused by the vehicle's motion and can lead to instability and perturbations in the flight **trajectory**.
[See also **anti-slosh baffle**, **pogo**]

slow-scan TV A television system which features a reduced number of picture 'frames' per second (i.e. less than the standard 25). Since the amount of picture information is reduced, this system requires a lower transmission **bandwidth**, which can reduce transmission charges (via satellite or otherwise). In fact, a typical **bit rate** for slow-scan TV is 64 kbits s^{-1} compared with 68 Mbits s^{-1} for 625-line colour TV. The main drawback to the system is that any significant motion produces a blur, due to the low 'refresh-rate' of the picture. The system is, however, useful for applications such as **videoconferencing** where subject movement is kept to a minimum. A typical bit-rate for videoconferencing is about 2 Mbits s^{-1}.

slow-wave structure The component part of a **travelling wave tube (TWT)** which reduces the velocity of an **RF wave** carried along it so that it travels slightly slower than the **electron beam**. This allows an interaction between the electron beam and the RF wave, resulting in an energy transfer from the beam to the wave – the RF amplification function. A slow-wave structure may take the form of a helix or of coupled cavities. The helix, made of copper, or of tungsten or molybdenum wire, is supported by three or four ceramic rods to isolate the RF fields on the helix from the metallic walls of the surrounding vacuum envelope [see figure T7]. The coupled-cavity type comprises accurately sized and shaped RF cavity sections, usually of copper, brazed together to form a structure of many cavities 'in cascade'.

slush hydrogen A potential **cryogenic propellant**: a mixture of solid and liquid hydrogen which is even colder than **liquid hydrogen**.

SLWT An abbreviation for super light weight tank, a version of the Space Shuttle **external tank (ET)** developed in the late 1990s.

SM An abbreviation for **structure model** or **service module**.

small self-contained payload (SSCP) The formal name for a **Getaway Special (GAS)** payload.

smallsat A colloquial term for a 'small satellite', which may be defined as having a **launch mass** between about 500 kg and 1000 kg. The exact definition is open to discussion, but it is common to define satellite mass ranges as follows:

- smallsat   500–1000 kg
- minisat   100–500 kg
- microsat   10–100 kg
- nanosat   <10 kg.

On the same scale, the range for 'medium-sized satellites' would be 1000–2000 kg, and 'large satellites' would have masses greater than 2000 kg. It is evident, however, that as satellite technology has developed, satellites have become larger and heavier, with a consequent shift in the definition of what constitutes a 'large satellite'.

[See also **in-orbit mass**, **mass budget**]

SMATV An abbreviation for satellite master antenna television, television distributed to cable **head-end**s by satellite.

[See also **CATV**, **cable TV**]

SNAP An acronym for systems of nuclear auxiliary power, a **radioisotope thermoelectric generator (RTG)** package used on spacecraft chiefly in the 1960s and early 1970s. Different versions of SNAP were used on the **Transit** navigation satellites (SNAP-3) and the **Apollo** Lunar Surface Experiments Package (**ALSEP**) (SNAP-27).

SNG An abbreviation for satellite news gathering, a satellite communications **application** which involves the transmission of a TV **signal** from a remote camera, via **satellite**, to a TV studio. The remote end of the **link** typically makes use of a transportable **earth station (TES)**.

snoopy cap A colloquial term for the 'communications carrier assembly', a close-fitting head cap, complete with headphones and a microphone, typically worn by **astronaut**s inside a **spacesuit** helmet. Named after the flying helmet worn by the cartoon character 'Snoopy'.

snubber A protective spacer which eliminates damage to folded **solar array** panels during **launch**.

soft dock – see **dock**

soft fail Jargon: the failure of a component, device or **subsystem** which does not seriously degrade the performance or **capacity** of a **system**.

[See also **degradation**, **graceful degradation**]

software Any form of coded information used in the operation of a computer system; a computer program; the non-physical part of a computer-based system (cf. **hardware**).

[See also **firmware**]

software patch A piece of computer software used to repair or enhance the

operation of an existing program which has either ceased to function
correctly or requires updating to enable it to perform a revised mission.

SOHO An acronym for:

    (i) small office/home office, a target market for **Internet** service providers
(ISPs), among others.

   (ii) Solar and Heliospheric Observatory, a **spacecraft** launched in 1995 as part
of **ESA**'s Solar-Terrestrial Science Programme (STSP) and the first of four
Cornerstone missions in the ESA Horizon 2000 programme.

Sojourner The first **teleoperated** **rover** to operate on a **planetary body** other than
the **Earth** and the **Moon**. It was carried to Mars by **NASA**'s Mars Pathfinder,
which landed on 4 July 1997.

[See also **Lunokhod**]

solar absorber – see **surface coatings**

solar array An area of **solar cells** designed to provide electrical **power** to a
spacecraft. Individual cells are connected in series to provide the voltage
required for a particular power subsystem design, and the series strings are
connected in parallel to form an 'array section'. This construction helps to
ensure that single cell failures do not 'short out' and deactivate whole
sections of the array.

    Owing to the large area of cells required for most applications, the majority
of arrays are stored folded against the spacecraft for launch and **deploy**ed in
space. Mechanically speaking, arrays are of two different types: rigid or
flexible.

    The rigid design is currently the most common. In the **three-axis stabilised**
spacecraft design, a hinged array of flat panels is connected to the spacecraft
by a light-weight yoke structure attached to a **solar array drive mechanism**.
Most three-axis satellites have two arrays, comprising several panels,
mounted perpendicular to their north and south faces. In order to remain
pointing at the Sun, the arrays rotate once per day about the satellite's
north–south axis (the **pitch axis**) [see figures M3, T4, U1]. The arrays on the
lower-power **spin-stabilised** spacecraft form the outer skin of the spacecraft
body [see figures E7, S14]; spinners with higher power capabilities have an
additional cylindrical 'skirt', mounted over the prime array, which is
deployed in a telescopic fashion.

    Flexible arrays or 'array blankets', for use on three-axis stabilised
spacecraft, are typically larger and of lower density [see figure S5]. They can
be stored in a box which is swung out from the side of the spacecraft on a

S5: **Solar array** for the **International Space Station (ISS)**. [Lockheed Martin]

hinged arm. An extendable truss structure, for example an '**astromast**', then pulls the array blanket from its box. The **NASA** Solar Array Experiment (SAE) flown on the STS 41-D mission was the largest example at the time: it was 32 m long, 4 m wide and comprised 84 panels together producing some 12.5 kW. The array was folded, concertina-fashion, into a stack only 19 cm thick, 0.6% of its extended length [see figure A12]. Flexible arrays, such as those carried by the **Hubble Space Telescope**, can also be rolled and unrolled on command [see figure H5].

The power-generation capability of solar arrays is dependent upon certain astronomical mechanisms. Owing to a slight eccentricity of the Earth's orbit, the Sun–Earth distance is greatest at the summer **solstice** and least at the winter solstice. This produces a seasonal variation in solar flux density [see figure D1]. More important though is the **declination** of the Sun with respect to the solar array surface which, for the usual orientation of an array surface parallel to the Earth's axis, is 0° at the **equinox**es and 23.5° at the solstices. Maximum power can only be derived if the array-surfaces are perpendicular to the solar vector.

[See also **degradation (of spacecraft materials), BAPTA, fuel cell, radioisotope thermoelectric generator, spacecraft axes**]

solar array drive mechanism (SADM) – see BAPTA

solar battery An archaic term for a **solar array** (and accompanying rechargeable **battery**), used in the early years of the **Space Age** when most **launch vehicle** and **spacecraft** power was derived from non-rechargeable batteries. The novelty of the **solar cell** led people to distinguish between a battery charged on the ground and one continually recharged by an array of solar cells, the solar battery.

solar cell A device for the conversion of solar energy to electrical energy. Also known as a photovoltaic cell, since it operates in accordance with the photovoltaic effect, whereby **electromagnetic radiation** incident on a thin film of n-type semiconductor material, on the surface of a p-type semiconductor, produces a potential difference between the two [see figure S6].

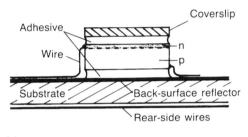

(a)

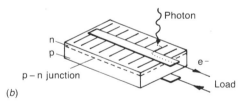

(b)

S6: Structure of a typical **solar cell**.

Silicon is the predominant semiconductor used in solar cells. The addition of phosphorus (a process known as 'doping') results in an n-type semiconductor; doping silicon with boron produces a p-type material ('n' stands for negative and 'p' for positive). Solar photons liberate excess **electrons** from the upper layer and their movement to the electron-deficient lower layer constitutes a current.

Silicon cells are typically made from ingots of pure silicon doped with boron. The ingot is cut into disks and phosphorus is diffused into the top surface to make a semiconductor (p/n) junction across the surface of the cell. For spacecraft applications the disks are then cut into rectilinear shapes to increase the packing density of the resultant **solar array**. Other semiconductors used for solar cells include gallium arsenide (GaAs) and indium phosphide. In the 1990s, particularly, GaAs cells began to replace silicon for some satellite applications, largely because of their greater conversion efficiency.

The first practical solar cell was developed at Bell Telephone Laboratories (USA) in 1954 and the first satellite to make use of it was **Vanguard** 1, launched by the USA in March 1958.

[See also **coverslip**]

solar constant A measure of the incident solar **energy** at a given distance from the Sun (measured in W m^{-2}); $S = 1.37$ kW m^{-2} (± 0.02 kW) at 1 **astronomical unit** (AU).

[See also **insolation**]

solar flare A sudden eruption on the surface of the Sun resulting in an increased flux of high energy particles (e.g. **electron**s, protons and, occasionally, heavier **ion**s). The flares cause radio and magnetic disturbances on the Earth, referred to as solar storms, and increase the activity of the **magnetosphere** and the **ionosphere** (so-called **geomagnetic storm**s). As a result, they threaten the crews of **manned spacecraft** with potentially damaging ionising radiation and can cause problems with the operation of satellites [see **space radiation, single event effect**].

The structural shells of spacecraft, and especially **space stations** and future **interplanetary** vehicles where exposure may be more prolonged, must be designed to be sufficiently thick to reduce radiation to tolerable levels, though it may be necessary for crews to retreat to specially shielded accommodation in the most severe solar storms. An increase in solar activity also tends to increase the average height of the **atmosphere**, thereby increasing **atmospheric drag** at a particular height.

[See also **Van Allen belts, solar wind, hardening (of spacecraft), space weather**]

solar generator An alternative name for **solar array**.

solar panel An individual element of a flat-panel **solar array**.

solar radiation pressure – see **solar wind**

solar reflector – see **surface coatings**

solar sailing The use of the **solar wind** in controlling the position and orientation of **spacecraft** in orbit or on **interplanetary** trajectories.

The **solar array**s of a **three-axis stabilised** satellite, mounted perpendicular to its north and south faces, rotate once per day about the satellite's north–south (**pitch**) axis to remain pointing at the Sun. This allows them to be used as an **actuator** for the **attitude and orbital control system**, providing adjustments in **roll** and **yaw**. If one of the array panels is driven to a position parallel to a line drawn from the Sun, the radiation pressure on the other will exert a moment on the spacecraft and it will begin to rotate about an axis orthogonal to the N–S axis; swapping the orientation of the arrays would rotate it in the opposite direction. Further, by orienting the arrays in the manner of an aircraft propeller, the spacecraft can be rotated about the third orthogonal axis, producing a '**windmill torque**'. Flaps attached to the arrays allow array offsets to be minimised, which limits the power loss due to array mispointing to about 1%.

Solar sailing has also been suggested as a method of interplanetary propulsion, mainly in the inner solar system where the solar wind has an appreciable magnitude. The techniques used would be analogous to those used in terrestrial sailing.

[See also **spacecraft axes**]

solar spectrum The distribution of **electromagnetic radiation** emitted by the Sun. Figure S7 shows how the Sun's output is spread over the frequency spectrum: approximately 7% is at ultraviolet **wavelengths** (<0.38 μm); 45.5% is in the visible (0.38–0.76 μm); and 47.5% is in the infrared and beyond (>0.76 μm). The 'spectral energy density' of the Sun is important in the design of spacecraft **surface coatings**, among other things.

[See also **thermal control subsystem**]

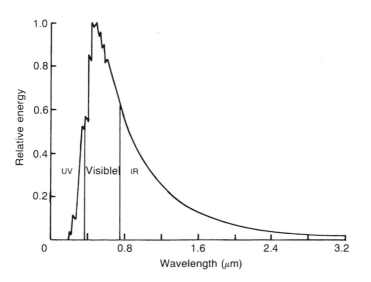

S7: Graphical representation of the **solar spectrum**.

solar storm – see **solar flare**

solar wind A continuous stream of atomic and other particles ejected from the
Sun; the cause of 'solar radiation pressure', one of the mechanisms that
results in the perturbation of a spacecraft's **orbit**. Depending on solar
activity, the velocity of the particles, or **plasma**, varies between about 300 and
500 km s^{-1}, the force imparted being sufficient to affect spacecraft with large
surface areas, for instance those with large **solar array** panels. The solar wind
is thus one of the **perturbations** affecting the **station keeping** of satellites in
geostationary orbit [see **libration period**]. It can, however, be used to
augment a spacecraft's **attitude and orbital control system** [see **solar
sailing**].

It is, incidentally, the solar wind that causes the tails of comets to orient
themselves in a direction away from the Sun wherever they might be in their
orbit about the Sun. It also causes an elongation of a planet's **magnetosphere**
[see figure V1].

[See also **solar flare, equilibrium point, triaxiality, luni-solar gravity,
atmospheric drag**]

solid propellant The combustible substance which provides **thrust** in a **rocket
motor** (note that rockets which use liquid propellant are known as **rocket
engines**). There are two main types of solid propellant: the homogeneous or
double-base propellants (nitrocellulose plasticised with nitroglycerine plus
stabilising products) which are limited in power and not widely used in space
applications; and the heterogeneous or composite type.

Composite propellants consist of a mixture of **fuel** and **oxidiser**. An
oxidiser in common use is crystalline ammonium perchlorate (NH_4ClO_4;
abbreviated to AP). It is mixed with an organic fuel, such as polyurethane or
polybutadiene, which also binds the two components together. The
inclusion of a plastic 'binder' produces a 'rubbery' material, making the
propellant relatively easy to handle. Performance is enhanced by the addition
of finely ground (10 μm) metal particles of aluminium, for example, which
increase the heat of the reaction owing to the formation of metal oxides, and
increase the overall density of the propellant. As an example, the propellant
used in the Space Shuttle **solid rocket booster**s comprises 14% polybutadiene
acrylic acid acrylonitrile (binder/fuel), 16% aluminium powder (fuel), 69.93%
NH_4ClO_4 (oxidiser), and 0.07% iron oxide powder (catalyst).

Solids are generally easy to handle and store, and do not corrode their containers as some liquid propellants do. The solid motor is simpler, but operationally less flexible [see **rocket motor**].

[See also **propellant grain, propellant, liquid propellant**]

solid rocket booster (SRB)

    (i) A propulsive component of a **launch vehicle** which uses **solid propellant**. An SRB added to a **core vehicle** to augment its **thrust** is otherwise known as a '**strap-on**' booster.

    (ii) Specifically, one of two boosters strapped to the **external tank** of the Space Shuttle launch vehicle. The SRBs burn with the **Space Shuttle main engine**s for about two minutes into the **flight** and are then jettisoned, parachuted into the ocean and recovered for refurbishment and re-use. Each booster is 38.5 m long and 3.7 m in diameter, contains about 500 tonnes of **propellant** [see **solid propellant**], and provides a thrust of approximately 12 million newtons (*in vacuo*) with a **specific impulse** of about 267 s. The **nozzle** can be gimballed for steering (±8° from the booster axis). These boosters were the first solid rocket boosters designed to be used with a manned vehicle; the failure of an 'O-ring' seal between two SRB segments was identified as the cause of the '**Challenger** accident' (mission STS 51-L) in January 1986.

    (iii) Another name for a **solid rocket motor (SRM)**.

[See also **rocket motor**]

solid rocket motor (SRM) – see **rocket motor**

solid state power amplifier (SSPA) A device which uses **semiconductor** components – typically gallium arsenide field effect transistors (**GaAsFETs**) – for the amplification of **radio frequency (RF)** signals. SSPAs are utilised in spacecraft **communications payload**s as an alternative to the **travelling wave tube amplifier (TWTA)**.

SSPA developments have improved the performance and reliability of the **transmit chain** of a **transponder**. However, since the solid-state medium is a much poorer conductor of heat than the copper (say) in a **travelling wave tube (TWT)**, SSPAs require large radiators for higher output powers. The excessive mass tends to limit SSPAs to relatively low-power applications, while TWTAs are specified for higher-power amplification. Typical late-1990s GaAsFETs could, individually, provide up to about 45 W at C-band and 15 W at Ku-band, while, using power combination techniques, total output powers of 500 W and 100 W, at C and Ku-band respectively, were available for certain applications.

solid state recorder A type of data recorder which utilises **semiconductor** storage devices (colloquially known as 'memory chips') as opposed to magnetic tape. Since it has no moving parts, it has a higher reliability than its mechanical counterpart and is being specified increasingly, for example, for **Earth observation** satellites.

solstice Either of two annual occasions when the Sun is overhead at the tropic of Cancer (summer solstice, June 21) or the tropic of Capricorn (winter solstice, December 21); the longest and the shortest day of the year, respectively.

[See also **equinox, eclipse, solar array**]

sonde – see **sounding rocket**

sound-in-syncs (SIS) A type of TV audio transmission in which the sound is coded into the line-synchronisation pulse of the **television** waveform.

sound suppression system A **launch pad** facility which protects a **launch vehicle** and its **payload**s from acoustic energy produced by the vehicle's **propulsion system**, at and following **ignition**, and reflected from the surface of the **launch platform**. Typical systems comprise volumes of standing water and water-sprays, the latter of which also serve to constrain the temperature of the structure and avoid damage. It is the evaporation of this water which produces the large clouds of steam often visible at rocket launches.

sounder Any instrument for the observation and/or measurement of atmospheric conditions (named by analogy with measurements of water depth: 'soundings'). Sounders which operate in the infrared or microwave **waveband**s are typically carried by **meteorological satellite**s and **Earth observation** satellites.

[See also **microwave sounder**]

sounding rocket A **rocket** capable of carrying a small **payload** on a **sub-orbital** trajectory, used mainly for the investigation of the Earth's upper atmosphere. Rocket or balloon-borne instruments are sometimes called 'sondes' (named by analogy with measurements of water depth: 'soundings').

South Atlantic anomaly A high concentration of radiation trapped by the Earth's magnetic field over the south Atlantic Ocean. See **space radiation**.

[See also **Van Allen belts**]

SOW – see **statement of work**

Soyuz The name given to a series of Soviet **manned spacecraft** and **launch vehicle**s.

The Soyuz **spacecraft** was developed in the 1960s and used in improved forms into the early twenty-first century [see figure S8]. It comprises three

S8: **Soyuz** spacecraft in Earth orbit on the **Apollo–Soyuz Test Project** mission in July 1975. Part of the docking mechanism is visible at the front of the spacecraft. [NASA]

modules: a spherical orbital compartment at the forward end, containing **life support** equipment and used for scientific experiments, etc.; a **re-entry** module, containing the cosmonaut's couches and flight control equipment; and an equipment module (similar in function to the **Apollo** spacecraft's service module) with two **deploy**able **solar array** panels.

The first manned Soyuz mission was launched on 23 April 1967. This first **variant** of the spacecraft was used increasingly for docking experiments and missions which placed multiple crews in orbit, and thereafter to deliver crews to the Soviet space station, **Salyut**. It originally carried a crew of three, but this was reduced to two following the Soyuz 11 mission in April 1971, in which the crew, who were not wearing **spacesuit**s, died on return from orbit when the spacecraft depressurised. The second Soyuz variant, Soyuz-T, first launched on 27 November 1980, was capable of carrying a crew of three (with spacesuits); it was later updated further to the Soyuz TM, which made its first flight on 21 May 1986. After being used to ferry crews to and from the **Mir** space station, the Soyuz continued in a similar role with the **International Space Station (ISS)**, in addition to acting as a temporary crew return vehicle (**CRV**).

The Soyuz launch vehicle was based on the R-7 booster, which launched **Sputnik** 1 and was developed as the basis of the **Vostok** and **Molniya** launch vehicles. The basic Soyuz has performed over 700 launches since its introduction in 1966 and is capable of lifting over 5000 kg to **low Earth orbit** (**LEO**). It is used, for example, to transport crews to the Mir space station and the ISS. Soyuz and its **upper stage**s (IKAR and Fregat) were commercialised in 1996 by an international joint venture.

[See also **Progress, Vostok, Voskhod, Apollo–Soyuz Test Project**]

space

   (i) The three-dimensional expanse in which all material objects are located.

  (ii) The region beyond the limits of the Earth's **atmosphere** which constitutes the operating environment of a **spacecraft** (e.g. 'near-Earth space', 'interplanetary space', **deep space**). There is no legal agreement as to where space begins, but it is generally defined to begin at about 100–110 km above sea level (since spacecraft cannot complete an **orbit** below this altitude because friction with the atmosphere causes them to re-enter – see **atmospheric drag**). The altitude record of the **X-15** rocket plane was in this region and some believe it should therefore be classed as a spacecraft.

[See also **space environment, NEO, space-time, degree of freedom, frame of reference**]

Space Age The period in which **space exploration** has been possible (i.e. since the launch of **Sputnik** 1, on 4 October 1957). Preferably written with initial capitals, to signify its importance alongside other 'ages of man', such as the Stone Age.

space astronomy Astronomy conducted from space, using observation equipment in Earth **orbit**, or in orbit about or on the surface of a **planetary body**, such as the Moon. The observation platform may be a dedicated **astronomical satellite**, or a **module** or **payload** of a **space station**. The main advantage for optical astronomy is the position of the **observatory** above the Earth's atmosphere, which perturbs and attenuates the **image**.

[See also **atmospheric attenuation**]

space capsule – see **capsule**

space colony A conceptual habitat in space, usually in Earth orbit (as opposed to on a planetary body) and usually self-sufficient (in contrast to a **space station**, which typically requires supplies of **consumables** to be transported from Earth).

Space Command A US military space organisation, headquartered at Peterson Air Force Base near Colorado Springs, which became operational in October 1985. It combined the US Air Force Space Command and Naval Space Command with elements of the Army Space Planning Group, but is dominated by the USAF. It operates a command and control centre in

Cheyenne Mountain, shared with **NORAD**. Since 1986, the US Space Command has operated the US **Space Surveillance Network (SSN)** to maintain the official 'Satellite Catalog'.

space debris A blanket term for any man-made artefact discarded, or accidentally produced, in **space**: in orbit around a **planetary body** [see **orbital debris**] or on a **trajectory** between planetary bodies. Although **spacecraft debris** on the surface of a planetary body is not technically 'in space', it may be included in the definition of space debris. While orbital debris is widely recognised as a major problem for the preservation and future use of the **space environment**, this is not yet the case for spacecraft debris on planetary surfaces.
[See also **NEO, micrometeoroid**]

space environment The region beyond the limits of the Earth's **atmosphere** considered, *inter alia*, as a place to live and work, conduct scientific investigations and operate spacecraft and other space systems. In a scientific sense, it constitutes the physical (e.g. temperature, **radiation** and **microgravity**) environment of **space** or, by extension, the environment of a **planetary body**; in an engineering sense, it also includes the application of space (as in the 'orbital environment' of a **space station** or the 'lunar environment' of a Moon base); and in an 'environmental' sense, it recognises that space, along with other planetary environments, is worthy of protection.
[See also **space radiation, space weather, space debris, orbital debris, NEO**]

space exploration The act or process of exploring **space**. In general usage, the term covers almost any aspect of **space science** and **space technology**; to those more closely involved with the subject, it is confined to the scientific and physical exploration of space by either **unmanned spacecraft** or **manned spacecraft**.

space insurance The business of providing financial protection against specified risks encountered in the course of manufacturing, handling, launching and operating space **systems** or **hardware**. An insurance policy is arranged by an insurance broker, acting as an intermediary between the client (which may be a satellite owner, operator or manufacturer, or a launch provider) and the underwriters (who underwrite the risk and make payments in the event of a successful claim). The majority of space insurance policies are written for commercial **communications satellites**, though the business is expanding into other space **applications**.

Traditionally, a **satellite** is insured according to a sequence of events in its life, which can be divided into four main phases: manufacturing, pre-launch,

launch and in-orbit operation. The last phase may be subdivided into in-orbit commissioning and in-orbit life. In practice, coverage for the in-orbit commissioning and initial in-orbit life phases can be included in the launch policy, which may cover a period of several years from launch vehicle **ignition**. Policies may thereafter be negotiated on an annual or other regular basis for the remaining in-orbit life.

Other space insurance policies cover, among others, third-party launch liabilities, delay risks, political risks and *force majeure*, **transponder** coverage and service interruption. The first space insurance policy was written in 1965 for **Intelsat** I, the first commercial geostationary communications satellite – otherwise known as **Early Bird**. It covered only pre-launch and third-party liability (for US$5 million and US$25 million, respectively). The first **astronaut** life insurance policy was written in 1969 for the **Apollo** 11 crew – Neil Armstrong, Edwin Aldrin and Michael Collins – by the Greek Insurance Company Inc: it covered "loss of life or disappearance in space and permanent total disability of the insured astronauts" (for US$10,000 each, which was intended to match the average life policy then current in the US). [See also **launch**, **pre-launch**, **lifetime**, **intentional ignition**, **clamp release**, **partial loss**, **total loss**, **constructive total loss**, **TLO**, **delivery in orbit**, **space law**, **Liability Convention**]

space law The body of law appertaining to space, specifically that embodied in a number of space treaties – see **Outer Space Treaty**, **Rescue Agreement**, **Liability Convention**, **Registration Convention**, **Moon Agreement** – which have been adopted by the United Nations General Assembly. However, since some space-faring nations are not signatories to all treaties, there is no fully international agreement to abide by this body of law.
[See also **space object**]

space object A legal term for a man-made object which has reached space (although there is no agreed definition of the term **space**). The use of the term 'space object' usually infers the ability to complete an **orbit**, since once the object is in orbit it is fully within the scope of Article 1 of the **Liability Convention**, the relevant legal document for re-entering **spacecraft** with the potential to cause third-party damage.
[See also **satellite**]

space probe A colloquial term for an **unmanned spacecraft**, usually designed for scientific investigation of a **planetary body** or **interplanetary** space. Sometimes written as a single word, spaceprobe, or shortened to **probe**.

space-qualified An attribute of any component or device considered able to perform its function in the **space environment**. The term is understood on two different levels, such that a space-qualified component is (i) one which has flown (and performed to specification) in space; (ii) one which has been tested on Earth in a simulated space environment to a point where it is considered capable of performing in space.

space radiation The natural **radiation** environment of space. From the point of view of a **spacecraft** and its **crew**, there are three main sources of radiation: the Sun, extra-solar system sources, and radiation trapped by the Earth's magnetic field. Solar radiation comprises clouds of charged particles released during eruptions on the surface of the Sun [see **solar flare**], and is the least predictable radiation mechanism. Radiation from outside the solar system comprises high-energy **electron**s and nucleons (chiefly protons), known as 'cosmic rays' or 'galactic cosmic radiation'; this is potentially the most harmful source since structural shielding is relatively ineffective. Geomagnetically trapped radiation comprises mainly low-energy protons, which do not penetrate shielding, although high concentrations exist over the south Atlantic Ocean (an effect known as the South Atlantic anomaly). [See also **electromagnetic radiation, electromagnetic spectrum, space environment, radiation hardening**]

space science That branch of science concerned with the features and processes of the **space environment**. It includes 'space physics', '**space astronomy**', etc., and can be extended to disciplines with applications in **space** (e.g. 'space medicine' and **microgravity** research).
[See also **space technology, space exploration**]

space segment
(i) That part of a **satellite**-based communications system or any **spacecraft** communications link which resides in space. The space segment of a communications system may comprise more than one satellite.
(ii) Any space-based hardware, when it is necessary to differentiate it from complementary hardware on the ground, the **earth segment**.

Space Shuttle In general, a reusable **spacecraft** or **launch vehicle** designed to carry personnel and/or **payload**s into orbit, return for refurbishment and repeat the exercise many times throughout its **operational** lifetime. In particular, the **NASA** Space Shuttle, otherwise known as the **space transportation system (STS)**, which comprises an **orbiter, external tank** (ET) and two **solid rocket booster**s (SRBs) [see figure S9]. In the case of the STS the

353

S9: **Space Shuttle** launch showing three **Space Shuttle main engines** and two **orbital manoeuvring system** engines (on the **orbiter**), **external tank** (beneath orbiter) and two **solid rocket boosters**. [NASA]

orbiter alone is considered a 'spacecraft'; the orbiter with its ET and SRBs is a 'launch vehicle'.

Although many design concepts for a reusable space transportation system appeared in the 1960s, the present Shuttle concept dates from the 1970s: the contracts to design and develop the Shuttle orbiter were announced in July

1972, and those for the ET and SRBs in 1973 and 1974 respectively. Approach and landing tests of the orbiter **Enterprise** were made in the late 1970s and the first spaceflight (designated STS-1 and conducted by **Columbia**) began on 12 April 1981.

[See also **Atlantis, Challenger, Discovery, Endeavour, payload bay, flight deck, Space Shuttle main engine (SSME), auxiliary power unit (APU), orbital manoeuvring system (OMS), remote manipulator system (RMS), thermal protection system (TPS), solid propellant, specific impulse, reusable launch vehicle (RLV)**]

Space Shuttle abort alternatives – see **abort**

Space Shuttle main engine (SSME) One of three **rocket engine**s used, along with two **solid rocket booster**s, to launch the **Space Shuttle** [see figures N2, S9]. The engines are mounted aft of the **orbiter**'s **payload bay** in a triangular configuration: with the orbiter horizontal and viewed from the rear the uppermost engine is designated 'No.1', the left-hand engine 'No.2' and the right-hand engine 'No.3'. The SSME uses the **cryogenic propellants liquid oxygen** and **liquid hydrogen** (in a 6:1 LOX/LH_2 mixture ratio) and develops a nominal (100%) **thrust** of 2.1 million newtons (470,000 lb) *in vacuo* (1,670,000 N/375,000 lb at sea level) [see **specific impulse**]. The engines are typically operated at the 104% thrust level in the latter stages of a flight [see **thrust profile**]. The engines are individually gimballed to steer the Shuttle on its ascent and regeneratively cooled by the liquid hydrogen [see **gimbal, regenerative cooling**].

[See also **combustion chamber, nozzle, ignition system, countdown**]

space station

   (i) In general, a **manned spacecraft** designed to remain in **orbit** and provide permanent, or semi-permanent accommodation and associated **life support system**s for a **crew** which may inhabit the spacecraft for long periods [see **Skylab, Salyut, Mir**].

  (ii) In particular, the **International Space Station (ISS)** [see figure I2].

 (iii) In the terminology of official regulatory bodies (e.g. the **ITU**), a 'space station' is any **spacecraft** 'stationed in **space**' (i.e. a **satellite**), as opposed to an **earth station**.

Space Surveillance Network (SSN) A system for **tracking** and cataloguing Earth-orbiting **satellites**, other space **hardware** and **orbital debris** operated by the US **Space Command** (and formerly by **NORAD**). As of 1 January 2000, the official 'Satellite Catalog' maintained by Space Command contained more than 26,000 objects, which had been catalogued since the launch of **Sputnik**

1 on 4 October 1957. Approximately 8600 were still in Earth **orbit**. Another several hundred objects were being tracked but had not been officially catalogued.

space technology The application of the physical and, to a lesser extent, biological sciences to the investigation, exploration and industrial/commercial utilisation of **space**; the methods, theory and practices governing such application.

[See also **space science**, **space exploration**]

space telescope In general, any telescope in Earth orbit, but the term usually refers to the **Hubble Space Telescope (HST)**.

space tourism A space **application** for which serious study and preparation began in the 1990s. Although terrestrially based space tourism (e.g. adventure holidays featuring experience of **astronaut** training facilities) already exists, the advent of commercial tourism in **space** requires the development of a relatively low-cost **reusable launch vehicle** and appropriate residential facilities on a **space station**.

space tracking and data network (STDN) A worldwide network of **tracking** stations for the provision of **communications** (e.g. **telemetry** and voice links) with **spacecraft** in Earth **orbit**. The network is operated by NASA's Goddard Space Flight Center (GSFC) in Greenbelt, Maryland and supplemented by the tracking and data relay satellite system (**TDRSS**).

[See also **deep space network (DSN)**]

space transportation system (STS) The formal name for the **Space Shuttle**. Following the first four 'test-flights', designated STS-1 to STS-4, the STS mission designation took the form (for example) STS 41-D. The first figure represented the final digit of the fiscal year for which the launch was planned (1984 in this example); the second figure represented the **launch site**, where 1 was **Kennedy Space Centre (KSC)** and 2 was **Vandenberg Air Force Base (VAFB)***, and the letter represented a particular flight in that fiscal year, 'D' being nominally the fourth flight. This system allowed payloads to be allocated to a particular **orbiter** as opposed to a specific position in the flight order, which meant that if a **payload** or orbiter suffered a delay, another mission which *was* ready could be launched instead. This system of flight

* Note: in fact no Space Shuttle launches were made from Vandenberg; after the Challenger accident the VAFB Shuttle launch complex was 'mothballed' and the **launch pad** was later converted to launch **expendable launch vehicle**s.

designation was retained until the **Challenger** accident in January 1986, whereupon it reverted to the simple STS-26, STS-27, etc., designation, owing largely to the significant reduction in the flight-rate.

space treaty – see **space law**

space tug – see **orbital manoeuvring vehicle**

space vehicle – see **spacecraft**

space weather An inclusive term for the various physical conditions in the Earth's **upper atmosphere** and **magnetosphere** which influence terrestrial systems, such as power grids, and **satellite**s in **orbit**. See, for example, **geomagnetic storm**, **space radiation**, **single event effect**, **solar flare**.

spacecraft A self-contained vehicle designed for **spaceflight**; alternatively known as a 'space vehicle' [but see **vehicle**]. The term functions as both singular and plural. Any space vehicle which is also capable of 'flight' within the atmosphere is termed an '**aerospace vehicle**'.

To qualify as a 'spacecraft', a vehicle typically has, as a minimum, a **propulsion subsystem**, a **power supply** and a **payload**. Exceptions include the simplest of (mainly scientific) satellites which use **magneto-torquer**s rather than propulsion systems for **attitude control**, and the historical **passive communications satellite**. Although **launch vehicle**s meet these requirements, they are not designed for spaceflight and do not qualify as spacecraft. Similarly, although launch vehicle **upper stage**s operate only in **space**, they are not generally considered to be spacecraft.

[See also **satellite**, **smallsat**, **manned spacecraft**, **unmanned spacecraft**, **man-tended spacecraft**, **manned manoeuvring unit (MMU)**, **attitude and orbital control system**, **thermal control subsystem**, **telemetry tracking and command (TT&C)**, **life support system**]

spacecraft axes The three orthogonal axes of rotation: roll, pitch and yaw. If the spacecraft has a recognisable longitudinal axis or a specified 'forward direction of flight', the axes are analogous to those of an aircraft, where the roll axis is the longitudinal axis; the pitch axis is in the plane of the wings; and the yaw axis is the 'vertical' axis, orthogonal to both the roll and pitch axes. The axes are mutually perpendicular, with an 'origin' at the vehicle's **centre of mass**.

For a winged spacecraft, such as a **Space Shuttle**, the similarity with an aircraft is obvious. For **expendable launch vehicle**s the roll axis is the axis which is vertical at launch and the other axes are more-or-less arbitrarily assigned since the vehicle rotates about the roll axis in flight [see **roll**

programme, but see also **pitch-over**]. The axes of a cylindrical spacecraft (e.g. **Apollo**, **Soyuz**, etc.) are similar to those of an ELV at launch, but once in orbit assume the axis-definition of an aircraft (i.e. defined relative to the pilot's seat).

The axes of a **satellite** mirror those of an aircraft 'flying along the orbital arc': the roll axis is aligned with the direction of travel; the yaw axis passes through the **sub-satellite point**; and the pitch axis is orthogonal to the other two [see figures S10, T4]. For a satellite in an **equatorial orbit**, the pitch axis is aligned approximately with the Earth's spin axis. The pitch axis is also the spin axis for the **spin-stabilised** satellite.

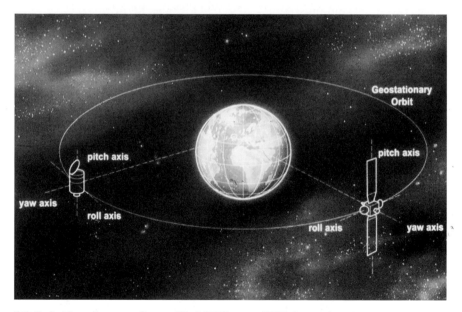

S10: Definition of **spacecraft axes**. [Mark Williamson/Willis Inspace]

In the compilation of engineering drawings the three orthogonal axes are often labelled in Cartesian fashion: $x =$ roll, $y =$ pitch, $z =$ yaw. For the **three-axis stabilised** spacecraft, the x-axis and y-axis are otherwise known as the east–west and north–south axes, respectively; the z-axis passes through the sub-satellite point. This leads to the definition of the box-shaped satellite's faces as follows: the 'plus-x face' faces east; the 'minus-x face' faces west; $+y$ faces south; $-y$ faces north; $+z$ is the Earth-pointing face; and $-z$ is the 'anti-Earth face'.

[See also **degree of freedom, barbecue roll, Dutch roll, actuator**]

spacecraft bus – see **bus**

spacecraft cabin – see **cabin**

spacecraft debris Any part of a **spacecraft**, intentionally jettisoned or accidentally produced in the course of the spacecraft's operation, which constitutes a potential hazard to other spacecraft (or to individuals), or an undesirable addition to the **space environment**.
[See also **space debris, orbital debris**]

spacecraft integration The assembly of the components of a **spacecraft** prior to final ground testing and **launch**. For a **satellite** it typically involves the joining of the **payload module, service module, antenna**s and **solar arrays**.
[See also **payload integration, assembly, integration & test (AIT)**]

spacecraft platform – see **bus**

spacecraft simulator – see **simulator**

spaceflight The act of travelling through space; a particular journey or mission in space. Sometimes split into two words – space flight – especially in older sources. See **spacecraft**.

Spacehab A **habitable module** designed to fit and operate within the **payload bay** of the Space Shuttle.

Spacelab A manned laboratory developed by the European Space Agency (**ESA**) for flights aboard the Space Shuttle [see figure S11]. It first flew in November 1983 (on STS-9) and its 22nd and final mission, designated Neurolab, was in April 1998 (STS-90). A Spacelab **payload** could include various sizes of 'pressurised module' and a number of '**pallet** segments', dependent upon the specific **mission**. When used, the pressurised module was connected to the **cabin** of the **orbiter** by a tunnel, which could be fitted with an **airlock** for **extra-vehicular activity (EVA)** if required. Materials and equipment designed to be exposed to space were mounted on the pallets. When configured as a pallets-only mission, the power, experiment control and **data**-handling equipment was protected within a pressurised and temperature-controlled housing called an 'igloo'.

Spacelab pallet – see **pallet**

spaceport In general, a facility engaged in the launching of **spacecraft** and other connected activities. In particular, in the 1990s, the term came to be applied to new facilities developed for commercial launches, mainly of relatively small launch vehicles (e.g. that developed by the Spaceport Florida Authority at **Cape Canaveral Air Station**, which conducted its first commercial launch – of an Athena 1 carrying **NASA**'s Lunar Prospector – in January 1998).

S11: **Spacelab** 1 in preparation for a **Space Shuttle** mission. Note **habitable module** at top and **pallet** carrying scientific instruments below. [NASA]

[See also **launch complex, launch vehicle, Alcantara, Baikonur Cosmodrome, Guiana Space Centre, Jiuquan, Kapustin Yar, Kagoshima, Kennedy Space Center (KSC), Northern Cosmodrome, Palmachim, Plesetsk, San Marco platform, Sea Launch, Sriharikota, Svobodny, Taiyuan, Tanegashima, Vandenberg Air Force Base (VAFB), Wallops Flight Facility, Woomera, Xichang**].

spaceprobe – see **probe**

spacesuit A pressurised suit worn by **astronaut**s as protection against the **space environment** (**vacuum**, thermal extremes and **radiation**). Also called a 'pressure suit' or 'extra-vehicular pressure garment'. **NASA** spacesuits are known as extra-vehicular mobility units (EMUs). Spacesuits for use inside the **cabin** (e.g. in case of cabin **decompression**), which are pressure-tight but have fewer protective layers, are known as 'intra-vehicular pressure garments'.

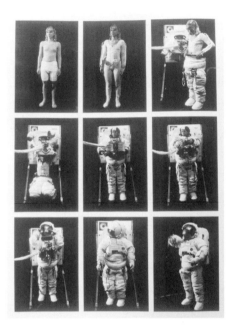

S12: Technician dons Space Shuttle **spacesuit**. [NASA]

The EMUs used on the **Space Shuttle**, for example, comprise a zip-on, one-piece undergarment and a two-piece pressurised spacesuit [see figure S12]. The former is known as a liquid cooling and ventilation garment (LCVG). It includes urine collection facilities, a 'drink bag' and a '**snoopy cap**', with headphones and a microphone, which fits over the head and chin. The spacesuit itself consists of upper and lower torso sections, joined at the waist in a hard ring, and a helmet. A visor assembly, which offers further protection against **micrometeoroid**s and ultraviolet and infrared radiation, fits over the helmet. The total weight is about 30 kg without **life support systems** and **consumables** (and about 120 kg with). The suit has several bonded layers: a polyurethane-on-nylon pressure bladder, several layers of **Kevlar composite**, and an outer abrasion-resistant layer of Kevlar, Teflon and **Dacron**. The hard upper torso has an **aluminium** shell. The torso sections are made in a variety of sizes for use by different crewmembers, unlike previous American spacesuits which were tailored for individual astronauts. An upgraded version of the Shuttle suit is used on **International Space Station (ISS)** missions.

By contrast, the Russian 'Orlan' spacesuit is a one-size (adjustable), one-piece suit, donned by climbing through a door at the back of the aluminium upper torso. It is designed to be stored in orbit, as opposed to being launched on individual missions, which means that it is simpler and more easily maintained. The slightly higher operating pressure of 40 kPa (5.8 psi), compared with 27 kPa (4.3 psi) for the Shuttle suit, reduces the time required for pre-breathing oxygen before an EVA. Since Russian EVAs have typically been shorter than Shuttle-based EVAs, the basic Orlan design does not incorporate a drink bag, food, or a **urine collection device (UCD)**.
[See also **portable life support system, pre-breathe, extra-vehicular activity (EVA)**]

space-time In physics, the four-dimensional continuum having three spatial coordinates and one time coordinate which completely specify an event. Also called 'space-time continuum'. Anything existing in both space and time, or concerned with space-time, is referred to as 'spatiotemporal'.
[See also **space**]

spacewalk – see **extra-vehicular activity (EVA)**

spatial diversity In **satellite communications**, a technique whereby **signals** received simultaneously from more than one **satellite** are combined to mitigate losses due to **aperture blockage** and **multipath** effects; the ability to switch transmission paths to avoid **rain attenuation**. The technique is employed in satellite **constellation**s, for example, to ensure the continuity of communications links or navigation signals.
[See also **navigation satellite**]

spatial frequency re-use – see **frequency re-use**

spatial resolution – see **resolution**

spatiotemporal – see **space-time**

SPCN An abbreviation for **satellite personal communications network**, a network for the provision of services via a **personal communications system (PCS)**.
[See also **GMPCS**]

spec A colloquial abbreviation for **specification**.

specific energy A figure-of-merit for a power source (e.g. a battery), measured in watt-hours per kilogram (Wh kg^{-1}). Since this is a measure of the usable

energy stored per unit mass, it is sometimes called the 'usable energy density'. Nickel–cadmium batteries typically have a specific energy of about 25 Wh kg^{-1} while nickel–hydrogen returns about 30 Wh kg^{-1}, lithium-ion 85–160 Wh kg^{-1} and silver–zinc 110–130 Wh kg^{-1}. See **battery**.

[See also **specific power**]

specific impulse

(i) The **impulse** per unit mass of **propellant** consumed; a measure of the efficiency and performance of a **rocket engine** or **rocket motor**. Specific impulse is given by the expression:

$$I_{sp} = I/M$$

where impulse, I, is in N s, and mass, M, is in kg. I_{sp} is therefore measured in N s kg^{-1}, which can be reduced to m s^{-1}.

(ii) An alternative, but equivalent, definition is: the ratio of **thrust** (N) to rate of propellant consumption (kg s^{-1}), which gives the same units for I_{sp} (m s^{-1}).

(iii) Specific impulse can also be defined in terms of **thrust coefficient** (C_F) and **characteristic velocity** (C^*), such that:

$$I_{sp} = C_F C^*/g$$

where g is the acceleration due to **gravity** (approx. 9.81 m s^{-2}). This implies that specific impulse is the figure of merit for the **nozzle** design and propellant, though in practice I_{sp} is regarded as a measure of the efficiency of the propellant alone.

Historical note: originally I_{sp} was defined as impulse per unit weight of propellant consumed, I/Mg, with I in newton-seconds and Mg in newtons. This gave the second (s) as the unit of specific impulse. In terms of definition (ii), when thrust was measured in pounds-force and rate of consumption in pounds per second, the units [lb/(lb/s)] also gave the second as the unit for I_{sp}. This was a convenient concept, since the specific impulse could be said to be the time (in seconds) over which the combustion of 1 kg of propellant produced a thrust of 1 kg: the longer the time, the less fuel used to produce an equivalent thrust and the more efficient the engine; the longer the time, the greater the specific impulse. Specific impulse is still often quoted in seconds; in fact the expression in definition (iii) can be reduced to s (an I_{sp} quoted in seconds can be converted to m s^{-1} by multiplying by g).

The different types of propellant used in the Space Shuttle system are indicative of contemporary achievable values: the **solid rocket boosters** develop on average 250 s dependent upon the altitude; the **Space Shuttle main engines**, which use LOX and LH_2, are rated at about 363 s at sea level and 453 s *in vacuo*; the orbital manoeuvring system (OMS) and reaction control system (RCS), which use the common **bipropellant** combination of **nitrogen tetroxide** and **MMH**, give around 315 s and 270 s respectively. In comparison, the **Saturn V**'s second and third stages (burning LOX and LH_2) had an I_{sp} of 425 s, somewhat less than the more advanced SSMEs, and its first-stage engines, burning LOX and **kerosene**, produced an I_{sp} of 265 s. An unaugmented **hydrazine thruster** has an I_{sp} of about 220 s, an electrically heated thruster (**hiphet**) increases this to around 300 s, and an **arcjet thruster** improves it further to over 500 s.

[See also **exhaust velocity**]

specific power A quantitative measure of the mass-efficiency of a **power** subsystem or device (units: $W\,kg^{-1}$). For example, for a **solar array** it is expressed as the ratio of array power output to array mass.

[See also **specific energy**]

specification A detailed description of all aspects of a **system, subsystem,** component, etc.; the act of specifying such aspects.

[See also **mil-spec, hi-rel, performance, reliability, COTS**]

spectral resolution A measurement of the ability of an instrument or recording medium to 'resolve' or distinguish between separate parts of a spectrum. For example, a **hyperspectral** sensor has a greater spectral resolution than a **multispectral** sensor.

[See also **resolution, electromagnetic spectrum, waveband, sensor, detector**]

spectrometer Any instrument designed to produce a spectrum, especially one in which **wavelength, energy**, intensity, etc., can be measured.

[See also **mass spectrometer, TOMS**]

spectrum

   (i) A range of **electromagnetic radiation**.

[See also **electromagnetic spectrum, frequency bands, waveband, radio spectrum, solar spectrum**]

   (ii) Any similar distribution or record of energies, velocities, masses, etc., of atoms, **ions, electrons**, etc. (e.g. an 'energy spectrum' or 'mass spectrum').

[See also **mass spectrometer**]

speed of light – see **light (velocity of)**

speedbrake A device on the tail fin of the Space Shuttle **orbiter**, used as a rudder for the atmospheric part of the descent from orbit and split vertically into two parts on landing to increase **drag** and assist deceleration.

Spektr A module of the **Mir** space station.

SPELDA A French acronym for Structure Porteuse Externe pour Lancements Doubles Ariane (external support structure for Ariane dual-launches); a container mounted on the third stage of the **Ariane** 4 launch vehicle, below the **payload fairing**, which allows two independent spacecraft to be carried and launched by a single vehicle. One spacecraft is contained within the SPELDA, another is mounted on top. It differs from the **SYLDA** in that it is part of the external structure or **fairing** of the vehicle, and not simply a container mounted inside the shroud. To accommodate triple **payload**s a SYLDA can be mounted on top of a SPELDA.
[See also **SPELTRA**]

SPELTRA A French acronym for Structure Porteuse Externe pour Lancements TRiples Ariane (external support structure for Ariane triple-launches). A **payload** container designed to be mounted on the **upper stage** of the **Ariane** 5 launch vehicle, below the **payload fairing**, which allows three (or more) independent spacecraft to be carried and launched by a single vehicle.
[See also **SPELDA, SYLDA**]

SPF – see **single point failure**

spike The wedge-shaped surface of a linear **aerospike engine**, against which combustion gases are discharged; also known as a ramp.

spillover The part of an **antenna** feed's radiation pattern not intercepted by the antenna reflector or **subreflector**.
[See also **overspill, antenna radiation pattern**]

spin-stabilised (spacecraft) A spacecraft which maintains its stability by rotating about its longitudinal axis [see **spacecraft axes** and figures S13, S14].
[See also **three-axis stabilised, gyroscopic stiffness, de-spun platform, spin-up**]

spin table A device mounted on a **launch vehicle** which 'spins up' its spacecraft payload prior to release, in order to provide spin stabilisation in the **transfer orbit** (used for example on the Space Shuttle, **Delta** and **Titan** launch vehicles). By contrast **Ariane** 4 provides spin stabilisation by spinning the whole third **stage**. (NB some satellites are **three-axis stabilised** in the transfer orbit).
[See also **spin-up, spin-stabilised**]

S13: 1960s **Syncom** model compared with 1990s **Intelsat** VI **spin-stabilised** communications satellite. Note **deployable antennas** at top, **travelling wave tube amplifiers** around centre and **propellant tanks** and **bipropellant thruster** below. Figure E7 shows Intelsat VI complete with **solar arrays**. [Hughes]

spinner A colloquial term for a **spin-stabilised** spacecraft.

spinning solid upper stage – see **payload assist module (PAM)**

spin-up

    (i)   The rotation of a **spin-stabilised** spacecraft prior to its release from a launch vehicle [see **spin table**].

S14: Cluster II **spin-stabilised** spacecraft showing **solar arrays**, deployed **omnidirectional antennas** and **thermal insulation**. [ESA/J.L. Atteleyn]

   (ii) The increase in rotation rate of a spin-stabilised spacecraft (to its **operational** spin-rate) at the beginning of its life in orbit. Spin-stabilised **communications satellites** typically rotate at a rate of approximately 60 rpm.

   (iii) The undesirable rotation of a spin-stabilised spacecraft's **de-spun platform** resulting in a loss of 'Earth-lock'.

splashdown The moment at which a **spacecraft** returning to Earth lands in a body of water; the equivalent term for a vehicle returning to a body of land is 'touchdown'. Early American **manned spacecraft** (i.e. **Mercury**, **Gemini** and **Apollo**) returned to a water-landing, mainly in the Pacific Ocean; Russian manned spacecraft (i.e. **Vostok, Voskhod** and **Soyuz**) were designed to land on solid surfaces, largely because there are no convenient oceans adjacent to Russian territory.

[See also **recovery**]

splashplate A type of **antenna feed** system incorporating a generally flat **sub-reflector**, rather than one with a precise mathematical profile (e.g. hyperbolic – see **conic sections**). Also called a 'ring-focus' feed, because the phase centre of the wave takes the form of a ring between the splashplate and the feed device.

split-spool device A non-pyrotechnic device designed to separate two items of **hardware** with minimum shock levels. It comprises two sections, held together by a spool of wire, which are separated by the action of a plunger forced between them. As the sections separate, the wire unspools in a controlled fashion, constraining the separation velocity and reducing the shock to adjacent equipment.
[See also **pyrotechnic cable-cutter, thermal knife**]

SPOT An acronym for Satellite Pour l'Observation de la Terre (**Earth observation satellite**), the name given to a series of French satellites for commercial **remote sensing**. During the system's development, the acronym stood for Satellite Probatoire pour l'Observation de la Terre, in recognition of its probationary nature.
[See also **Landsat**]

spot beam A **beam** formed by a **satellite** communications **antenna** which provides coverage of a narrowly defined area. Spot beams concentrate the satellite's **radiated power** in relatively small areas, thus increasing the **power flux density** of the **signal** at the ground. This may be used, in designing a communications **link budget**, to decrease the size of receiving antennas or, alternatively, to provide higher traffic **capacity** and/or wide **bandwidth** services. An antenna which produces a number of spot beams is sometimes called a spot beam antenna – see **multiple beam**.
[See also **beamwidth, global beam, hemispherical beam, zone beam, traffic**]

spread spectrum In **telecommunications**, a transmission technique which spreads the **data** over a **bandwidth** many times that required for the original data [see **code division multiple access (CDMA)**]. Distributing the **radiated power** reduces the **amplitude** of the **signal** to such an extent that it disappears into the **noise**, allowing the residual capacity of an allocated **transponder** to be used as well as dedicated transponders. This technique is useful for low data rates and small receiving antennas. It is used primarily for military communications since it offers a degree of protection against **jamming**.

spreading loss A loss in power density experienced by a **radio frequency (RF)** signal as it radiates from its point of origin. If **power** is radiated isotropically from a point source, forming a spherical wavefront, the power density on the wavefront diminishes progressively with distance (obeying an inverse-square law). In calculating the loss of **signal** power in a practical

telecommunications system, this natural tendency is incorporated with frequency to give an expression for free space loss.
[See also isotropic, equivalent isotropic radiated power (EIRP)]

spurious signal A signal in a communications system which is not intended; one which has arisen by spurious means, not always understood. Sometimes referred to in the plural as 'spurii' (pronounced 'spuri-eye').

Sputnik The name given to a series of Soviet unmanned spacecraft (from the Russian for 'fellow traveller'); the word was also used, in the former Soviet Union, as a general term for spacecraft. Sputnik 1, launched on 4 October 1957, was the world's first artificial satellite and marked the opening of the Space Age; Sputnik 2, the world's second satellite, was launched on 3 November 1957 carrying the dog Laika; later spacecraft designated Sputnik were prototypes for the Vostok series of manned spacecraft.

spy satellite A colloquial term for a reconnaissance satellite or surveillance satellite.

squint A term describing the angular offset between the geometrical axis (boresight) of an antenna and its radio frequency (RF) boresight. The existence of a squint angle may be accidental – due to poor manufacturing standards or damage, which can lead to inaccuracies in pointing – or intentional, as part of the antenna design.

SRB – see solid rocket booster

Sriharikota The location of an Indian launch site, known as SHAR (Sriharikota High Altitude Range), operated by the Indian Space Research Organisation (ISRO). The facility, which conducted its first launch, of the Rohini-1 satellite, on 19 July 1980, is on Sriharikota Island, at approximately 14° N, 80° E, in the Bay of Bengal.

SRM An abbreviation for solid rocket motor. See rocket motor.

SSM – see second surface mirror

SSME – see Space Shuttle main engine

SSN – see Space Surveillance Network

SSO – see sun-synchronous orbit

SSPA – see solid state power amplifier

SSTDMA – see satellite-switched time division multiple access

SSTO – see single stage to orbit

SSUS An abbreviation for spinning solid upper stage. See payload assist module (PAM).

stabilisation system A system or subsystem designed to maintain the stability of a **spacecraft** in **orbit** or on a **trajectory** (between one **planetary body** and another, for example). See **three-axis stabilised**, **spin-stabilised**, **attitude control**, **gravity gradient stabilisation**, **magneto-torquer**.

stable platform – see **inertial platform**

stack Jargon: an assembly of **stage**s which, when stacked together, constitute a **launch vehicle** (e.g. usage: American **astronaut**s, during launch, are said to be 'riding the stack').

stacking The process of assembling the **stage**s of a **launch vehicle** one on top of the other. See **stack**.

stage

(i) A propulsive segment of a **launch vehicle**; a self-contained sub-assembly of a **multi-stage rocket** [see figure S15]. Most practical launch vehicles used to deliver **payload**s to **orbit** (as opposed to **sub-orbital** trajectories) have more than one stage, each with its own **engine**s and **propellant tank**s. The most common design has three stages, although both two- and four-stage rockets exist. The lowermost stage is known as the first stage, since it is the first to provide propulsion. The main reason for having several stages is that the whole mass of the launch vehicle, including

S15: **Ariane** 5 main cryogenic **stage** being prepared for shipment to the **Guiana Space Centre**. [Aerospatiale]

empty propellant tanks and associated structure, does not have to be carried all the way to orbit, with the result that a larger payload can be placed in a given orbit [see **mass ratio**].

(ii) Part of an **amplifier**, TWT **collector**, etc.

[See also **staging, single stage to orbit, upper stage, apogee stage, perigee stage, live stage, dummy stage, core vehicle, strap-on, solid rocket booster**]

stage separation The moment at which a used **stage** separates from the rest of a **launch vehicle**; the process of **staging**.

staging The process of **jettison**ing one **stage** of a **launch vehicle** and preparing to fire the next. This may include the firing of **retro-rocket**s on the used stage to avoid a collision with the subsequent stage, the jettisoning of an **interstage**, and the firing of **ullage rocket**s on the uppermost stage in preparation for its **ignition**. Also called stage separation.

Star City The location (north-east of Moscow) of the **Gagarin Centre** for **cosmonaut** training.

star network A **VSAT** network with a central 'hub' station and a number of remote terminals linked independently to the hub. The main alternative is a **mesh network**.

star sensor A device used to establish a spacecraft's **attitude** relative to a number of 'fixed stars', the positions of which are stored within an onboard database. **Sensor**s with wide fields-of-view are used in a **transfer orbit** to provide initial orientation, and subsequently should a **pointing** problem occur, and narrow FOV sensors are generally used for normal, in-orbit operations.

[See also **sun sensor, earth sensor**]

Stargazer – see **Pegasus**

Start A commercial **launch vehicle** converted from the former-Soviet SS-25 'Sickle' intercontinental ballistic missile (**ICBM**) and operated under a Russian joint venture. It is available in two versions: the 'Start-1', a five-**stage** vehicle capable of lifting 390 kg to **low Earth orbit** (**LEO**), and the 'Start', a six-stage vehicle with a **payload capability** of about 600 kg. The Start family began commercial operations in March 1995.

[See also **Rockot, Dnepr, Cosmos**]

START An acronym for Strategic Arms Reduction Treaty. See **ICBM**.

statement of work (SOW) Industrial terminology for the part of a contract which specifies precisely what is required from the contractor.

[See also **prime contractor, subcontractor**]

static discharge – see **electrostatic discharge**

static firing A test-firing of a **rocket engine** or **rocket motor** during which the device is constrained from moving – usually conducted in a **test stand**. [See also **flight readiness firing**]

station keeping – see **orbital control**

stationary orbit An **orbit** in which a **spacecraft** appears stationary with respect to the **planetary body** it is orbiting. For example, if the body is the Earth, the orbit is known as **geostationary orbit (GEO)**. Although not widely used, the equivalent for Mars is areostationary orbit (from Ares, the Greek counterpart to the Roman god of war, Mars).

stationary plasma thruster (SPT) A type of Hall effect thruster – see **ion engine**.

STDN – see **space tracking and data network**

steerable antenna

   (i) A spacecraft **antenna** which can be steered by ground command to enable its **beam** to illuminate an alternative **coverage area**.

   (ii) An **earth station** antenna with the capability to point at different **spacecraft**, or remain pointing at a single **satellite** with greater accuracy [see **box, free-drift strategy**].

   [See also **antenna pointing mechanism, phased array antenna**]

Stennis Space Center – see **NASA**

step-rocket An archaic term for a **launch vehicle** with several **stage**s; a multistage rocket.

step-track A type of **earth station** system for tracking satellites which moves an **antenna** in predetermined 'steps' in a direction dependent on the strength of the **signal** received. The signal is sampled at regular intervals and compared with the previous sample; the antenna continues to step in a particular direction until the signal power is lower than the previous sample, at which time it reverses its direction. The process then repeats.

stoichiometric ratio The ratio of **fuel** and **oxidiser** required for complete **combustion** of both **propellant**s. True stoichiometric combustion leaves no excess of either component.

stopband A range of frequencies not passed by a **filter**. Frequencies which the filter allows to pass are in the 'passband'. [See also **bandwidth, low-pass filter, high-pass filter, band-pass filter, channel filter, input filter, output filter**]

storable propellant – see **propellant**

store-and-forward A type of **satellite communications** link in which **data** is stored temporarily on board a low Earth orbiting satellite and **downlink**ed to a receiving station when it comes into view. In the early days of satellite communications, a satellite operating in this manner was known as a delayed **repeater** satellite.

strapdown gyro system An **inertial platform** in which the **gyroscope**s are connected directly to the vehicle structure ('strapped down') as opposed to being isolated from the structure in a set of **gimbal**s.
[See also **laser gyro**]

strap-on (booster) A propulsive component of a **launch vehicle** attached to a **core vehicle** to augment its thrust. The addition of either solid or liquid propellant strap-ons allows a core vehicle to be tailored to the payload and gives rise to a number of **variant**s of a basic vehicle. See **Ariane**.
[See also **solid propellant**, **liquid propellant**, **Long March**]

strategic defense initiative (SDI) A development programme for a space-based defensive 'shield' against nuclear weapons, proposed in March 1983 by US president Ronald Reagan and colloquially termed 'Star Wars' after the film of the same name. The programme was cancelled before any **operational** flight hardware was **deploy**ed.

stratopause – see **atmosphere**

stratosphere – see **atmosphere**

stress (engineering) The branch of spacecraft design engineering concerned with the behaviour of a structure under mechanical loads. Stress is measured in terms of force per unit area ($N\ m^{-2}$). Where this force is applied as a result of a change in temperature, it is known as 'thermal stress' which can, if sufficiently severe or prolonged, lead to structural failure.
[See also **load path**, **finite-element modelling**, **thrust structure**, **thermal cycle**, **thermal model**]

stress-corrosion cracking The fracture of a material due to the combined action of tensile **stress** and a corrosive environment. Stress-corrosion is the corrosion of a metallic surface enhanced by local stresses.

stretched (propellant tank) Jargon: extended or elongated. To **uprate** or improve a **launch vehicle** or spacecraft **propulsion system**, it is quite common to adapt an existing design by simply increasing the length of a **propellant tank**. This allows an increase in **propellant** capacity and overall **vehicle** performance while minimising the complexity of the design change. A launch vehicle **stage** with a stretched propellant tank is often called a 'stretched stage'.

stringer A longitudinal structural brace used to reinforce the bodies of **launch vehicles** and **spacecraft** and maintain the shape of a **skin** or other panel; a smaller version of the **longeron**.
[See also **airframe**]

structure model The version of a spacecraft built for the purposes of structural tests, and verification that all components and subsystems will fit in and about the proposed structure. The structure model undergoes vibration tests: if any component is 'out of spec' the **flight model** is altered accordingly. Figure S16 shows the **primary structure** of a **three-axis stabilised** satellite under construction.
[See also **thermal model, engineering model, vibration facility, acoustic test chamber**]

STS An abbreviation for **space transportation system**, the formal name for the **Space Shuttle**.

subcarrier An electromagnetic wave of fixed amplitude and **frequency** which, in order to convey information through a radio transmission system, is modulated by a **signal** and in turn modulates a main **carrier** wave. Subcarriers are used for colour information and sound in a TV system and for **data** transmission, for example.
[See also **modulation**]

subcontractor Industrial terminology for a supporting contractor in a project or programme; a contractor responsible for delivering a product to the **prime contractor**, not directly to the customer.

sub-orbital hop A colloquial expression for the journey taken by a **launch vehicle** or **spacecraft** on a **sub-orbital** or **ballistic trajectory**. For example, the early **Mercury** flights were described as sub-orbital hops.

sub-orbital trajectory The **flight path** of a **launch vehicle** or **spacecraft** in which the **vehicle** has insufficient energy to reach orbital velocity and attain **orbit**. Typically used to describe the flight path of a launch vehicle or spacecraft which follows a parabolic or **ballistic trajectory**.
[See also **trajectory, sub-orbital hop**]

subreflector A subsidiary reflector in a 'reflector antenna' system, interposed between the **feedhorn** and the main reflector in a so-called 'folded-optics' configuration. See, as examples, **Cassegrain reflector** and **Gregorian reflector**. An advantage of the subreflector design is that the feedhorn and associated hardware can be positioned near the base of the main reflector. This not only reduces **waveguide** runs (and thus **RF** losses) but also reduces

S16: **Primary structure** of a **three-axis stabilised** satellite under construction [see **structure model**]. Note **thrust cylinder** in centre and **honeycomb panels** around base. [Space Systems/Loral]

the system **noise temperature**, since the **receiver** is 'looking at' cold sky as opposed to warm earth.

[See also **antenna**]

sub-satellite point The point of intersection, on the surface of a **planetary body**, of a line drawn between a satellite in **orbit** and the body's centre.

subsonic flow – see **Mach number**

substitution heater – see **simulation heater**

subsystem Literally, a constituent part of a **system**. Although 'system' and

'subsystem' tend to be used interchangeably, 'system' is the more general term. For example, a **communications system** would include all **hardware**, **software** and procedures used to operate the system (both **space segment** and **ground segment**); a 'communications subsystem' (e.g. on a **communications satellite**) would include only the 'hardware' on board the satellite – it would be synonymous with **communications payload**.
[See also **propulsion system, propulsion subsystem**]

Sun The star about which the **Earth** orbits. Usage: when used as a proper name (when referring directly to the Earth's sun), Sun should have an initial capital (e.g. 'in **orbit** around the Sun'); when referring to any other star or when used as a common noun (to describe an object or system), it should have a lower case initial (e.g. **sun sensor, sunshield**). A related adjective is 'solar' ('of the Sun', from sol, Latin for Sun) – see, for example, **solar cell**.

Physical data: equatorial diameter 1,392,000 km; mass 2×10^{30} kg; **escape velocity** 617.5 km s^{-1}; surface **gravity** 273 m s^{-2}; average distance to Earth 149,597,870 km (1 **astronomical unit**); average light travel time to Earth 8 m 19 s.
[See also **light (velocity of)**]

sun sensor A device used to establish a spacecraft's **attitude** relative to the Sun. **Sensor**s with wide fields-of-view are used in a **transfer orbit** to provide initial orientation, and subsequently should a **pointing** problem occur, while narrow FOV sensors are generally used for normal, in-orbit operations.

A common direction-sensing technique involves an arrangement of sensors (e.g. **solar cell**s) grouped around the pointing axis so that each receives the same illumination when the sensor is exactly aligned with the target source. When the sensor is directly in line with the Sun the output from each of the four cells is the same; if the sensor is tilted with respect to the Sun the outputs change accordingly. This type of differential output can be used to control a satellite's **reaction wheels**, **reaction control thruster**s, etc., which make adjustments until the sensor outputs are equal and **pointing** is regained. This is described as a 'null-seeking' technique.
[See also **earth sensor**]

sunshield
  (i) Any device on a **spacecraft** positioned to avoid sunlight falling on a **sensor**, instrument or other sensitive hardware.
  (ii) A shield used to protect a spacecraft in the Space Shuttle's **payload bay** from the heating effects of the Sun. The shield is mounted on the **cradle**

carrying the spacecraft and operates like a 'clamshell', its two doors rotating back to allow the spacecraft to leave the payload bay [see figures M1, O3].

sun-synchronous orbit (SSO) – see **heliosynchronous orbit**

SUPARCO An acronym for Space and Upper Atmosphere Research Commission, Pakistan's national space agency and operator of the SUPARCO Flight Test Range, a **launch site** north-west of Karachi (at approximately 25° N, 66° E) used mainly for **sounding rockets**.

super light weight tank (SLWT) A version of the Space Shuttle **external tank (ET)** developed in the late 1990s.

supersonic flow – see **Mach number**

supersynchronous orbit An elliptical orbit with an **apogee** higher than that of geosynchronous orbit, into which a satellite is injected before its transfer to the circular **geostationary orbit (GEO)**; an alternative to **geostationary transfer orbit (GTO)**.

[See also **parasitic station acquisition**]

surface coatings (of a spacecraft) Any of a variety of paints or 'finishes' applied to the surface of a spacecraft as part of its **thermal control subsystem**. In a definitive sense there are four main types of surface available to the thermal design engineer, which assist in the maintenance of a spacecraft's temperature between certain limits. The 'solar reflector' and 'solar absorber' respectively reflect or absorb most of the incident energy in the **solar spectrum**. The 'flat reflector' and the 'flat absorber' reflect or absorb practically all the incident energy, both within and beyond the solar spectrum – they have a 'flat' response across the frequency band.

[See also **second surface mirror, thermal insulation**]

surveillance satellite A military satellite, usually based in **geostationary orbit**, concerned primarily with high **resolution** imaging of 'enemy territory', although the definition could be extended to included the detection of ballistic missile launches. Colloquially termed a 'spy satellite' and similar to a **reconnaissance satellite**, except that reconnaissance implies a periodic collection of **data**, whereas surveillance implies continual data collection.

[See also **remote sensing**]

Surveyor The name given to a series of American spacecraft designed for lunar research. Surveyor 1 was the first US spacecraft to execute a controlled landing on another **planetary body** (on 2 June 1966). It and later Surveyors

S17: **Swath** and adaptive coverage of Envisat **Earth observation** satellite. [ESA]

took photographs and soil samples, confirming the constitution of potential landing sites for **Apollo**.

sustainer engine – see **Atlas**

Svobodny The location of a Russian military **launch site** (at approximately 51.2° N, 128° E), which conducted its first commercial launch on 4 March 1997.

[See also **Baikonur Cosmodrome**, **Northern Cosmodrome**, **Kapustin Yar**]

swath A narrow strip defining the coverage of an **Earth observation** satellite's **sensor** or **payload** [see figures G5, S17].

swath-width The width of the narrow strip which defines the coverage of an **Earth observation** satellite's **sensor** or **payload**; one of the key parameters of a **remote sensing** payload.

swing-arm A projection from a launch vehicle **service structure** which can be rotated towards the vehicle for access, **propellant**-loading, etc. and swung away from the vehicle at the moment of **launch**, when any **umbilical**s are automatically detached.

[See also **access arm**]

swing-by The action of a spacecraft using the **gravitational field** of a **planetary**

body to re-direct its **trajectory** (e.g. Voyager performed a gravitational swing-by at Jupiter). See **gravity propulsion**.

SYLDA A French acronym for SYsteme de Lancement Double Ariane (Ariane dual-launch system). A container mounted inside the **payload fairing** of the **Ariane** launch vehicle to allow two independent spacecraft – one within the SYLDA, the other on top – to be carried and launched by a single vehicle. A similar structure designed for the **Titan** III launch vehicle was called an 'extension module'.

[See also **SPELDA, SPELTRA**]

Syncom The name – an abbreviation of 'synchronous communications' – given to a series of American **communications satellite**s [see figure S13]. Syncom 1 was launched into a 33° inclination **geosynchronous orbit** on 14 February 1963, but communication was lost at orbital **injection**; Syncom 2 was launched to a 28° inclination orbit on 26 July 1963; Syncom 3, launched on 19 Aug 1964, was the first **spacecraft** to attain a good approximation to **geostationary orbit** (**apogee** 34,191 km; **perigee** 36,271 km; inclination 0.1°, compared with a nominal 35,786 km, 35,786 km and 0.0° for GEO). An unrelated series designated Syncom IV (or 'Leasat') was launched in the 1980s by the Space Shuttle.

[See also **orbital inclination**]

synodic period An orbital period defined in the **frame of reference** of the **celestial body** being orbited. See **orbit**.

synthetic aperture radar (SAR) A **radar** which produces an **image** by integrating **data** from successive **antenna** positions, thereby constructing an effective antenna aperture much larger than the 'real', physical aperture [see figures S17, S18]. SAR payloads are carried on some **Earth observation** satellites and some planetary **probe**s [see figure R1].

While most spacecraft **sensor**s or remote sensing **payload**s are passive (simply receiving incident **radiation**), the synthetic aperture radar is an active device (transmitting **microwave** energy towards an objective and measuring the reflected radiation). The **energy** source, in most cases, is a high-power **travelling wave tube amplifier (TWTA)**. The degree to which energy is reflected depends on the instrument's illumination parameters, such as incidence angle, **wavelength** and **polarisation**, and on attributes of the reflecting surface, such as moisture content, vegetation cover and overall surface geometry. An understanding of these and other parameters allows

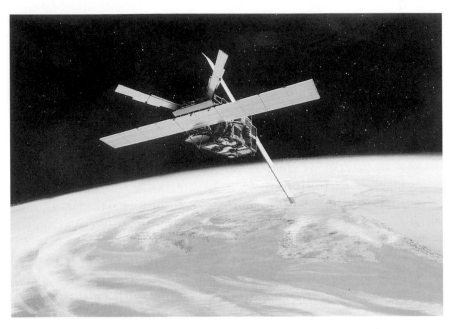

S18: ERS **Earth observation** satellite showing deployed **synthetic aperture radar** antenna and **wind scatterometer** antennas above. [Matra Marconi Space]

interpretation of the radar image. A typical instrument has a surface **resolution** measured in metres or tens of metres and a swath-width of tens of metres to several hundred kilometres.

[See also **swath-width, side-looking radar**]

system In general, a group or combination of interrelated or interacting elements forming a collective entity. In space technology (and other branches of technology):

   (i)   any assembly of electronic, electrical and/or mechanical components ['**hardware**'] forming a self-contained unit [e.g. **communications system, attitude and orbital control system, propulsion system, thermal protection system, life support system, space transportation system**];

   (ii) any coordinated assemblage of facts, concepts, **data**, etc. [e.g. **software**].

[See also **control system, subsystem**]

T

tail service mast A support structure on a **launch pad**, which carries **propellant** and/or electrical supplies to a **launch vehicle**. A simple service mast may carry a single **umbilical**, which is connected to the vehicle until the moment of **launch** and then ejected and carried clear of the vehicle by the mast. More complex masts, like those mounted on the Space Shuttle's **mobile launch platform**, supply **liquid oxygen** and **liquid hydrogen** to the vehicle's **external tank** [see figures F2, S3]. At launch, they retract the umbilicals and close over them to protect them from the engine exhaust.
[See also **swing-arm, service structure**]

Taiyuan A Chinese **launch site**, in Shanxi Province south-west of Beijing (at approximately 38° N, 112° E), which became operational in 1987. The first launch conducted from the site (on 6 September 1988) was that of Feng Yun 1, a **meteorological satellite**.
[See also **Jiuquan, Xichang**]

Tanegashima An island off the south coast of Kyushu, Japan (at approximately 30° N, 131° E); the location of **NASDA**'s Tanegashima Space Center, which includes the Takesaki and Osaki **launch sites**, used respectively for small and large **launch vehicles**. The latter conducted its first launch, of ETS-1 (Kiku-1), on 9 September 1975 and has been used since then mainly for the **N** and **H**-series launch vehicles.
[See also **Kagoshima**]

tank – see **propellant tank**

tanker A colloquial name given to space vehicles which carry supplies to an orbiting **space station**, particularly the '**Progress** tankers' used to supply the former Soviet Union's **Mir** station.

taper The variation in electric field produced by a **feed** across an **antenna** surface.

tape-wrapping A method of manufacturing low-density spacecraft structures, which involves wrapping a continuous length of '**pre-preg**' tape around a former to produce a hollow body, such as a propellant tank. Tape-wrapping is a coarser version of **filament-winding**.

Taurus An American four-stage solid propellant **launch vehicle** developed in the early 1990s as a small satellite launcher. It has a **payload capability** of about

1400 kg to **low Earth orbit (LEO)** and 450 kg to **geostationary transfer orbit (GTO)**. The vehicle, which was first launched in 1994, uses derivatives of the **Pegasus** rocket motors with a first stage based on the Thiokol Castor 120, a solid **rocket motor** that has been used on several launch vehicles.

[See also **Athena, Atlas, Delta, Titan, Scout, solid propellant**]

TC An abbreviation for **telecommand**.

TC&R – see **telemetry, command and ranging**

TCP An abbreviation for transmission control protocol, a **software** procedure concerned with **data** transmission via the **Internet**. Often combined with the term Internet protocol in TCP/IP.

TDM – see **time division multiplexing**

TDMA – see **time division multiple access**

TDRS An acronym for Tracking and Data Relay Satellite [see figures A4, D4].

TDRSS An acronym (pronounced "teedriss") for Tracking and Data Relay Satellite System, a series of **NASA** satellites based in **geostationary orbit (GEO)**, designed to provide communications links between spacecraft in **low Earth orbit** (e.g. **Space Shuttle, Hubble Space Telescope**, etc.) and terrestrial ground stations. The baseline system comprises one TDRS satellite over the Pacific Ocean (at the 171° W **orbital position**), referred to as TDRSS-West, one over Brazil (at 41° W) referred to as TDRSS-East, and an **in-orbit spare**; TDRS-1 was launched in April 1983. The reason for its implementation was that spacecraft in LEO can communicate directly with a given **earth station** for only a few minutes on each **orbit**; allowing them to communicate via a satellite in GEO increases the access time.

[See also **inter-orbit link (IOL), inter-satellite link (ISL)**]

technology demonstrator A **spacecraft, launch vehicle** or **payload** designed to demonstrate a given technology as opposed to providing a commercial or **operational** service.

[See also **experimental**]

telecommand An instruction transmitted to a spacecraft from a controlling **earth station** or **mission control centre**. See **telemetry, tracking and command**.

telecommunications The science and technology of **communications** at a distance, by means of **telephony, television, data communications**, etc., using the **radio spectrum, transmission line**s or **optical fibre** as the medium.

teleconference A meeting between two physically separated parties whereby audio signals are transmitted between them to simulate the audio-presence of the other party. At its simplest it is little more than a two-way telephone call, although it is usually held between more than two people. The better systems use wider **bandwidth** audio channels and feature background noise-reducing circuits, among other things.

[See also **videoconference**]

telemetry Information or measurements transmitted by **radio frequency** waves from a remote source to an indicating or recording device (e.g. from a **spacecraft** to an **earth station**). A telemetry **data**-stream is usually transmitted separately from any other communications **channel**, since it concerns only the status of the spacecraft subsystems. The data is used both to control the spacecraft and to establish the performance of the **subsystem** equipment. Historically, telemetry has been allocated its own **frequency band** and has been transmitted by dedicated subsystem equipment through a **TT&C antenna** on the spacecraft, but during the **transfer orbit** phase it may be transmitted by part of a **communications payload** using frequencies within the payload band.

[See also **telemetry, tracking and command (TT&C)**]

telemetry, command and ranging (TC&R) – see **telemetry, tracking and command (TT&C)**

telemetry, tracking and command (TT&C) A spacecraft **subsystem** incorporating the three functions: **telemetry**, which is **downlink**ed from the spacecraft to the ground; **tracking**, whereby an **earth station** tracks the telemetry **carrier** or a **beacon** carried on the spacecraft; and **command**, whereby instructions are **uplink**ed from the ground to the spacecraft. Sometimes called a telemetry, command and ranging subsystem (TC&R).

telephony A **telecommunications** system by which speech or other sounds, converted into an electronic **signal**, can be transmitted via **transmission lines**, **microwave link**s, **satellite communications** links, etc., and reproduced at a distant receiver.

[See also **personal communications system (PCS), GMPCS**]

teleoperate To operate at a distance (i.e. by remote control). See **telepresence**.

telepresence Presence at a distance: a technique which allows a task or operation to be undertaken from a remote position. It generally involves the use of real-time video communications and a form of **remote manipulator system**, which allow an operator to carry out the task from a remote position, in relative comfort and safety. For example, an operation may be conducted in space or on a planetary surface by an operator in a **spacecraft**, **space station** or on Earth.

telescope A device for collecting, focusing and detecting electromagnetic radiation from astronomical sources.

[See also **Hubble Space Telescope**, **Newtonian telescope**, **Cassegrain reflector**, **Gregorian reflector**, **Schmidt telescope**, **Coudé focus**]

teletext A system of data **broadcasting** which uses the spare lines of the field blanking interval, the 'space' in the TV waveform between successive TV pictures. Bursts of digital signals carry a computer-based **videotext** service of written or graphical information to be transmitted to the consumer with the TV signal. To make use of the signals a separate decoder within the TV set is required. Teletext is, in effect, broadcast videotext and is a non-interactive system. Viewdata, on the other hand, is interactive videotext, whereby the consumer is linked to the computer by telephone and can select information as required.

televideo The system or process concerned with video communication (including an audio link), over a distance, between a small number of participants (typically two). The system used by a 'videophone'. The term may be used to describe a small-scale **videoconference**.

television The system or process concerned with transforming a visual scene, usually with an accompanying audio **signal**, into an electrical waveform which can be transmitted over a distance to and displayed on a cathode ray tube, or other display, within a TV receiver.

telex An international telegraphic system whereby text is transmitted between teleprinters via **transmission line**s, **microwave link**s, **satellite communications** links, etc.

Telstar The name given to a series of **communications satellite**s. Telstar 1, launched into a **low Earth orbit** on 10 July 1962, was the first privately owned (non-government) communications satellite, the first satellite to provide an instantaneous (as opposed to **store-and-forward**) communications link and the first to carry **television** across the Atlantic Ocean.

temporal resolution A measurement of the ability of an instrument or recording medium to 'resolve' or distinguish between different points in time.

[See also **resolution**]

terrestrial Of or related to the Earth. See **Earth**.

terrestrial tail A link between a satellite **earth station** and the local terrestrial **telecommunications** network; also known as a 'backhaul link'.

[See also **single hop**, **double hop**]

TES An abbreviation for transportable **earth station**, typically an earth terminal small enough to be towed by a road-going vehicle or transported on board an aircraft. For example, an earth station used in satellite news gathering (**SNG**).

test stand A ground-based facility for testing **rocket engine**s and **rocket motor**s. Also called a proving stand.

tether

    (i) A safety line which links an **astronaut** on an EVA (**extra-vehicular activity**) to a **spacecraft**. In the early days of **spaceflight** (e.g. on 'spacewalks' made from the **Gemini** capsules) the tether incorporated an **umbilical** which supplied oxygen and cooling water from the spacecraft; nowadays astronauts wear a **portable life support system** 'back-pack' and the tether is simply a safety line. EVAs conducted using the Space Shuttle-based **manned manoeuvring unit** (**MMU**) are termed 'untethered EVAs'.

    (ii) A physical link between two orbiting spacecraft, which takes the form of a long cable (several kilometres in length). If the tether is made of a conducting material, it can be used to generate **power** as it cuts the magnetic field lines of the **planetary body** it is orbiting. Tethers can also be used for **gravity gradient stabilisation** and to change a spacecraft's orbital altitude: see **tethered satellite**.

tethered satellite A satellite attached to a larger orbiting spacecraft by a cable known as a **tether** [see figure T1]. Using a tether to place the two, linked orbiting masses far apart effectively elongates the spacecraft, which can then be stabilised by **gravity gradient stabilisation**, since the difference in gravitational attraction between the two masses acts to align the tether with the local vertical. If a tether joining a tethered satellite to a larger spacecraft, **space platform** or **space station** is severed, the spacecraft's altitude is changed in proportion to the mass of the tethered satellite and the length of the tether. This method can be used to raise or lower the orbital altitude of a spacecraft without the expenditure of **propellant**.

thermal balance – see **thermal energy balance**

thermal barrier – see **thermal insulation**

thermal blanket – see **thermal insulation**

thermal control subsystem The collection of equipment on a spacecraft which maintains its temperature within the limits dictated by the design. Methods of thermal control are either passive or active.

    Passive control is the simplest and usually cheapest method and requires no intervention, from man or machine, in its normal operation. It includes the use of reflective and absorptive **surface coatings**, **thermal insulation** and

T1: **Tethered Satellite** System (TSS) leaving capture device in **Space Shuttle** (STS-46) **payload bay**. [NASA]

radiators. Early spacecraft required only passive control techniques to maintain their temperature between certain operating limits, but as spacecraft became more complex and more powerful, active methods of thermal control evolved to accommodate the increase in complexity.

Active control involves more complex devices which are controlled either by direct ground command or by automatic systems working in a 'feedback loop' mode. These range from the simple, thermostatically controlled **heater** to the more advanced forms of variable conductance **heat pipe**.
[See also **optical solar reflector**, **second surface mirror**, **heat sink**, **louvre**, **thermal doubler**, **heat spreader plate**]

thermal cycle The cyclic variation in temperature experienced by a spacecraft in **orbit** due, predominantly, to the relative position of the Sun and its **eclipse** by the Earth.

A spacecraft in **low Earth orbit** undergoes a large number of thermal cycles during its **lifetime** due to its short orbital period. It also experiences an eclipse on every orbit. This can have a detrimental effect on spacecraft equipment, particularly the batteries which have a correspondingly short-period **duty cycle** (charging while in sunlight, discharging in shadow). A spacecraft in **geostationary orbit** is affected mainly by the diurnal variation of the Sun's position and the biannual eclipse. The differing aspects of thermal cycling are simulated during ground-testing in a **thermal-vacuum chamber**.

[See also **stress**]

thermal doubler A local increase in the thickness of a spacecraft structural panel used to improve the thermal conduction path and distribute heat more effectively.

[See also **heat sink, thermal control subsystem**]

thermal energy balance A quantitative expression of a body's thermal **energy** inputs and outputs. For a **spacecraft** orbiting the Earth, the thermal energy balance may be expressed as:

$$Q_{sol} + Q_{int} = Q_{rad} + Q_{str}$$

where Q_{sol} is the rate of solar energy absorption, Q_{int} is the rate of internal energy generation, Q_{rad} is the rate of thermal energy radiation and Q_{str} is the rate of energy storage by the spacecraft.

If the spacecraft is in thermal equilibrium (energy input = energy output) then $Q_{str} = 0$ and the equation can be simplified to:

$$Q_{sol} + Q_{int} = Q_{rad}$$

which effectively means that both the heat absorbed from the Sun and that generated by the spacecraft itself must be radiated to space in order to maintain the thermal energy balance of the spacecraft. For spacecraft in **low Earth orbit**, heat input due to **albedo** and **earthshine** may also be important.

[See also **solar spectrum, thermal control subsystem**]

thermal gaiter – see **gaiter**

thermal grease – see **interface filler**

thermal infrared – see **waveband**

thermal insulation A material or structure which prevents the transmission of heat; a thermal barrier which limits both heat input and output. The thermal barriers used in spacecraft are of two main types: 'thermal blankets' and 'thermal stand-offs'.

The material colloquially termed a thermal blanket, but more correctly known as multi-layer insulation (MLI) because of its layered construction, is the material that gives a **satellite** its foil-wrapped, chocolate-box appearance [e.g. see figure S14]. The simplest examples of MLI consist of layers of **Mylar** foil, aluminised on one side and crinkled to produce insulating voids. **Kapton** may be used as an alternative foil, and low conductivity interlayers (of fibre-glass, silk, **Nomex** or **Dacron** net) can improve the insulation [see figure T2]. One example of a high-temperature superinsulating material, particularly useful for rocket **nozzle**s, comprises stainless steel or titanium foil skins with silica fibre interlayers. Thermal blankets have many uses on and within **spacecraft**: they maintain the temperature of isolated components, insulate cryogenic **propellant tank**s to minimise **boil-off**, and can be wrapped around even larger components, such as astronomical telescopes, to reduce thermal distortion which could otherwise render a precision instrument useless.

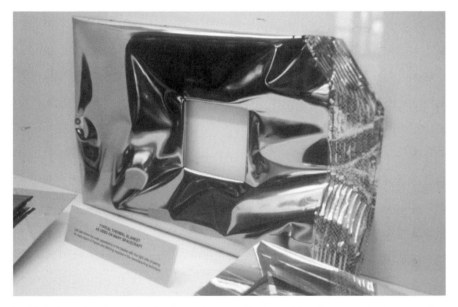

T2: Multi-layer insulation (MLI), a type of spacecraft **thermal insulation** with criss-crossed interlayers to improve insulating properties. [Mark Williamson]

A 'thermal stand-off' is an insulating spacer, as opposed to a blanket or continuous mat, placed between a unit of spacecraft hardware and the panel or surface to which it is attached. This type of thermal barrier can be used to prevent heat-loss from the spacecraft by isolating items such as **antenna** reflectors, which might otherwise act as **radiator** panels.

[See also **thermal control subsystem, cryogen**]

thermal knife A device designed to sever a cable connecting two items of **hardware** by means of the application of high temperature (e.g. for releasing **solar array** panels folded against the side of a **satellite**).

[See also **pyrotechnic cable-cutter, split-spool device**]

thermal louvre – see **louvre(s)**

thermal model The version of a spacecraft built to test the **thermal control subsystem** and verify that all components and subsystems will operate in the thermal environment of space. If any component is 'out of spec' the **flight model** is altered accordingly.

[See also **engineering model, thermal-vacuum chamber**]

thermal noise A source of **noise** in a communications system which is due to the intrinsic random motion of electrons. The recognised method for reducing this noise is to cool the **receiver** or other relevant equipment. The greatest benefit is gained by cooling to cryogenic temperatures, close to **absolute zero**: at these temperatures molecular (and therefore electron) motion is reduced and thus so is the noise. Also called Johnson noise.

[See also **shot noise, noise power, noise temperature, noise figure, cryogen**]

thermal protection system (TPS)

(i) In general, any system that provides protection to a **spacecraft** or **launch vehicle** from the effects of heat, particularly that due to friction with the Earth's atmosphere [see **aerodynamic heating, heat shield, ablation**], but also due to solar heating [see **insolation**].

(ii) The formal name for the layer of heat-resistant **materials** applied to the outer surface of the Space Shuttle **orbiter** [see figure T3]. About 70% of the vehicle's surface was originally covered with some 31,000 silica-fibre tiles of varying shape and thickness; some were later replaced with thermal blankets. The white tiles on the upper surfaces, designed to protect the orbiter from temperatures between 400 and 650 °C, are composed of a silica compound with added alumina oxide to improve their reflectivity. The tiles on the underside and other areas exposed to temperatures up to 1260 °C are coated with a black borosilicate glass. Each of the tiles is

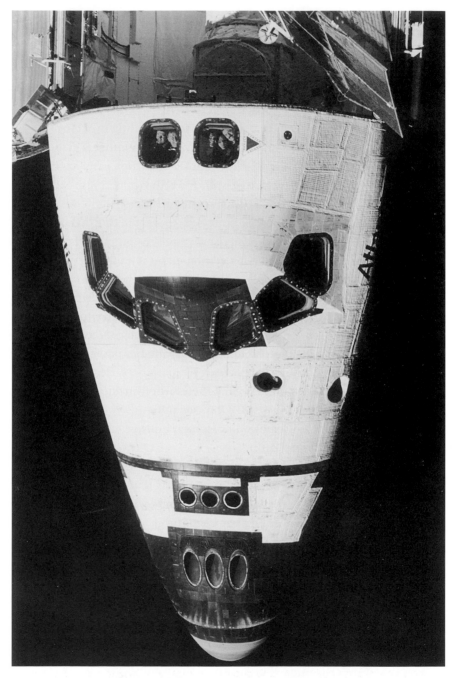

T3: **Thermal protection system (TPS)** of the Space Shuttle **orbiter** showing thermal blankets round top windows and down sides, and tiles round forward windows and **reaction control thruster**s on nose. [NASA]

bonded first to a synthetic **Nomex** felt 'strain-isolator' pad, which allows expansion and contraction of the tile without cracking, and then to the aluminium **skin** of the orbiter. The **payload bay** doors and some upper wing surfaces, which are heated to less than 400 °C, are covered with coated Nomex felt; the nose and wing leading-edges, which experience temperatures exceeding 1260 °C, are clad in reinforced carbon–carbon **composite**.

[See also **thermal insulation, thermal control subsystem**]

thermal stand-off An insulating device used to separate instruments, or other devices, from a spacecraft structure to avoid heat transfer. Also called a 'thermal barrier'. See **thermal insulation**.

thermal stress – see **stress**

thermal subsystem – see **thermal control system**

thermal tiles – see **thermal protection system**

thermal-vacuum chamber A ground-based test facility which simulates the thermal and vacuum environment of space. The larger chambers admit the whole spacecraft. Thermal-vacuum tests simulate a number of **eclipse seasons** to test the operation of all on-board equipment. For most satellite programmes a special **thermal model** version of the spacecraft is built to test the **thermal control subsystem**.

[See also **acoustic test chamber, vibration facility, anechoic chamber, space environment**]

thermoelectric effect The phenomenon whereby a current is produced in a circuit made by two different conductors joined at two junctions which are maintained at different temperatures. Also called the Seebeck effect after its discoverer, the Russo-German physicist, Thomas Johann Seebeck [1770–1831]. Used in the operation of the **radioisotope thermoelectric generator (RTG)**.

The reverse effect, whereby a current produces a temperature difference, is known as the Peltier effect (as in 'Peltier cooling'); and the effect whereby heat is generated or absorbed when a current flows across a temperature gradient is called the Thomson effect (after William Thomson – see **kelvin**).

thermosphere – see **atmosphere**

third space velocity – see **escape velocity**

Thor An American single-**stage** liquid **propellant** intermediate range ballistic missile (IRBM) developed by the US Air Force in the 1950s (effectively as an equivalent to the US Army's **Redstone**). Propellant: **liquid oxygen/kerosene**. Various **upper stages** (e.g. Able and **Agena**) were added to the Thor first stage to form the Thor-Able and Thor-Agena, etc., which were used largely for

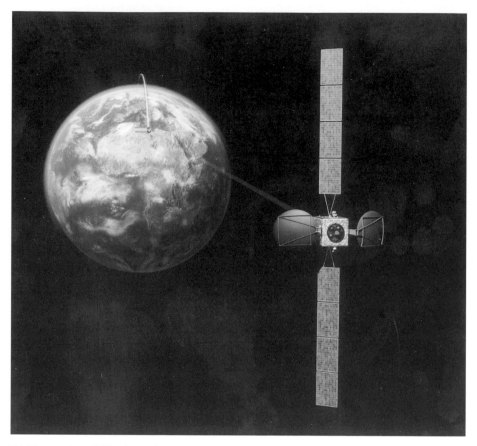

T4: **Three-axis stabilised** satellite in **geostationary orbit** showing **deployable antennas, solar arrays** and an **inter-orbit link** with a satellite in **polar orbit**. [ESA]

launching scientific and weather satellites, etc. The Thor is no longer **operational** in these forms, but over a period of years the vehicle evolved into the three-stage Thor-Delta, which is now known simply as the **Delta**.
[See also **Jupiter**]

three-axis stabilised (spacecraft) A spacecraft which maintains its stability using an **attitude control** system comprising **sensors** and **actuators**, as opposed to spin-stabilisation [e.g. see figures T4, S10, M3]; sometimes called 'body stabilised'.
[See also **spin-stabilised, spacecraft axes**]

threshold A limit or starting point (e.g. the minimum strength of a **signal**, etc., which will produce a specified response).

throat The narrowest part of a rocket exhaust nozzle; the region where the exhaust gases make the transition from subsonic to supersonic flow. See **nozzle**.

[See also **Mach number**]

throttling The act of increasing or decreasing the flow of **propellant** to a **rocket engine** (typical usage: 'throttle up' to increase **thrust**; 'throttle down' or 'throttle back' to decrease).

[See also **thrust profile, dynamic pressure, rocket motor**]

thrust A force produced by any **propulsion system**: the product of the mass of gas ejected per unit time and the **exhaust velocity** ($t = \dot{m}v_0$), where t is measured in newtons (N), $\dot{m}$ in kg s^{-1} and v_0 in m s^{-1}. For more precise determinations a pressure term may be added: $t = \dot{m}v_0 + A_e(P_e - P_a)$, where A_e is the area of the **nozzle** exit, P_e is the exit pressure and P_a is the ambient pressure.

[See also **lift, drag, reaction control thruster**]

thrust coefficient The figure of merit for a rocket **nozzle** design, given by the expression:

$$C_F = t/P_c A_t$$

where t is the **thrust** (N), P_c is the **combustion chamber** pressure (N m^{-2}) and A_t is the nozzle **throat** area (m^2).

[See also **characteristic velocity**]

thrust cone – see **thrust structure**

thrust cylinder – see **thrust structure**

thrust frame A structure which supports a **rocket engine** or **rocket motor** and, when it is fired, protects the rest of the **spacecraft** or **launch vehicle** by absorbing and dissipating the mechanical **loads** due to the **thrust**.

[See also **load path, thrust structure, diffusion member, propulsion bay**]

thrust profile A graphical representation of the **thrust** developed by a **rocket engine** or **rocket motor** throughout the period of **combustion**; a thrust versus time characteristic. As an illustration, a typical thrust profile for a **Space Shuttle main engine** is as follows: ignition at $T - 6.6$ s; thrust increases to 100% within 4.5 s (**lift-off** occurs at $T - 0$ when the **solid rocket boosters** are ignited); at $T + 25$ s the SSMEs are 'throttled down' to 84% for 11 s, then to 66% for 37 s; at about $T + 74$ s they are 'throttled up' to 104% and will operate at this level until about $T + 466$ s; thrust will then gradually be reduced to 65%, followed by engine cut-off at $T + 522$ s (8.7 minutes). The reduction in thrust shortly after launch corresponds with the period of **maximum dynamic**

pressure. The ability to 'throttle' the SSMEs also limits the vehicle's acceleration to 3g during ascent.

[See also **throttling, T-minus** . . .]

thrust structure

    (i) A **spacecraft** structure designed mainly to protect the spacecraft from the forces imparted by a **launch vehicle**. In many satellites the basis of the **service module** is a central 'thrust tube' or 'thrust cylinder', often of corrugated or honeycomb construction, which provides stiffness in a direction parallel to the launch vehicle's thrust axis. It often houses elements of the **propulsion system** (e.g. **propellant tank**s, **pressurant** tanks, **apogee kick motor**). The remainder of the satellite's **primary structure** is connected to the thrust structure [see **load path**]. In spin-stabilised satellites the 'thrust tube' may comprise a lower conical shell, an intermediate cylindrical shell and an upper cone.

        An alternative design for a **three-axis stabilised** spacecraft has no central thrust tube and relies on a rigid-box construction of panels. This type of structure is typically used for satellites with a **combined (bipropellant) propulsion system**, since there is no need for a central tube structure in which to house an apogee motor.

    (ii) As part of a **launch vehicle**, any structural component which distributes the thrust applied to the vehicle by the **rocket motor**(s) or **rocket engine**(s).

[See also **thrust frame, propulsion bay, diffusion member, airframe**]

thrust tube – see **thrust structure**

thrust vector

    (i) A mathematical quantity which defines the magnitude and direction of a **thrust**.

    (ii) The axis along which the thrust acts.

[See also **vector steering**]

thruster – see **reaction control thruster**

THz An abbreviation for terahertz – see **hertz, frequency bands**.

tiles (covering Space Shuttle) – see **thermal protection system**

time division multiple access (TDMA) A coding method for information transmission between a potentially large number of users, whereby all users utilise the same **frequency band** but transmit and receive at different times (measured in fractions of a second, and known as **burst**s). The satellite **transponder** can thus be accessed by many users who require a relatively wide **bandwidth**, one at a time but in extremely quick succession, which makes for efficient use of the available bandwidth. TDMA is used for digital **signals**.

[See also **frequency division multiple access**]

time division multiplexing (TDM) In **telecommunications**, the process in which several **signals** share the same **carrier** but at different times. Each signal is sampled in a pre-determined sequence according to a closely defined repeating time schedule. In such a way, a large number of data **channels** can modulate the same RF carrier.

[See also **modulation**, **frequency division multiplexing**, **radio frequency (RF)**]

Tiros The name given to a series of 19 American **meteorological satellites** launched in the 1960s; an acronym for television and infrared observation satellite. Tiros 1, launched on 1 April 1960, was the first dedicated weather satellite: its **payload** comprised a pair of half-inch **vidicon** TV cameras capable of producing **images** in both the visible and **infrared** parts of the spectrum, a capability which led to the satellite's name. The first ten Tiros satellites were **experimental** in nature; the remainder were part of an **operational** system designed to guarantee the provision of weather **data** on a day-to-day basis.

Titan An American **launch vehicle** developed in the late 1950s as a back-up to the Atlas intercontinental ballistic missile (**ICBM**). The two-stage Titan I used the **liquid propellants liquid oxygen** and **kerosene**. The two-stage Titan II, which launched the **Gemini** spacecraft, used the storable liquid propellants **nitrogen tetroxide** and **aerozine-50**. The three-stage Titan III had two **solid propellant** strap-on boosters; with the addition of the **Centaur** upper stage, Titan IIIs were used to launch the **Viking** spacecraft to Mars and the **Voyagers** to Jupiter and Saturn, for example. The Titan III became a commercially operated launch vehicle in the late 1980s, but ceased operation in the early 1990s. The Titan IV, in two **variants** (IVA and IVB), was developed chiefly as a military heavy-lift launcher.

[See also **Atlas**, **Delta**, **Scout**, **upper stage**, **strap-on**]

Titan-Centaur A Titan **launch vehicle** with a Centaur **upper stage**. See **Titan**.

titanium (Ti) A low-density metal, widely used (when alloyed with other metals) in the aerospace industry. Typical applications: spacecraft body-panel **face-skins**, **antenna** reflector face-skins, structural components.

[See also **materials**]

TLO An abbreviation for total loss only; a term used in **space insurance** to indicate that a **spacecraft** is insured only against the risk of a **total loss**, not a **partial loss**.

TM

   (i) An abbreviation for **telemetry**.

   (ii) An abbreviation for **thermal model**.

T-minus ... A verbal announcement preceding the **lift-off** of a **launch vehicle** indicating the time remaining prior to lift-off, where 'T' refers to the time of lift-off (e.g. 'T-minus 30 seconds and counting ...'; also written $T-30$ s). The time after lift-off is indicated by 'T-plus ...' etc.

[See also **countdown, countdown sequencer, 'hold', launch, launch window**]

TOMS An acronym for total ozone mapping spectrometer, a **payload** carried on some **remote sensing** satellites to monitor changes in the ozone content of the Earth's stratosphere.

[See also **spectrometer, mass spectrometer**]

torque coil A device which makes use of a planet's magnetic field to stabilise a **satellite**; also known as a magneto-torquer or magnetic torquer. It is based on the principle that a freely suspended, current-carrying coil aligns itself with the local magnetic field. Small disturbances in the satellite's **attitude** will be damped out as a battery-fed coil, mounted in the structure of the satellite, acts to realign itself with the planet's field. Since the effect varies with the strength of the field, the method is best suited to applications in low and medium, rather than high altitude orbits.

[See also **stabilisation**]

total loss A term used in **space insurance** to indicate the occurrence of an incident or an amount of degradation in the capability of a **spacecraft** sufficient to allow a payment of the insured amount as specified in the insurance policy. Typically, a total loss is declared if the spacecraft is destroyed or fails to separate from the **launch vehicle**, or fails to attain its designated **orbital position**, for example.

[See also **partial loss, constructive total loss, TLO, deductible**]

touchdown The moment of contact between a **spacecraft** and the surface of a **planetary body** (used equally for a planetary **probe** landing on Mars, for instance, or a manned spacecraft returning to Earth).

[See also **splashdown**]

TPS – see **thermal protection system**

tracking The determination of the **orbital parameters** of a **spacecraft**, or the **trajectory** of a **launch vehicle** from a ground installation; a function of a spacecraft's **telemetry, tracking and command** subsystem.

Using a spacecraft's telemetry **carrier**, ground stations can determine its direction (so-called 'angular tracking') and its distance. The determination of distance is also known as 'ranging' (e.g. as in a telemetry, command and ranging system, TC&R). The determination of range variations is termed 'range-rate measurement'. Some spacecraft also carry a radio frequency **beacon**, which can also be tracked. Launch vehicles are generally tracked using both **radar** and optical means [see **kinetheodolite**].

[See also **orbital debris, Space Surveillance Network (SSN)**]

traffic

(i) In a **communications system**, the volume of 'messages' (telephone calls, **data** transmissions, etc.) handled by the system in a given period.

(ii) A general term for the messages themselves.

[See also **capacity**]

trajectory The path described by a body moving in three dimensions under the influence of external forces. Typically used to describe the path of a **launch vehicle**, as a synonym for **flight path**. Also used for a spacecraft moving in a path which is not an **orbit**: orbits tend to be 'closed curves'; trajectories tend to be 'open'.

[See also **ballistic trajectory, sub-orbital trajectory, free-return trajectory, trans-lunar trajectory**]

transducer Any device, such as a microphone or an electric motor, that converts one form of energy into another.

transfer orbit An elliptical Earth orbit used to transfer a spacecraft to a higher altitude orbit (i.e. usually from **low Earth orbit (LEO)** to **geostationary orbit (GEO)**. See **geostationary transfer orbit (GTO)**.

[See also **Hohmann transfer orbit**]

Transit A **constellation** of six **navigation satellites** established in the early 1960s and operated by the US Navy, primarily to determine the positions of Polaris missile-carrying submarines. It was replaced, in the 1980s, by the **Global Positioning System (GPS)**, which improved accuracy by a factor of five.

[See also **Glonass, GNSS, SNAP**]

transit The passage of a **satellite** or **celestial body** across the face of a larger body. In astronomy, the apparent passage of a celestial body across the local meridian.

trans-lunar injection (TLI) The point at which a spacecraft bound for the Moon leaves its **parking orbit** on a **trans-lunar trajectory**.

trans-lunar trajectory A **trajectory** used to transfer spacecraft from Earth orbit to lunar orbit (used, for example, during the **Apollo** lunar programme). [See also **transfer orbit, free-return trajectory**]

transmission line A conducting entity designed to provide a controlled and protected propagation path for electrical signals and **carrier** waves, i.e. as opposed to propagation through **free space**. The two commonly used types are **waveguide** and **coaxial cable** [see figure T5]; others include 'microstrip' and 'stripline'.

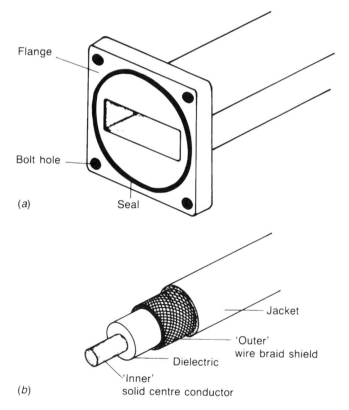

T5: Two types of **transmission line**: (a) **waveguide** and (b) **coaxial cable**.

transmit chain A collection of electronic and microwave equipment which amplifies, filters and outputs a signal to a transmit-antenna. In a spacecraft **communications payload** the transmit chain generally comprises an **IF**

amplifier, upconverter, TWTA or SSPA, output filter or multiplexer, and associated RF switches and hybrids. Whether or not to include the transmit-antenna in the definition of the transmit chain is largely one of personal choice.

[See also receive chain, amplifier chain, input section, output section]

transmitter A device which generates and amplifies a radio frequency (RF) carrier, modulates the carrier with information and feeds it to an antenna for transmission.

[See also modulation, transponder, transmit chain]

transparent repeater A repeater in which all operations, such as filtering and frequency conversion, are performed on the carrier as opposed to the baseband signal, the type and characteristics of the signal being of little consequence. It is sometimes called a transparent transponder, and is also known as a 'bent-pipe repeater' (following the analogy of an angled pipe simply channelling an input from one direction and outputting it in another). No satellite switching or on-board processing is performed.

[See also regenerative repeater]

transponder Generally, a type of radio or radar transmitter-receiver that transmits signals automatically when it receives predetermined signals (derived from 'transmitter + responder'); on board a satellite, a chain of electronic communications equipment which receives, filters, amplifies and transmits a signal, part of the communications payload [see figure C5]. The term 'transponder' is often used, in error, synonymously with travelling wave tube (TWT) or travelling wave tube amplifier (TWTA), but a transponder comprises several devices, typically including a number of separate amplifiers, and as a result of redundancy measures, there is usually more than one TWTA in a transponder.

If a spacecraft's communications payload comprises a single transponder chain, the term 'transponder' includes all the devices of that payload with the exception of the receive and transmit antennas. Usually, however, there are many transponder chains in a payload which, when grouped together (without the antennas), are known as a repeater.

From the point of view of a service provider (e.g. Intelsat, Eutelsat, etc.) or satellite user a 'transponder' represents a certain amount of frequency bandwidth available for telecommunications services.

[See also half-transponder, upconverter, saturated output power, back off (from saturation)]

transponder back off – see **back off (from saturation)**

transponder chain – see **transponder**

trapped radiation Radiation trapped by the Earth's magnetic field. See **space radiation**.

travelling wave tube (TWT) A device for the high-power amplification of **radio frequency (RF)** signals, colloquially termed a 'tube' and manufactured in two main types (helix and coupled cavity). When matched with its power supply unit, or **electronic power conditioner (EPC)**, the combination is referred to as a **travelling wave tube amplifier (TWTA)** [see figure T6].

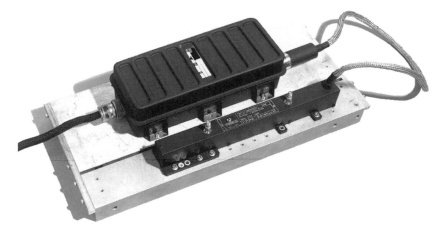

T6: **Travelling wave tube amplifier (TWTA)** – comprising **travelling wave tube (TWT)** and **electronic power conditioner (EPC)** – for **Earth observation** satellite. Note the TWT's **signal** input and output connectors and power cord, and compare with figure T7. [Thomson Tubes Electroniques]

The travelling wave tube owes its name to its mode of operation: it is designed to cause an **RF wave** to travel along its length in a carefully predetermined manner. The **energy** for amplification is derived from a high-powered **electron beam**, which is made to interact with the RF wave carried on a **slow wave structure**, usually in the form of a helix [see figure T7]. The interaction is such that **electrons** are, on average, decelerated by the electric fields of the RF wave and thereby lose energy to it, amplifying the **signal** carried by the RF wave. The remaining energy of the electron beam is dissipated as heat in a **collector**. An alternative design of slow wave structure is based on a series of coupled cavity sections (coupled by a slot in the wall of each cavity); the cavities in a **klystron** are similar but independent.

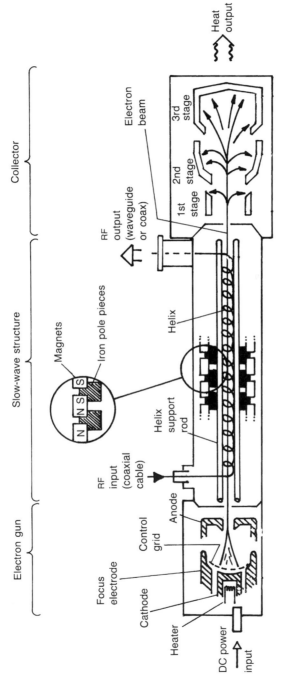

T7: Component parts of a **travelling wave tube** (TWT), from left to right: **electron gun**, **slow wave structure** and **collector**. Note collector of high-powered version depicted radiates directly into space.

The TWT is a **broadband** device, which can handle a signal **bandwidth** up to about 800 MHz at **Ku-band** and even wider bandwidths at higher **frequency bands**. Ground-based tubes deliver higher output **powers** than space-based tubes, typically up to 700 W at Ku-band and 3 kW at **C-band** for helix tubes and up to 10 kW for coupled-cavity tubes. Space-based devices, which tend to be of the helix type, can deliver up to about 300 W at Ku-band. Individual TWTs have been built with power-gains of more that 10,000,000 (70 dB).

Historical note: the TWT was invented in 1943 by Dr Rudolf Kompfner, an Austrian refugee working for the British Admiralty, and the first practical device was developed at Bell Telephone Labs in 1945 by J. R. Pierce and L. M. Field (a detailed theory of its operation (small signal theory) was published by John Pierce in 1947). TWTs have been widely used for space communications (in **spacecraft** and **earth station**s alike) since the first privately owned **communications satellite**, **Telstar** 1, was launched in 1962.

[See also **electron gun, repeater, transponder, gain**]

travelling wave tube amplifier (TWTA) An assembly of devices for the high-power amplification of **radio frequency (RF)** signals [see figure T6]. A TWTA comprises a **travelling wave tube** and an **electronic power conditioner** (i.e. TWTA = TWT + EPC). The acronym TWTA is sometimes pronounced 'tweeta'.

[See also **solid state power amplifier (SSPA), klystron**]

triaxiality An attribute of an ellipsoid with three dissimilar orthogonal axes (e.g. a **planetary body**); one of the mechanisms that results in the perturbation of a spacecraft's **orbit**.

The Earth, for example, is not perfectly spherical: its rotation has caused a bulging of the equator and a flattening of the poles; its equatorial radius varies such that a cross-section through the equator approximates to an ellipse; and the southern hemisphere is larger than the northern, placing the planet's **centre of mass** south of the equator. The Earth's oblateness (the polar radius is some 21 km less than the average equatorial radius) produces perturbations in a **polar orbit**, but has little effect on **equatorial orbit**s. The difference between the minimum and maximum equatorial radii is only about 70 m, but this is sufficient to perturb a satellite in equatorial orbit: the gravitational attraction varies around its circumference, which causes the satellite to accelerate and decelerate along the orbit direction. In fact, whereas the equatorial bulge has been known since Newton's day, the ellipticity of the equator was only detected through its effect on the orbital paths of the early satellites.

[See also **orthogonal axis, perturbations, orbital control, equilibrium point, luni-solar gravity, solar wind**]

tri-axis stabilised – see **three-axis stabilised**

triggered lightning Lightning triggered by a **launch vehicle** (or any **rocket**) flying through electrically charged clouds. This can damage sensitive electronic components carried by the vehicle or, at worst, lead to its destruction.

tropopause – see **atmosphere**

troposphere – see **atmosphere**

Tsukuba Space Center – see **NASDA**

Tsyklon A Russian **launch vehicle** developed in the 1970s and first flown in 1977; alternative spelling: Cyclone. It had a **payload** capability of about 4000 kg to **low Earth orbit (LEO)**.

[See also **Proton, Energia, Angara, Molniya**]

TT&C – see **telemetry, tracking and command**

TT&C antenna A spacecraft **antenna** designed specifically for use with a **telemetry, tracking and command** subsystem; an **omnidirectional antenna**.

tube A colloquial abbreviation for **travelling wave tube (TWT)**.

tumbling An end-over-end rotation of a body about its **centre of mass**. Some **launch vehicle** stages or components are given a tumbling motion prior to **re-entry** to discourage any aerodynamic tendencies the stage may have (which could cause it to fly a non-**ballistic trajectory**) and to make the **splashdown** point more predictable. For example, by firing a **pyrotechnic** valve in the nose-cap of the Space Shuttle **external tank**, and releasing **pressurant** gas from the **liquid oxygen** tank, the ET can be made to tumble at a rate of about 10 degrees/second.

tundra orbit An **elliptical orbit** with a 46,000 km **apogee** and a 25,000 km **perigee** (approx.) [see figure O1]. It is similar to the **Molniya orbit** in that it offers coverage of high latitudes, and similar to **geostationary orbit** in that it has a period of 24 hours, which gives it one apogee as opposed to the Molniya's two. The tundra orbit's high perigee means that it avoids both **atmospheric drag** and the **Van Allen belts**.

[See also **orbit**]

turbopump A device in a liquid **propulsion system** (i.e. a **rocket engine**) which raises the pressure of the **propellant** taken from the tanks and delivers it to the engine(s) at the required pressure and flow-rate. There are typically two pumps per engine, one for the **fuel** and one for the **oxidiser** [see figure R6]. They are each driven by a turbine (hence 'turbopump') which is in turn driven by a jet of gas from a '**gas generator**'.

turnkey Industrial terminology for a contract, project or **system** for which a single contractor has complete responsibility (typically from the time of contract signature to the time operational **hardware** and/or **software** is handed over to the client). A turnkey contract for a satellite **communications system**, for example, might include a **satellite** (delivered in **orbit**), an **earth station** facility (including **telemetry, tracking and command**) and the appropriate software to operate and maintain the system.
[See also **delivery in orbit**]

TV distribution The transfer of **television** programming, via **satellite**, from an originating source (e.g. a TV company) to a point of distribution into the public network, or private cable network. Although such transmissions have been intercepted by unauthorised users, they are not intended to be received directly by the public, as they are with DBS (**direct broadcasting by satellite**) or **DTH** (direct-to-home) transmissions.

TVRO An abbreviation for television receive only. Generally used with reference to the small **antenna** installations for **direct broadcasting by satellite (DBS)**.

tweeta Engineering slang for a TWTA (**travelling wave tube amplifier**).

twit Engineering slang for a TWT (**travelling wave tube**).

TWT – see **travelling wave tube**

TWTA – see **travelling wave tube amplifier**

Tyuratam The location of the **Baikonur Cosmodrome**, in Kazakhstan.

U

UCD – see **urine collection device**

UDMH An abbreviation for unsymmetrical dimethylhydrazine, $(CH_3)_2N.NH_2$, a storable liquid **fuel** used in **rocket engines**. See **liquid propellant**.
[See also **UH25**]

UH25 A storable liquid **fuel** used in **rocket engines**; a mixture comprising 75% **UDMH** and 25% **hydrazine** hydrate. See **liquid propellant**.

UHF (ultra high frequency) – see **frequency bands**

ullage

(i) The volume by which a liquid container (e.g. a **propellant** tank) falls short of being full. Also called 'ullage space'.

(ii) The quantity of liquid lost from a container by leakage or evaporation.
[See also **ullage rocket, ullage vapour**]

ullage pressure – see **ullage vapour**

ullage rocket A rocket **thruster** on a **liquid propellant** launch vehicle **stage**, which is fired before **ignition** of the stage's engines to position the **propellant** over the outlet valve(s). Between the **boost phases** of successive stages the launch vehicle is **coasting** and, since there is no applied **thrust**, the propellant is 'weightless' and can float freely within its tank. Ullage rockets return the propellant to the lower end of the tank in preparation for the next boost phase.
[See also **ullage vapour**]

ullage space – see **ullage**

ullage vapour The evaporated **liquid propellant** or any other **pressurant** which occupies the 'ullage space' in a **propellant** tank. The pressure of the vapour is called the 'ullage pressure'. See **ullage**.
[See also **ullage rocket, vent and relief valve**]

ultraviolet – see **waveband**

umbilical A line carrying electrical or fluid services from a service tower or other structure to a **launch vehicle**, or between a **spacecraft** and a **spacesuit**.
[See also **service structure, tail service mast, tether**]

umbilical tower – see **service structure**

UNCOPUOS An acronym for United Nations Committee for the Peaceful Uses of Outer Space.

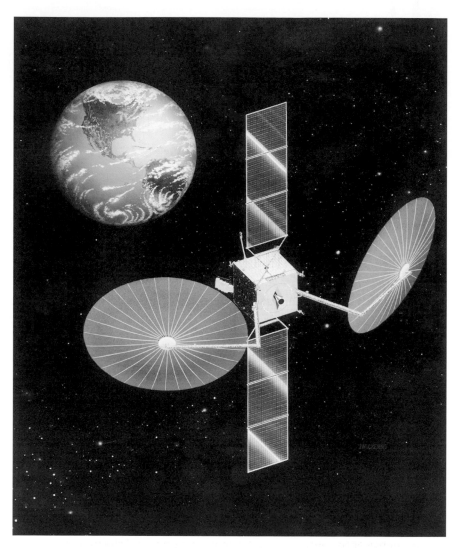

U1: **Unfurlable antenna**s and **solar arrays** deployed from a **three-axis stabilised** HS601 satellite. [Hughes]

underexpanded – see **expansion ratio**

unfurlable antenna A spacecraft **antenna** which unfolds (often in the manner of an umbrella) from a position of storage once the **vehicle** reaches its planned **orbit** or **trajectory** [see figures U1, A4, D4]. Using this method, an extremely large-diameter antenna reflector can be stored in a small container. An unfurlable antenna may also have to be deployed from the spacecraft body – see **deployable antenna**.

unicasting The transmission of **data** to a single recipient. A term, used mainly in connection with **Internet** communications, which has arisen as a logical extension of the terms **broadcasting**, **multicasting** and **narrowcasting**. The equivalent in **satellite communications** is 'point-to-point' [see **link**].

unified propulsion system – see **combined (bipropellant) propulsion system**

Unity A connection node of the **International Space Station (ISS)**, otherwise known as 'Node 1'.

unmanned spacecraft Any **spacecraft** which operates without a **crew** (e.g. a **satellite** or space **probe**).

[See also **manned spacecraft, man-tended spacecraft, astronomical satellite, communications satellite, Earth resources satellite, meteorological satellite, reconnaissance satellite**]

unregulated power supply A spacecraft **power supply** which provides a variable voltage, dependent upon the power source and electrical load. An unregulated system feeds the output of the **solar array**, **battery**, etc., direct to the spacecraft's **power-bus**, whatever the voltage. Payload and other subsystem equipment must therefore be designed to operate across a range of voltages.

[See also **regulated power supply**]

unsymmetrical dimethylhydrazine – see **UDMH**

untethered EVA – see **tether**

upconverter A device for increasing the **frequency** of a **signal**. It contains a circuit called a **mixer**, which 'mixes' the incoming signal with a **local oscillator** frequency. This process, known as the heterodyne process or 'heterodyning', produces frequencies corresponding to the sum and the difference of the two original frequencies. The output of the upconverter is the sum signal.

A 'dual-conversion' **transponder** contains a **downconverter** and an upconverter separated by a number of filter and amplifier stages. The frequency of the output from the downconverter and the input to the upconverter is known as the **intermediate frequency (IF)**. A 'single conversion' **transponder** contains a downconverter but no upconverter, since the conversion is directly from the **uplink** frequency to the **downlink** frequency.

uplink The communications path or **link** between an **earth station** and a **spacecraft**, 'from **Earth** to **space**'; also called 'feeder link'. Opposite: **downlink**.

upload The process of 'uplinking' data or **commands** to a **spacecraft** from an **earth station** or earth terminal. Opposite (for **telemetry**): **download**.

upper atmosphere The upper regions of a planetary atmosphere; for **Earth**, those regions of the **atmosphere** above the troposphere.

upper stage

(i) The uppermost stage in a **multistage rocket**. Some named stages (e.g. **Centaur**) are used, in different forms, on more than one **launch vehicle** (e.g. **Atlas**-Centaur, **Titan**-Centaur) [see figure U2].

U2: Centaur **upper stages** for **Titan** (foreground) and **Atlas** (background). [Lockheed Martin]

(ii) A rocket **stage** used to launch a spacecraft from the Space Shuttle (e.g. **payload assist module (PAM)**, **inertial upper stage (IUS)**). NB The PAM has also been used as the uppermost stage of the **Delta** launch vehicle.

uprate To improve the **performance** of a **launch vehicle**, rocket **stage**, engine or motor. A launch vehicle can be 'uprated' by the addition of an extra stage or **strap-on** boosters, or by 'stretching' existing propellant tanks. The 'rating' of a launch vehicle is based on its **thrust**, the **specific impulse** of its **propellants** and, ultimately, on its **payload** capacity.

[See also **man-rated**]

UPS An abbreviation for:

(i) unified propulsion system – see **combined (bipropellant) propulsion system**.

(ii) uninterruptable power supply, a back-up to a primary **power supply** that is activated immediately the primary fails (mainly ground-based).

urine collection device (UCD) A bag, and its associated tubes and connectors, attached to the inside of a **spacesuit** for the hygienic collection of urine, particularly during EVA (**extra-vehicular activity**).

US Space Command – see **Space Command**

V

V-2 The German World War II ballistic missile, A-4, which was renamed V-2 ('Vergeltungswaffe Zwei' or 'vengeance weapon two') for propaganda purposes. Its **propellants** were a mixture of ethyl alcohol and water (**fuel**) and **liquid oxygen** (**oxidiser**). It was developed at the Peenemunde Army Experimental Station under Wernher von Braun, who surrendered to the Allies at the end of the war and became one of America's leading rocket engineers. Examples of the V-2 were taken to the USA, tested and adapted to form the beginnings of America's missile – and eventually space **launch vehicle** – industry, which produced the **Atlas** and the **Saturn V** to name but two (Von Braun was instrumental in the development of the Saturn V).

VAB – see **vehicle assembly building**

vacuum A region completely devoid of matter; the theoretical concept of 'nothingness'. In practice, only an approximation to a vacuum is possible, although '**outer space**' represents a very good approximation for all practical purposes.
[See also **vacuum chamber**]

vacuum chamber A type of test chamber for **spacecraft** that simulates the **vacuum** properties of the **space environment**.
[See also **environmental chamber**]

VAFB – see **Vandenberg Air Force Base (VAFB)**

valve

   (i) Any device which controls, regulates or restricts the flow of a fluid (e.g. a **propellant** valve in a **rocket engine**).
   [See also **pyrotechnic valve**]

   (ii) An electrical device in which a flow of **electrons** between electrodes takes place (see **electron tube**).

Van Allen belts Two regions of charged particles in the Earth's **magnetosphere** named after the American physicist James Alfred Van Allen. He was instrumental in the provision of the Geiger-counter payload on the first American **Explorer** satellite, launched in 1958, which detected the belts. Launched later that year, **Pioneer** 3 recorded peaks in the particle flux at

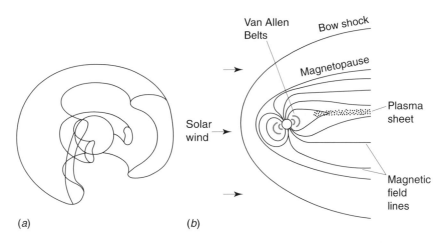

V1: **Van Allen belts**: (a) inner and outer belts compared with the Earth; (b) structure of the Earth's **magnetosphere**.

altitudes of 3218 and 16,090 km. The inner belt has since been determined to extend from about 2400 to 5600 km and the outer one from 13,000 to 19,000 km in the equatorial plane, where the belts are widest. They surround the Earth and tail off towards the poles as they follow the magnetic lines of force that constitute the Earth's magnetic field [see figure V1]. The inner belt consists predominantly of protons (hydrogen nuclei) with energies greater than 30 MeV, being at its densest within about 35 degrees of the geomagnetic equator. The outer belt comprises mainly high-speed **electron**s (with energies from 50 keV to 5 MeV) and low-energy protons of less than 1 MeV, and spreads between about 55 and 70 degrees north and south of the geomagnetic equator.

The belts present a potential hazard to both manned and unmanned space missions due to their degrading effect on materials and the effects of radiation on the human body. In practice, however, most spacecraft orbits and trajectories avoid these regions: **low Earth orbits** tend to be below the inner belt, at altitudes of less than 1000 km; **medium Earth orbits** tend to lie between the two belts; and **geostationary orbit**, and other **high Earth orbit**s, are above the outer belt. It was decided that the **Apollo** astronauts, en route to the Moon, travelled at a velocity sufficient to reduce the **radiation** exposure time to safe levels.

[See also **space radiation, ionisation, ionosphere, solar flare, South Atlantic anomaly**]

Vandenberg Air Force Base (VAFB) An American **ICBM** and space launch facility north of Los Angeles, California (at approximately 34.6° N, 120.6° W), part of which is also referred to as the Western Test Range (WTR) (prior to May 1964 called the Pacific Missile Range). VAFB is particularly suited for launches into **polar orbit**, since the nearest land mass due south is Antarctica. The first space-related launch from the WTR was that of Discoverer 1 on 28 February 1959. The Space Shuttle Launch Complex (SLC-6), known colloquially as 'slick-6', was 'mothballed' in 1986 following the **Challenger** accident and the **launch pad** was later converted to launch the **Athena** vehicle.

[See also **Kennedy Space Center (KSC), Wallops Flight Facility**]

Vanguard The name given to a series of American **unmanned spacecraft** launched in the late 1950s. Vanguard 1, America's second **satellite**, was launched on 17 March 1958 by a **launch vehicle** of the same name which was based on the **Viking** sounding rocket. The 1.5 kg sphere was nicknamed 'grapefruit' by Soviet Premier Khrushchev, but studies of its orbital **perturbations** produced an accurate figure for the flattening of the **Earth**'s poles and showed that the southern hemisphere was larger than the north. It was also significant in being the first satellite to derive its **power** from **solar cells**.

[See also **Explorer, Sputnik**]

variable conductance heat pipe (VCHP) – see **heat pipe**

variant Something which differs from a standard type, design or **configuration**. Typically used in connection with a **launch vehicle**, which may be available in a number of variants (with or without **strap-on** boosters, for example).

[See also **core vehicle, upper stage, solid rocket booster**, Ariane]

vector

(i) In mathematics, a variable quantity that has magnitude and direction, and can be resolved into components.

(ii) More loosely, the axis along which a force acts (e.g. **thrust vector**).

(iii) To direct a force or **thrust** (e.g. to steer a **launch vehicle**). Usage: 'to vector a **thruster**'; 'vectored thrust', etc.

vector steering A **launch vehicle** steering method in which one or more **rocket engine**s or **nozzle**s are **gimbal**-mounted to enable the **thrust vector** to be rotated with respect to the vehicle to produce a turning moment.

[See also **vernier engine**]

vehicle In space technology, any conveyance of any type of **payload** (e.g. a **spacecraft, launch vehicle**, etc.). As an alternative for 'spacecraft', the term 'vehicle' is used predominantly for **manned spacecraft**, particularly with regard to **extra-vehicular activity (EVA), dock**ing, etc.

[See also **lunar roving vehicle (LRV), manned manoeuvring unit (MMU), expendable launch vehicle (ELV), heavy lift vehicle (HLV), orbital transfer vehicle (OTV), automated transfer vehicle, vehicle assembly building (VAB)**]

vehicle assembly building (VAB) The building at **Kennedy Space Center** in which the Space Shuttle is prepared for flight [see figure V2]. It is divided into a 'low bay' and a 'high bay', the latter of which is further subdivided into four bays. The low bay is used for assembling the parts of the **solid rocket booster**s (SRBs) which do not contain **propellant**; high bays 2 and 4 are for the final assembly of the SRBs and preparation of the **external tank** (ET); and high bays 1 and 3 are for the **integration** of the **orbiter** (transferred from the **orbiter processing facility**) with the ET and SRBs on the **mobile launch platform**, upon which the Shuttle combination is transported to the **launch pad**.

The VAB was built in the 1960s for the **Apollo** programme and was originally designed to hold four **Saturn V** launch vehicles, one in each of the four high bays. At 158 m wide, 218 m long and 160 m tall, it remains one of the most capacious buildings in the world.

[See also **Crawler**]

V2: **Vehicle assembly building (VAB)** at **Kennedy Space Center** with **launch control centre** in front. [Mark Williamson]

413

vehicle equipment bay – see **instrument unit**

velocity of light – see **light (velocity of)**

Venera The name given to a series of Soviet planetary exploration **probes** launched towards the planet Venus: Venera 7 made the first radio transmissions from the surface of another planet on 15 December 1970.

vent and relief valve A device in a **liquid propellant** system, usually as part of a **launch vehicle**, which combines the functions of the vent valve and the relief valve; also called a 'boil-off valve'. A vent valve relieves the excess pressure in a propellant system when it receives an external command; a relief valve opens automatically when the pressure reaches a predetermined value. The vent function is only available before **launch** (e.g. to vent **ullage vapour** during tank-filling); the relief function is available both before and after launch.
[See also **beanie cap**, **ullage**]

vent valve – see **vent and relief valve**

vernier engine A small **rocket engine** used to make fine adjustments to the **thrust** (magnitude and direction) of a **launch vehicle**, allowing adjustments in velocity and **trajectory** to be made. Mounted at an angle to the main thrust axis, the vernier provides an alternative to gimballing the main engines.
[See also **gimbal**, **vector steering**]

vertical polarisation – see **polarisation**

VfR An abbreviation for Verein fur Raumschiffahrt, a **spaceflight** society established in Germany in 1927 by a group of enthusiasts including rocket-car builder Max Valier and author Willy Ley. It published a journal, '*Die Rackete*', and in the early 1930s (by which time **Wernher von Braun** was a member) began experiments with **liquid propellant** rockets. In 1933, its personnel and records were taken over by the German army and the VfR was disbanded.
[See also **ARS**, **BIS**]

VHF (very high frequency) – see **frequency bands**

vibration facility A ground-based test facility which simulates the vibrations transmitted to a **spacecraft** from a **launch vehicle**. The amplitude and frequency-range of the vibrations differ between launch vehicles: the spacecraft mechanical design must avoid resonant frequencies which could damage it.
[See also **vibration table**, **acoustic test chamber**, **thermal-vacuum chamber**, **anechoic chamber**]

vibration table A test facility which subjects components or assemblies to vibrations similar to those that will be encountered during a launch. The vibration environment differs between **launch vehicle**s: data derived from actual launches can be used to test the spacecraft. Sometimes called a 'shake-table' or 'shaker-table'.

videoconference A meeting between two physically separated parties whereby audio and video signals are transmitted between them to simulate the presence of the other party. The better systems use full-**bandwidth** video to produce standard **television** pictures with accompanying hi-fi audio. Inferior systems utilise reduced-bandwidth **slow-scan TV** and standard telephone-quality lines. There are also systems of intermediate quality.

[See also **teleconference**, **televideo**]

videotext A system for providing a written or graphical representation of computerised information on a television screen. When transmitted or broadcast to the consumer it is referred to as **teletext**.

vidicon A type of television camera tube incorporating a photoconductive surface, on which an electric **charge** pattern forms in response to incident light. The camera's optical system projects an **image** onto the tube, forming a pattern of charge which corresponds to the image. The image is converted into a TV **signal** by scanning the pattern with an **electron beam**. The generic term for such a device is an iconoscope, vidicon being a contraction of 'video iconoscope'.

[See also **charge-coupled device (CCD)**]

Viking

(i) A series of two American planetary exploration **probes**, each comprising an 'orbiter' and a 'lander', launched towards Mars on 20 August and 9 September 1975.

(ii) The name given to a Swedish **spacecraft**, designed to measure space **plasma** phenomena in the Earth's **magnetosphere**, launched in February 1986.

(iii) The name of an engine used in the **Ariane** launch vehicle.

(iv) An American **sounding rocket** developed in the late 1940s which formed the basis of the **Vanguard** launch vehicle.

visible spectrum – see **electromagnetic spectrum**

VLBI An abbreviation for very long baseline interferometry. For certain astronomical observations, particularly in radio astronomy, the baseline for the observations can be extended beyond the Earth by using a **spacecraft** in **orbit**.

VLF (very low frequency) – see **frequency bands**

VLS A Brazilian small satellite launch vehicle (Veiculo Lancador de Satellites), launched from the **Alcantara** launch site. Its first flight, which carried the Brazilian SCD-2A data-messaging satellite on 2 November 1997, was a failure.

voltage – see **bus voltage**

Vomit Comet A nickname given to the aircraft used by **NASA** for zero-g simulations, a modified version of the Boeing 707 designated KC-135. 'Comet' is a reference to the DeHavilland Comet, the world's first jet airliner. See **weightlessness**.

Voskhod One of two Soviet **manned spacecraft** launched in the mid-1960s as a follow-on to the **Vostok** programme. Since the Voskhod was little more than a 'stripped-down' Vostok, with no room for either **spacesuit**s or ejection seats but with meagre accommodation for three **cosmonaut**s, it is considered to be a spacecraft designed for political purposes: Voskhod 1, launched on 12 October 1964, carried a three-man **crew** on a 24-hour **mission** before even a two-man spacecraft had flown; and Voskhod 2, launched on 18 March 1965, pre-empted the first launch of America's two-man **Gemini** by only five days. Voskhod 2 carried a crew of two, leaving room for a flexible telescopic **airlock** which enabled Alexei Leonov to make the first 'spacewalk' [see **extra-vehicular activity (EVA)**].

[See also **Soyuz**]

Vostok The name given to the world's first **manned spacecraft** and its **launch vehicle**.

The Soviet Vostok **spacecraft**, a series of six one-man **capsule**s, were launched into **low Earth orbit** in the early 1960s. Vostok 1, for example, carried the first man in space, Yuri Gagarin, into orbit on 12 April 1961 (**flight** duration: 1 h 48 m) and Vostok 6 carried the first woman in space, Valentina Tereshkova, on 16 June 1963 (flight duration: 70 h 50 m). The Vostok spacecraft comprised two main sections: a spherical **re-entry** capsule containing a rocket ejection seat; and an instrument compartment which contained service and **propulsion** subsystems (including **retro-rocket**s).

The Vostok launch vehicle was based on the R-7 booster, which launched **Sputnik** 1 and was later developed as the basis of the **Soyuz** and **Molniya** launch vehicles. The Vostok was operational from 1957 until the late 1980s.

[See also **Voskhod, Soyuz**]

Voyager A series of two American planetary exploration **probe**s launched towards the outer planets of the solar system – Jupiter, Saturn, Uranus and

Neptune – on 20 August 1977 (Voyager 2) and 5 September 1977 (Voyager 1) [see figures G3, R2]. On 17 February, 1998, Voyager 1 overtook Pioneer 10 to become the most distant of mankind's creations: it was 70 astronomical units (AU) – some 10.4 billion kilometres – from Earth. In the so-called 'Voyager Extended Mission', it is hoped that the **deep space network (DSN)** will be able to use the spacecraft to detect the heliopause, the point where the Sun's electromagnetic influence gives way to interstellar space [see **heliosphere**].

VSAT An acronym for very small aperture terminal, an **earth station** with an **antenna** diameter between that of a **gateway station** and a terminal for **direct broadcasting by satellite (DBS)**; a typical VSAT antenna is approximately 3 m in diameter, although some may be as small as 60 cm or as large as 7 m, depending on the **application**. VSATs are used by private industry, educational and institutional users for a wide variety of voice and **data** applications, including **Internet** communications, and operate as either one-way (receive-only) or two-way installations. A typical VSAT network comprises a central 'hub' station and a number of remote terminals linked independently to the hub; it is known as a 'star network'. If the terminals are interlinked with all other terminals, there is no hub and it is termed a 'mesh network'.

Vulcain The name given to an engine used in the core stage of the **Ariane** 5 launch vehicle [see figure R6].

W

Wallops Flight Facility A **NASA** facility on Wallops Island, Virginia (at approximately 37.9° N, 75.4° W) for the launch of **sounding rockets**, scientific balloons, etc. Its first space-related launch was that of **Explorer** 9 on 16 February 1961.

[See also **Kennedy Space Center (KSC)**, **Vandenberg Air Force Base (VAFB)**]

WARC An acronym for World Administrative Radio Conference, a conference convened under the auspices of the International Telecommunication Union (ITU) to plan **communications** services (particularly with regard to the allocation of radio frequencies). Following the reorganisation of the ITU in 1992, such meetings were known as World Radiocommunication Conferences (**WRC**).

The first world space radiocommunication conference was held in Geneva in 1963, and the first allocation of **frequency bands** to the **broadcasting-satellite service (BSS)** was made at the WARC for space telecommunications (WARC-ST) in 1971. A plan for the BSS in 'Region 1' and 'Region 3' [see **ITU**] was devised at WARC 1977 and ratified at WARC 1979, a plenary session which considered the whole of the **radio spectrum**; 'Region 2' was covered at the Regional Administrative Radio Conference, RARC 1983. WARC 1985 (ORB '85) and WARC 1988 discussed, among other things, the use of **geostationary orbit**.

WASS An acronym for wide area augmentation system; a US network of regional ground-based reference stations intended to augment the service provided by the **Global Positioning System (GPS)**.

[See also **GNSS**]

waste management system – see **environmental control and life support system (ECLSS)**

wave number The reciprocal of **wavelength**. The number of waves per unit distance in the direction of propagation.

waveband A continuous range of wavelengths of **electromagnetic radiation** between two limits; an alternative description of this range is embodied in the term **frequency band**, which is based on **frequency** as opposed to **wavelength**. By convention, satellite **payloads** for **Earth observation** tend to be specified in terms of wavelength, while **satellite communications** frequencies are specified in terms of frequency [see **frequency bands**].

The **electromagnetic spectrum** is commonly divided into sections labelled radio, **microwave**, infrared (IR), visible, ultraviolet (UV), x-rays and gamma-rays, which cover the following wavebands:

- Radio $\qquad\qquad$ $3 \times 10^5\,\text{m}$–$0.3\,\text{m}$
- Microwave $\qquad$ $0.3\,\text{m}$–$3 \times 10^{-3}\,\text{m}$
- Infrared (IR) $\qquad$ $3 \times 10^{-3}\,\text{m}$–$7.6 \times 10^{-7}\,\text{m}$
- Visible $\qquad\quad$ $7.6 \times 10^{-7}\,\text{m}$–$3.9 \times 10^{-7}\,\text{m}$ (red–violet)
- Ultraviolet (UV) $\quad$ $3.9 \times 10^{-7}\,\text{m}$–$3 \times 10^{-9}\,\text{m}$
- X-rays $\qquad\quad$ $3 \times 10^{-9}\,\text{m}$–$3 \times 10^{-11}\,\text{m}$
- Gamma-rays $\quad$ $3 \times 10^{-11}\,\text{m}$–$3 \times 10^{-15}\,\text{m}$ ...

Sections are often subdivided and, for convenience, quoted in derived units: e.g. near IR (0.7–1.4 μm), middle IR (1.4–5.5 μm) and far IR (5.5–3000 μm); in addition, the lower part of the far IR band (10–15 μm approx.) is often termed 'thermal IR'. NB All divisions and ranges are approximate.

waveguide A metal or metal-coated tube, usually of rectangular cross-section, which confines and conveys (or 'guides') electromagnetic waves [see figures T5, W1]. In simple terms the wave can be thought of as being internally reflected along the tube in a similar way to light in an **optical fibre**; in reality the wave is propagated in one or more 'waveguide modes', described by electric and magnetic fields which induce currents in the surface of the guide. The size of the guide is dependent on the **frequency** of the radio waves, such that higher frequencies require smaller waveguides and tighter tolerances on other waveguide components. To reduce mass, metal-coated **carbon-fibre reinforced plastic (CFRP)** waveguides have been developed for certain applications.
[See also **waveguide switch**, **feed horn**, **beam waveguide**, **electromagnetic spectrum**]

waveguide switch A device designed to isolate part of a circuit and divert a signal carried in **waveguide** (one of two types of commonly used electrical **transmission line**). Devices are either switched magnetically using solenoids, causing a mechanical deviation of the waveguide run, or electronically using diodes [see figure W1].
[See also **coaxial cable**, **coaxial switch**]

wavelength The distance, measured in the direction of propagation, between two points of the same phase in consecutive cycles of an electromagnetic wave.

W1: **Waveguide switch** matrix for Telecom 2 **communications payload** comprising 15 switches (cylindrical items), showing **waveguide** runs on front face and coaxial connectors on top. [Teldix]

Wavelength is measured in metres (m) or multiples and sub-multiples of metres. Common sub-multiples used in technical circles (apart from cm and mm) are the micron ($1\mu = 10^{-6}$ m), the nanometre (1nm $= 10^{-9}$ m) and the angstrom (1 Å $= 10^{-10}$ m). The reciprocal of wavelength is wave number. [See also **frequency**, **frequency bands**, **waveband**, **electromagnetic spectrum**]

web The thickness of **solid propellant** in a **rocket motor** from the initial ignition surface to the insulated wall of the **motor case**; the total length of an end-burning charge [see **propellant grain**]. NB Not to be confused with **shear web**. Also an abbreviation for World Wide Web (**WWW**).

weightlessness The state of having little or no weight experienced in an orbiting **spacecraft** or an aircraft flying on a parabolic arc; alternatively called '**zero gravity**'; sometimes expressed as 'free fall'.

Weightlessness can be experienced in **orbit** because both the spacecraft and its occupants are in free fall in the Earth's gravitational field: since both are falling at the same rate, the spacecraft offers no resistance to the falling motion of the occupant. The effect can be reproduced in an aircraft flying a

ballistic trajectory: the aircraft is put into a dive to increase its velocity then 'pulled up', and 'pushed over' a parabolic arc. Using this technique in a light aircraft gives just a few seconds of 'zero gravity', but this can be increased to 15–45 s in a subsonic jet, about a minute in a supersonic jet and several minutes in a rocket-powered aircraft (e.g. the **X-15**). **NASA**, for example, generally conducts its zero-g simulations using a KC-135 aircraft, a modified version of the Boeing 707, which provides about 30 seconds of weightlessness per parabola (the aircraft has been nicknamed 'the **Vomit Comet**'!). **ESA** uses an Airbus aircraft for its zero-g simulations. Alternatively, several seconds of zero-g in a vacuum can be experienced by objects released into an evacuated **drop tower**.

[See also **gravity, low gravity, microgravity**]

Western Test Range – see **Vandenberg Air Force Base (VAFB)**

Whipple shield A shield mounted on a **spacecraft** to protect it from impacts, either random impacts from **space debris** or from particles produced by a comet (named after Fred Whipple, the American astronomer who hypothesised that comets were 'dirty snowballs'). Such a shield is known more generally as a 'bumper shield' or, if it is the outer of two or more shields, as a 'sacrificial shield' (since its integrity is sacrificed in atomising incident particles which then impact a second shield with significantly decreased energy). Typically made using light-weight ceramic cloth (e.g. **Kevlar composite**) in a similar way to the bullet-proof vest.

white room An air-conditioned **cleanroom** mounted on an **access arm** of a launch vehicle **service structure**, used to transfer a **crew** to a **manned spacecraft** [see figure W2].

White Sands Test Facility – see **NASA**

wideband A modifier indicating the wide **bandwidth** of the quantity under consideration (e.g. wideband communications, wideband **signal**, **wideband noise**, etc.). Essentially the same as **broadband**.

wideband noise – see **broadband noise**

wind scatterometer An **active** microwave **payload** designed to measure surface wind-speed and direction by detecting reflectivity perturbations of the sea surface; a type of **radar** carried by some **Earth observation** satellites. A typical instrument has a surface **resolution** measured in tens of metres and a **swath-width** of several hundred kilometres.

[See also **synthetic aperture radar (SAR)**]

windmill torque The rotation of a **three-axis stabilised** spacecraft about an axis

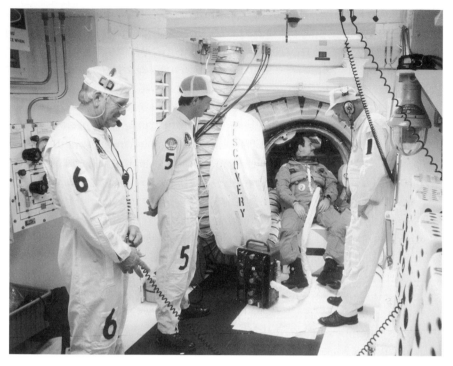

W2: **White room** attached to a **swing-arm** on the **service structure** of Space Shuttle **launch complex** 39B. Note **astronaut** leaning into access hatch of Discovery **orbiter** and **portable life support system (PLSS)** connected to **spacesuit**. [NASA]

in the **roll–yaw** plane [see **spacecraft axes**]. This can be caused by the undesirable impingement of a **thruster** plume on a **solar array** surface [see **plume impingement**], or by the intentional use of the **solar wind** in controlling the **attitude** of the spacecraft [see **solar sailing**].

window – see **launch window**

wind-shear A rapid change in wind direction with height, particularly important for a **launch vehicle** rising through the atmosphere since it places increased structural (dynamic) **load**s on the vehicle.

wire antenna – see **antenna**

wiring harness An **assembly** of electrical wiring bundles, otherwise known simply as a harness.

wireless Without wires (or **cable**); another word for **radio** (chiefly British and now rarely used).

wireless local loop A terrestrial **communications system** which links a customer's premises to a provider's national or international network using a **microwave link** as opposed to **cable**. The link is often referred to colloquially as the 'last mile' (chiefly US), acknowledging the difficulty in connecting customers in remote areas to a terrestrial network. **Direct broadcasting by satellite (DBS)** and **VSAT** systems do not suffer this disadvantage.

Woomera The location of an Australian **launch site** (at approximately 31° S, 136° E), which conducted its first orbital launch, of the Weapons Research Establishment Satellite WRESAT-1, on 29 November 1967. The launch of the first Skylark **sounding rocket**, for research into the upper atmosphere, was made on 13 February 1957. The Woomera facilities also provided a launch site for the UK's **Blue Streak** missile and Europe's Europa **launch vehicle**.

World Administrative Radio Conference – see WARC

WRC An abbreviation for World Radiocommunication Conference, a conference convened under the auspices of the International Telecommunication Union (**ITU**) to plan **communications** services (particularly with regard to the allocation of radio frequencies); formerly known as the World Administrative Radio Conference (**WARC**). The first WRC was held in Geneva in 1993.

WWW An abbreviation for World Wide Web, the collection of **data** stored on computers around the world and accessible via the **Internet**.

X

X-1 A rocket-propelled research aircraft conceived in 1943 and designed to reach and exceed Mach 1, the speed of sound, in horizontal flight: on 14 October 1947 Charles E. (Chuck) Yeager broke the sound barrier in the XS-1 (the full designation of the first flight model). Between them, the XS-1 and two later planes designated X-1, made 156 flights, setting a speed record of 1540 km h^{-1} (about Mach 1.45) and an altitude record of 21,916 m. The aircraft, which were dropped from the modified bomb-bay of a B-29 bomber for their flight tests, were designed and built by Bell Aircraft Corporation and sometimes known as the Bell X-1.

[See also **X-15, Mach number**]

X-15 A rocket-propelled research aircraft designed to investigate the effects of high speed at high altitude on future aerodynamic space vehicles (as opposed to ballistic **re-entry** vehicles) [see figure X1]. The main engine burnt the **propellant**s anhydrous ammonia and **liquid oxygen**, while **hydrogen peroxide** was used for the **attitude control** thrusters.

NASA's three X-15s were dropped from beneath the wing of a converted B-52 bomber in a series of 199 flights (the first and last powered flights were in September 1959 and October 1968, respectively). Designed to reach speeds of up to Mach 6 in horizontal flight, the X-15 set a record of 6605 km h^{-1} (about Mach 6.2), and an altitude record of 107,960 m. At one time there were plans for the X-15 to be flown into **orbit**, but this became the goal of the **Mercury** programme instead. The X-15 did, however, lead to winged **lifting body** research and eventually the **Space Shuttle**.

[See also **X-1, Mach number**]

X-33 A **NASA** advanced **technology demonstrator** designed to develop the technology for a **reusable launch vehicle (RLV)**. Its **propulsion system** is based on a pair of linear **aerospike engine**s, each supplying 20 combustion chambers burning **liquid hydrogen** and **liquid oxygen**.

X-band: 8–12.5 GHz – see **frequency bands**

xenon-ion A type of **ion engine** which uses xenon (Xe), a gaseous trace element in air, as a **propellant**.

[See also **XIPS**]

X1: **X-15** rocket-propelled research aircraft after landing, with B-52 carrier aircraft overhead. [NASA]

Xichang A Chinese **launch site** in the Sichuan Province of south-west China (at approximately 28° N, 102° E), formerly known in the West as Chengdu (or Ch'eng-tu) after a nearby town. It conducted its first launch, of SW-1, on 29 January 1984 and is now concerned, among other things, with commercial launches of the **Long March** vehicle.

[See also **Jiuquan**, **Taiyuan**]

XIPS An acronym for xenon ion propulsion system, a commercial gridded **ion engine** developed by US spacecraft manufacturer Hughes.

XPD An abbreviation for **cross-polar discrimination** – see **discrimination**.

X-ray spectrum – see **electromagnetic spectrum**

XS-1 – see **X-1**

Y

yaw A rotation about the yaw axis – see **spacecraft axes**.

yaw axis – see **spacecraft axes**

yo-yo A mass on a **tether** released from a spinning launch vehicle **stage** to increase its moment of inertia and thus reduce its rotation rate (usually prior to the release of its satellite-**payload**).

Z1: **Zvezda** service module for the **International Space Station (ISS)** under construction in Russia showing **docking module** on front. [NASA]

Z

Zarya A **module** of the **International Space Station (ISS)**, otherwise known as the **functional cargo block** (or FGB) [see figure F4]. 'Zarya' is Russian for 'daybreak'.

Zenit A Ukrainian **launch vehicle** developed in the 1980s and first launched in 1985. Its **payload capability** is about 15,000 kg to **low Earth orbit (LEO)** and 5900 kg to **geostationary transfer orbit (GTO)**. It was adapted in the 1990s to form the Zenit-3SL **variant** for use in the **Sea Launch** programme.

zenith The point on the celestial sphere directly above an observer and diametrically opposite the **nadir**.

zero gravity A commonly used alternative term for **weightlessness**, often abbreviated to 'zero-g'; a colloquial term referring to the conditions experienced inside a space vehicle in 'free fall' (when the force of gravitational attraction is apparently zero). The acceleration due to **gravity** in a freely orbiting **spacecraft** is zero for most practical purposes, but this is only an approximation since any force acting from outside or within the spacecraft will produce a gravitational acceleration – see **microgravity**. [See also **gravitational anomaly, perturbations**]

zero momentum system – see **reaction wheel**

zero pre-breathe suit – see **pre-breathe**

zone beam A **beam** formed by a **satellite** communications **antenna** which provides coverage of an area smaller than a hemisphere (see **hemispherical beam**) but larger than that covered by a **spot beam** (for example, a zone beam may provide coverage of the contiguous United States ('CONUS') or of Europe). [See also **beamwidth, global beam, multiple beam**]

zoom antenna A type of spacecraft **antenna** with the capability to vary its area of coverage, or **footprint** (by analogy with a 'zoom lens'). An antenna with a number of **feed** elements can be used to cover a relatively large area if all the elements are fed; feeding only the centre elements can produce a smaller **coverage area** with a higher **power flux density** – the 'zoom' effect.

Zvezda A **module** of the **International Space Station (ISS)**, otherwise known as the Service Module, designed to provide initial power, propulsion, life support and other functions [see figure Z1]. 'Zvezda' is Russian for 'star'.

Classified List of Dictionary Entries

1 Spacecraft Technology

AC power
accelerometer
access panel
acoustic test chamber
actuator
ADCS
aerial
aerobraking
aerocapture
AI
AIT
altimeter
amplifier
amplifier chain
anechoic chamber
antenna
antenna array
antenna efficiency
antenna farm
antenna feed
antenna module
antenna platform
antenna pointing mechanism (APM)
antenna radiation pattern
antenna taper
anti-Earth face
AOCS
aperture antenna
aperture blockage
aperture efficiency
aperture taper
APM
array
array antenna
array blanket
array degradation
array shunt regulator
artificial intelligence (AI)
ASIC
assembly, integration & test (AIT)

astromast
asymmetric feed
attenuator
attitude
attitude and orbital control system (AOCS)
attitude control
attitude determination and control system (ADCS)
autonomous control system
axes (of a spacecraft)
axial ratio
axisymmetric feed

baffle
balance mass
ballute
band-pass filter
BAPTA (bearing and power transfer assembly)
barbecue roll
bathtub curve
battery
battery reconditioning
beam-forming network
bearing and power transfer assembly (BAPTA)
beginning of life (BOL)
body stabilised
BOL
bolometer
bolt-cutter
boom
boresight
boresighting
box
bumper shield
burn-in
bus
bus voltage

C/A-code
cable-cutter
carrier generator
cavity
CCD
central processing unit (CPU)
centre-fed (antenna)
CEU
channel filter
charge-coupled device (CCD)
charge–discharge cycle
cleanroom
closure panel
CMG
coax
coaxial cable
coaxial switch
cold-soak
collector
command
communications module
communications package
communications payload
communications repeater
communications subsystem
computer
constant conductance heat pipe (CCHP)
constellation
control electronics unit (CEU)
control law
control moment gyro (CMG)
cooling system
corner-cube reflector
Cosmos
COSTAR
coupled cavity
coverglass
coverslip
CPU
cross-strapping
CRV
CTV
cupped dipole

DANDE
data bus

DC power
dead-band
deployable antenna
deployable battery
deployable radiator
depth of discharge (DOD)
despin active nutation damping
    electronics (DANDE)
de-spun platform
detector
dielectric lens
differential navigation
diffusion member
diplexer
dipole
directivity
dish
dock
docking module
docking port
docking probe
doubler
downconverter
drogue
drum-stabilised
dry mass
duty cycle

earth sensor
Earth-lock
Earth-pointing face
east–west station keeping
eclipse season
EGNOS
electron gun
electron tube
electronic power conditioner (EPC)
electronic steering
elliptical antenna
ELT
EMC
encounter
end of life (EOL)
energy density
engineering model
entry interface

entry-probe
environmental chamber
EOL
EPC
EPIRB
EPS
equilibrium point
expert system

f/D ratio
face-skin
failure rate
feed
feeder link
feed(er) losses
feedhorn
FET
field effect transistor (FET)
filter
finite-element modelling
fixed conductance heat pipe (FCHP)
flight hardware
flight model
FOV
free-drift strategy
free-flying pallet
frequency source
front end
fuel cell

GaAsFET
gaiter
gallium arsenide
Global Positioning System (GPS)
Glonass
GNC
GNSS
GPS
GPS rollover
gravity gradient stabilisation
ground spare
guidance (inertial)
gyro
gyrodyne
gyroscope
gyroscopic stiffness

hard dock
hardening (of spacecraft)
harness
heat exchanger
heat pipe
heat rejection
heat shield
heat sink
heat spreader-plate
heater
heat-soak
helix
heterodyning
HGA
high power amplifier (HPA)
high-pass filter
honeycomb panel
horizon scanner
horizon sensor
horn
horn antenna
housekeeping
HPA
hybrid
hyperspectral

iconoscope
IF amplifier
IFOV
image enhancement
image intensification
imaging
imaging radar
imaging team
Imux
inertia wheel
inertial guidance
inertial platform
inertial reference unit (IRU)
infrared sensor
in-orbit mass
in-orbit spare
input filter
input multiplexer
input section
insulation

interface filler
IOC
IOT
IRU

kinetheodolite
klystron
KPA

laser
laser ranging retroreflector
latch-up
launch mass
lens antenna
LEOP
libration period
lifting body
light baffle
line heater
line scanner
linearity
lithium-ion cell
LNA
LO
load path
local oscillator (LO)
louvre(s)
low noise amplifier (LNA)
low-pass filter

magnetometer
magneto-torquer
major axis
maser
mass budget
mass dummy
mass spectrometer
MEMS
microprocessor
microwave lens
microwave monolithic integrated
    circuit (MMIC)
microwave sounder
minor axis
minus-x face
minus-y face

minus-z face
mixer
MLI
MMIC
modem
momentum bias
momentum dumping
momentum off-loading
momentum wheel
monocoque
monopulse RF sensor
multi-layer insulation (MLI)
multipaction
multiplexer
multispectral
multispectral scanner
mux

nickel–cadmium (NiCd) cell
nickel–hydrogen (NiH$_2$) cell
nickel-metal-hydride (NMH) cell
NMH
north–south station keeping
nuclear reactor
nutation
nutation damper

offset-fed (antenna)
omni
omnidirectional antenna
OMT
Omux
ONET
on-station
optical bench
optical solar reflector (OSR)
optical tracking
orbital control
orbital relocation
orthomode transducer (OMT)
OSR
output filter
output multiplexer
output section

pad
panchromatic

parabolic antenna
parallel tubes
parametric amplifier
paramp
parasitic station acquisition
payload
payload module
P-code
PCU
phased array antenna
photovoltaic cell
pin-wheel louvre(s)
platform
plumbing
plume impingement
plus-x face
plus-y face
plus-z face
polariser
power
power budget
power bus
power subsystem
power supply
primary power supply
primary structure
prime spacecraft
probe
propulsion module
pumped fluid loop
pushbroom scanner
pyrotechnic cable-cutter
pyrotechnics

QM

radar altimeter
radiation pattern
radiator
radioisotope thermoelectric generator
   (RTG)
radiometer
ranging
reaction wheel
reaction wheel assembly (RWA)
reactor

receive chain
receiver
reciprocity (principle of)
recovery
re-entry
re-entry corridor
reflector antenna
regulated power supply
remote agent
rendezvous
repeater
retroreflector
re-visit capability
RF chamber
RF power
RF sensor
RF switch
ring redundancy
robotics
rover
RTG
RWA

sacrificial shield
SADA
SADE
SADM (solar array drive mechanism)
SAR
satellite
scanner
scatterometer
scatterometry
second surface mirror (SSM)
secondary power supply
secondary structure
SEE
selective availability
semiconductor
sensor
service module
SEU
shake table
shaped reflector
shear web
shunt dump regulator
shutter control

Si
side lobe
side-looking radar
silicon cell
silver–zinc cell
simulation heater
simulator
single event effect (SEE)
single event upset (SEU)
slow-wave structure
SM
SNAP
snubber
soft dock
solar absorber
solar array
solar array drive mechanism (SADM)
solar battery
solar cell
solar generator
solar panel
solar radiation pressure
solar reflector
solar sailing
solid state power amplifier (SSPA)
solid state recorder
sounder
space probe
space segment
space telescope
space vehicle
spacecraft
spacecraft axes
spacecraft bus
spacecraft integration
spacecraft platform
spacecraft simulator
spaceprobe
specific energy
specific power
spectrometer
spin-stabilised (spacecraft)
spinner
spin-up
split-spool device
SSM

SSPA
stabilisation system
stable platform
star sensor
station keeping
steerable antenna
stress (engineering)
structure model
subreflector
substitution heater
sun sensor
surface coatings (of a spacecraft)
swath
swath-width
synthetic aperture radar (SAR)

tanker
taper
TC
TC&R
TDRS
TDRSS
telecommand
telemetry
telemetry, command and ranging
    (TC&R)
telemetry, tracking and command
    (TT&C)
tether
thermal balance
thermal barrier
thermal blanket
thermal control subsystem
thermal cycle
thermal doubler
thermal energy balance
thermal grease
thermal insulation
thermal knife
thermal louvre
thermal model
thermal stand-off
thermal stress
thermal subsystem
thermal-vacuum chamber
three-axis stabilised (spacecraft)

thrust cone
thrust cylinder
thrust structure
thrust tube
TM
TOMS
torque coil
tracking
transducer
transmission line
transmit chain
transmitter
transponder
transponder chain
travelling wave tube (TWT)
travelling wave tube amplifier (TWTA)
tri-axis stabilised
TT&C
TT&C antenna
tube
tweeta
twit
TWT
TWTA

unfurlable antenna
unregulated power supply
upconverter
UPS

vacuum chamber
valve
variable conductance heat pipe (VCHP)
vibration facility
vibration table
vidicon
voltage

WASS
waveguide
waveguide switch
Whipple shield
wind scatterometer
windmill torque
wire antenna
wiring harness

zero momentum system
zoom antenna

2 Communications Technology

active satellite
altitude-azimuth mount
AM
amplitude modulation (AM)
angle modulation
antenna beam
antenna gain
anti-jam
AOS
asymmetric link
atmospheric attenuation
attenuation
azimuth

back off (from saturation)
background noise
backhaul link
band (frequency)
bandwidth
baseband
baud
beacon
beam
beam area
beam centre
beam edge
beam waveguide
beam-hopping
beamwidth
bent-pipe repeater
BER
binary phase shift keying (BPSK)
bit
bit error rate (BER)
bit rate
blockage
Boltzmann's constant
BPSK (binary phase shift keying)
broadband
broadband noise
broadcasting

broadcasting-satellite service
  (BSS)
BSS
burst
byte

cable
cable TV
capacity
carrier
carrier wave
carrier-to-interference ratio
carrier-to-noise power density ratio
  (C/N_o)
carrier-to-noise ratio
Cassegrain reflector
CATV
C-band
CDMA
channel
channel isolation
channel separation
channelisation
circular polarisation
coast earth station
co-channel interference
code division multiple access (CDMA)
codec
coding
collocate
common carrier
commonality
communication
communications
communications link
communications system
companding
compression
co-polar(ised)
coverage area
crosslink

cross-polar discrimination (XPD)
cross-polar(ised)

DAMA
data
data communications
data link
data rate
datacoms
datum
dBc ('dB relative to carrier')
dBHz ('dB-hertz')
dBi ('dB-isotropic')
dBK ('dB-kelvin')
DBS
dBW ('dB-watts')
decibel (dB)
declination
decoding
decryption
de-emphasis
delay
delayed repeater satellite
delta modulation
demand assignment multiple access
    (DAMA)
demodulation
demodulator
descrambler
descrambling
digital compression
digital video broadcast (DVB)
direct broadcasting by satellite (DBS)
direct-to-home (DTH)
discrimination
Doppler shift
double conversion transponder
double hop
downlink
download
drift tube
DTH
dual conversion transponder
dual-gridded reflector
ducting
DVB
DVS

earth segment
earth station
earth terminal
eclipse-protected
edge of coverage
EHF (extra high frequency)
EIRP
electron beam
elevation (angle)
elliptical polarisation
encryption
end user
energy dispersal
ENT (equivalent noise temperature)
equatorial mount
equivalent isotropic radiated power
    (EIRP)
equivalent noise temperature
Eurovision
extranet

facsimile
fax
FDM
FDMA
fibre optics
figure of merit (G/T)
fixed-satellite service (FSS)
FM
footprint
forward error correction (FEC)
free space loss
frequency
frequency allocation
frequency allotment
frequency assignment
frequency band
frequency bands
frequency coordination
frequency division multiple access
    (FDMA)
frequency division multiplexing (FDM)
frequency modulation (FM)
frequency re-use
frequency separation
frequency spectrum
frequency-hopping

MPEG
MSS
multicasting
multimedia
multipath
multiple access
multiple beam
multiplexed analogue component
  (MAC)
multiplexing
multipoint-to-point

narrowcasting
noise
noise factor (*F*)
noise figure (NF)
noise power
noise power density (NPD; N_o)
noise temperature
Nyquist frequency

OBDP
OBP
on-board data processing
on-board processing
optical fibre
orbital replacement unit (ORU)
ORU
outdoor unit
output
overspill

passband
passive intermodulation (PIM)
PCM
PCS
peak
personal communications system
  (PCS)
PFD
phase modulation (PM)
phase shift keying (PSK)
PIM
PM
pointing
pointing loss
point-to-multipoint

point-to-point
polar mount
polarisation
polarisation conversion
polarisation frequency re-use
polarity
power flux density (PFD)
pre-emphasis
prime focus
propagation delay
protection ratio
PSK
pulse code modulation (PCM)

QPSK (quadrature phase shift keying)
quasi-DBS
quaternary phase shift keying (QPSK)

radiated power
radio (band)
radio frequency (RF)
radio link
radio propagation
radio signal
radio spectrum
rain attenuation
regenerative repeater
RF
RF link
RF signal
RF wave
RHCP (right-hand circular polarisation)

satellite communications
satellite switching
satellite-switched time division
  multiple access (SSTDMA)
saturated output power
saturation
S-band
SCPC
scrambler
scrambling
service area
SHF (super high frequency)
ship earth station
shot noise

signal
signal delay
signal-to-noise ratio
single channel per carrier (SCPC)
single conversion transponder
single-entry interference
single hop
slow-scan TV
SMATV
SNG
SOHO
sound-in-syncs (SIS)
spatial diversity
spatial frequency re-use
SPCN
spillover
splashplate
spot beam
spread spectrum
spreading loss
spurious signal
squint
SSTDMA
star network
step-track
stopband
store-and-forward
subcarrier

TCP
TDM
TDMA
telecommunications
teleconference
telephony
teletext

televideo
television
telex
terrestrial tail
TES
thermal noise
THz
time division multiple access (TDMA)
time division multiplexing (TDM)
traffic
transparent repeater
transponder back off
TV distribution
TVRO

UHF (ultra high frequency)
unicasting
uplink
upload

vertical polarisation
VHF (very high frequency)
videoconference
videotext
VLF (very low frequency)
VSAT

wideband
wideband noise
wireless local loop
WWW

X-band
XPD

zone beam

3 Propulsion Technology

ABM
aerospike
aerospike engine
air-breathing rocket
AKM
AMF
annular nozzle
apogee boost motor (ABM)
apogee engine
apogee engine firing
apogee kick motor (AKM)
apogee motor firing (AMF)
apogee stage
arcjet thruster
area expansion ratio
ascent engine
ascent stage
axial thruster

bipropellant propulsion system
bipropellant thruster
blowdown system
booster
burn
burn-out
burst test

characteristic velocity
chemical propulsion
chilldown
chugging
cigarette combustion
cold-gas thruster
combined (bipropellant) propulsion
   system
combined cycle engine
combustion
combustion chamber

delta V (ΔV)
descent engine
descent stage

EHT
ejection velocity
electric propulsion (EP)
electrically heated thruster (EHT)
electromagnetic propulsion
electron bombardment ion thruster
electrostatic propulsion
electrothermal hydrazine thruster
   (EHT)
electrothermal propulsion
engine
engine cut-off
engine re-start (capability)
EP
exhaust
exhaust nozzle
exhaust plume
exhaust velocity
exit cone
expansion nozzle
expansion ratio

flame inhibitor
flow separation
fuel budget
fuel estimation

gas jet

Hall effect thruster
hiphet
hiphet thruster
hybrid rocket
hydrazine thruster
hypersonic flow

igniter
ignition
ignition system
impulse
inhibitor

ion engine
ion propulsion
ion thruster

LAE
linear aerospike engine
liner
liquid apogee engine (LAE)

magnetoplasmadynamic thruster
motor
motor case
MPD

NERVA
neutraliser
nozzle
NTR
nuclear propulsion
nuclear thermal rocket (NTR)

OMV
orbital manoeuvring vehicle
  (OMV)
orbital transfer vehicle (OTV)
OTV
overexpanded

PAHT
PAM
payload assist module (PAM)
percussion primer
perigee kick motor (PKM)
perigee stage
photon engine
PKM
plasma rocket
plasma thruster
plug nozzle
plume
plume shield
PMD
power-augmented hydrazine thruster
  (PAHT)
pressurant
pressure expansion ratio
pressurisation

propellant budget
propellant estimation
propellant liner
propellant management device
  (PMD)
propellant tank
propulsion subsystem
propulsion system
pulsed thrust
pyrotechnic valve

radial thruster
radio frequency ionisation thruster
ramp
reaction control system (RCS)
reaction control thruster
re-light
restartable upper stage
retro-pack
retro-rocket
rocket
rocket booster
rocket engine
rocket equation
rocket motor
rocket nozzle

safing
solid rocket motor (SRM)
space tug
specific impulse
spike
spinning solid upper stage (SSUS)
SRM
SSUS
stationary plasma thruster (SPT)
stoichiometric ratio
subsonic flow
supersonic flow

tank
thermal gaiter
throat
throttling
thrust
thrust coefficient
thrust profile

4 Launch Vehicle Technology

A-4
access arm
access tower
aerodynamic heating
aerodynamic stress
aerospace vehicle
Agena
airframe
Angara
anti-slosh baffle
Ariane
ASLV
Athena
Atlas
Atlas-Centaur
autopilot
avionics

baffle
balloon tank
Bell X-1
blast-off
blockhouse
Blue Streak
boost phase
built-in hold

Centaur
clamp band
coasting
coning
core vehicle
Cosmos
countdown
countdown sequencer
cowling
CSG
Cyclone

Delta
dispenser

Dnepr
downrange
dry mass
dry weight
dummy payload
dummy stage
Dutch roll
dynamic pressure

EELV
ELA
ELV
Energia
engine bay
envelope
equipment bay
escape tower
Europa
Evolved Expendable Launch Vehicle
    (EELV)
expendable launch vehicle (ELV)
extension module

fairing
fin
firing range
firing room
fixed service structure
flame bucket
flame deflector
flame trench
flight
flight controller
flight hardware
flight path
flight termination system (FTS)
FTS

gantry
gas generator
gimbal
GSLV

H (launch vehicle)
heavy-lift launch vehicle (HLLV)
heavy-lift vehicle (HLV)
HLV
hold (in countdown)
hold-down arm
Hotol

ICBM
inertial upper stage (IUS)
instrument unit
intercontinental ballistic missile
    (ICBM)
interstage
intertank structure
IRBM
IUS

jettison
Jupiter

kinetic heating

laser gyro
laser ring gyro
launch
launch campaign
launch capability
launch commit criteria
launch complex
launch control centre
    (LCC)
launch escape tower
launch pad
launch platform
launch profile
launch shroud
launch site
launch vehicle
launch window
launcher
launcher release gear
lift-off
live stage
LLV
LMLV
Long March

longeron
LRG

M-5
mass ratio
maximum dynamic pressure
max-Q
mobile service structure
Molniya
multistage rocket

N (launch vehicle)
N-1
nose cone
nose fairing

ogive
one-and-a-half stage
optical gyroscope

passivation
payload capability
payload envelope
payload fairing
payload fraction
payload integration
payload mass ratio
payload shroud
Pegasus
pitch-over
plugs-out test
pogo
pogo suppressor
propellant dispersal system
propellant mass fraction
propulsion bay
Proton
proving stand
PSLV
purge

ramjet
range
range safety system
ranging
recycle countdown
Redstone

regenerative cooling
relief valve
reusable launch vehicle
    (RLV)
ring laser gyro (RLG)
RLG
rocket stage
Rockot
roll programme

Sanger
Saturn 1B
Saturn V
Scout
scramjet
scrub
Sea Launch
sequencer
service structure
service tower
shock diamonds
shock suppression system
shroud
single stage to orbit (SSTO)
skin
skirt
slant range
sloshing
sonde
sound suppression system
sounding rocket
Soyuz
spaceport
SPELDA
SPELTRA
spin table
SSTO
stack
stacking
stage
stage separation
staging
Stargazer
Start
static firing
step-rocket
strapdown gyro system

strap-on (booster)
stretched (propellant tank)
stringer
sustainer engine
swing-arm
SYLDA

tail service mast
Taurus
test stand
Thor
thrust frame
thrust structure
Titan
Titan-Centaur
T-minus . . .
triggered lightning
Tsyklon
turbopump

umbilical
umbilical tower
uprate

V-2
variant
vector steering
vehicle equipment bay
vent and relief valve
vent valve
vernier engine
Viking
Vostok
Vulcain

white room
window
wind-shear

X-1
X-15
X-33
XS-1

yo-yo

Zenit

5 Space Shuttle

airborne support equipment
   (ASE)
approach and landing test
APU
ASE
Atlantis
auxiliary power unit (APU)

beanie cap
Buran

cargo bay
Challenger
chase-plane
Columbia
cradle
Crawler
crawlerway
crossrange capability

Discovery

elevon
Endeavour
end effector
Enterprise
ET
external tank (ET)

flight deck
flight readiness firing (FRF)
frisbee ejection system

GAS
gaseous oxygen arm
Getaway Special (GAS)
glass cockpit
glide path
glide slope
grapple fixture

manipulator arm
manned manoeuvring unit (MMU)
mate/de-mate device
MECO
mid-deck
mission specialist
MMU
mobile launch platform

OMS
OPF
orbital manoeuvring system (OMS)
orbiter
orbiter processing facility (OPF)

pallet
payload bay
payload canister
payload changeout room
payload specialist
personal rescue sphere (PRS)
PRS

remote manipulator system (RMS)
RMS
rotating service structure (RSS)
runway (for Space Shuttle)

Shuttle
Shuttle carrier aircraft (SCA)
Shuttle landing facility (SLF)
SLWT
small self-contained payload (SSCP)
solid rocket booster (SRB)
Space Shuttle
Space Shuttle abort alternatives
Space Shuttle main engine (SSME)
space transportation system (STS)
Spacelab pallet
speedbrake

6 Manned Spaceflight

activated charcoal canister
airlock
Alpha
ALSEP
anoxia
Apollo
Apollo-Soyuz test project (ASTP)
ASTP
astronaut
astronaut manoeuvring unit (AMU)
astronautics

bends

cabin
cabin pressure
capcom
capsule
capsule communicator
carbon dioxide absorber
carbon dioxide narcosis
centrifuge
closed-ecology life support system (CELSS)
CO_2 scrubber
COF
Columbus Orbital Facility (COF)
command and service module
command module
communications carrier assembly
constant-wear garment
cooling system (of a spacesuit)
cosmonaut
cosmonautics
crew
cupola

decompression
decompression sickness
dysbarism

ebullism
ECLSS
egress
EMU
environmental control and life support system (ECLSS)
EVA
extra-vehicular activity (EVA)
extra-vehicular mobility unit (EMU)
extra-vehicular pressure garment

FGB
flotation bag
flotation collar
foot restraint
Freedom
functional cargo block (FGB)

Gemini

habitable module
Hermes
hypoxia

ICM
igloo
ingress
Interim Control Module (ICM)
International Space Station (ISS)
intra-vehicular activity (IVA)
intra-vehicular pressure garment
ISS
IVA

JEM

Krystall
Kvant

LEM
life support

7 Unmanned Spacecraft

applications satellite
artificial satellite
astronomical satellite
ATV
automated transfer vehicle
   (ATV)

communications satellite

Early Bird
Earth resources satellite
Echo
Explorer

Galileo
Giotto
Gorizont

HST
Hubble Space Telescope (HST)
hybrid satellite

Kosmos

Landsat
Luna
Lunokhod

Mariner
Mars Pathfinder
meteorological satellite
Meteosat
metsat
microsat
minisat

Molniya
MSG
multi-purpose satellite

nanosat
navigation satellite
Navstar
NGST

passive communications satellite
picosat
Pioneer
Pioneer Venus

Ranger
reconnaissance satellite

smallsat
Sojourner
SPOT
Sputnik
spy satellite
Syncom
surveillance satellite
Surveyor

Telstar
tethered satellite
Tiros
Transit

Vanguard
Venera
Viking
Voyager

8 Materials

ablation
aluminium (Al)

beryllium (Be)

carbon composite
carbon fibre reinforced plastic (CFRP)
carbon–carbon composite
ceramic matrix composite (CMC)
ceramics
CFRP
CMC
cold-welding
composites

Dacron
degradation (of spacecraft materials)

fatigue
filament-winding
flat absorber
flat reflector
fracture mechanics

GaAs
graphite epoxy

interlayer
Invar

Kapton
Kevlar composite
KRP

materials
metal matrix composite
  (MMC)
MMC
Mylar

Nomex

outgassing

pre-preg

refractory metal

shear
silicon (Si)
stress-corrosion cracking

tape-wrapping
titanium (Ti)

9 Propellants

aerozine-50
ammonium perchlorate (NH_4ClO_4)
AP

binder
bipropellant
boil-off

charge
composite propellant
cryogenic propellant
CTPB

double-base propellant

end-burning

fuel

grain

heterogeneous propellant
homogeneous propellant
HTPB
hydrazine (N_2H_4)
hydrogen peroxide (H_2O_2)
hypergolic

kerosene

liquid hydrogen
liquid oxygen (LOX)

liquid propellant
LOX

mixed oxides of nitrogen (MON)
MMH
MON
monomethyl hydrazine (MMH)
monopropellant

NHMF
nitrogen tetroxide (N_2O_4)
NTO

oxidiser

polybutadiene
propellant
propellant charge
propellant grain
pyrophoric fuel

RFNA
RP1

slush hydrogen
solid propellant
storable propellant

UDMH
UH25
unsymmetrical dimethylhydrazine
   (UDMH)

web

10 Orbits

altitude
aphelion
apoapsis
apogee
areostationary orbit
ascending node
atmospheric drag

ballistic fly-by
ballistic trajectory

circular orbit
Clarke orbit

debris
debris mitigation
decay
de-orbit
descending node
direct orbit
drag compensation
drag make-up
drift
drift orbit

eccentricity
elliptical orbit
equatorial orbit
escape velocity

first space velocity
fly-by
free-return trajectory

GEO
geostationary arc
geostationary orbit (GEO)
geostationary transfer orbit
   (GTO)
geosynchronous orbit
graveyard orbit

gravitational attraction
gravitational boost
gravitational perturbation
gravity propulsion
ground track
GSO
GTO

halo orbit
heliosynchronous orbit
HEO
high earth orbit (HEO)
highly elliptical orbit (HEO)
Hohmann transfer orbit

ICO
inclination
inclined orbit
injection
insertion
intermediate circular orbit (ICO)

Lagrange point
LEO
libration
libration point
low Earth orbit (LEO)
luni-solar gravity

MCC
medium Earth orbit (MEO)
MEO
mid-course correction (MCC)
mission control
mission control centre (MCC)
Molniya orbit

nadir
NGSO
nodes
nominal orbital position

orbit
orbital debris
orbital decay
orbital elements
orbital environment
orbital inclination
orbital injection
orbital parameters
orbital period
orbital plane
orbital position
orbital slot
orbital spacing
orbital track

parabolic trajectory
parking orbit
periapsis
perigee
perihelion
perturbations
plane change
polar orbit
prograde orbit

retrograde orbit

second space velocity
semi-major axis
semi-minor axis
sling-shot (gravitational)
SSO
stationary orbit
sub-orbital hop
sub-orbital trajectory
sub-satellite point
sun-synchronous orbit
  (SSO)
supersynchronous orbit
swing-by

third space velocity
trajectory
transfer orbit
trans-lunar injection (TLI)
trans-lunar trajectory
tundra orbit

zenith

11 Physics and Astronomy

absolute temperature
absolute zero
acceleration due to gravity
albedo
ALH84001
arean
asteroid
astrogeology
astronomical unit (AU)
atmosphere
ATOX
AU

celestial body
centre of gravity
centre of mass
centrifugal force
centripetal force/acceleration
cislunar
corona
cosmic radiation
Coudé focus (telescope)
cryogen
cryogenics

deep space
dielectric
digital elevation model
drag
drop tower
dynamic pressure

Earth
earthshine
eclipse
ecliptic
electromagnetic pulse
  (EMP)
electromagnetic radiation
electromagnetic spectrum

electron
electrostatic discharge (ESD)
EM
EMI
EMP
energy
ephemeris
equinox
ESD
exosphere
extra-solar planet

false colour image
far infrared
focal plane
focal plane assembly (FPA)
FPA
frame of reference
free fall
free space

gamma-ray spectrum
geomagnetic storm
geometric resolution
g-force
gravitational anomaly
gravitational constant
gravitational field
gravity

hard vacuum
heliopause
heliosphere

image
inertial frame
infrared
insolation
interplanetary
ion

thermoelectric effect
thermosphere
transit
trapped radiation
triaxiality
tropopause
troposphere

ultraviolet
upper atmosphere

vacuum
Van Allen belts

vector
velocity of light
visible spectrum
VLBI

wave number
waveband
wavelength
weightlessness

X-ray spectrum

zero gravity

12 Space Centres and Organisations

Alcantara
Ames Research Center
Andoya
ARPA
ARS

Baikonur Cosmodrome
BIS
BNSC

Cape Canaveral
Cape Canaveral Air Station (CCAS)
Cape Kennedy
CCAS
CCIR
CCITT
CEPT
CNES
Comsat
Cosmodrome
COSPAR
COSPAS-SARSAT
CSG

DARPA
Deep Space Network (DSN)
DOD
Dryden Flight Research Center
DSN

Eastern Test Range
EBU
ECS
Edwards Air Force Base
ELDO
ESA
ESOC
ESRIN
ESRO
ESTEC
Eumetsat

European Space Agency (ESA)
Eutelsat

FCC

Gagarin Centre
Glenn Research Center
Goddard Space Flight Center
   (GSFC)
Guiana Space Centre

IAA
IAF
IFRB
IGY
IISL
Inmarsat
Intelsat
Intercosmos
Intersputnik
ISAS
ISRO
ISY
ITU

Jet Propulsion Laboratory (JPL)
Jiuquan
Johnson Space Center (JSC)
JPL

Kagoshima
Kakuda Propulsion Center
Kapustin Yar
Kennedy Space Center (KSC)
Kourou
KSC

Langley Research Center
Lewis Research Center

Marshall Space Flight Center

NACA
NASA
NASDA
National Space Technology
  Laboratories
NOAA
NORAD
Northern Cosmodrome

Palmachim
Plesetsk
PTT

RAKA
RARC
RASA
Regional Administrative Radio
  Conference (RARC)
RKA
Rosaviakosmos
RSA

San Marco platform
SHAR
Space Command
Space Surveillance Network (SSN)
Space Tracking and Data Network
  (STDN)
Sriharikota

SSN
Star City
STDN
Stennis Space Center
SUPARCO
Svobodny

Taiyuan
Tanegashima
Tsukuba Space Center
Tyuratam

UNCOPUOS
US Space Command

VAFB
Vandenberg Air Force Base
  (VAFB)
VfR

Wallops Flight Facility
WARC
Western Test Range
White Sands Test Facility
Woomera
World Administrative Radio
  Conference (WARC)

Xichang

13 Miscellaneous

abort
active
aerodynamics
aerospace
announcement of opportunity
   (AO)
AO
application
areocentric
assembly

back-contamination
back-up
BDR
beam-builder
block
boilerplate
breadboard

CDR
clamp release
commercial off-the-shelf
computer enhancement
configuration
conic sections
constructive total loss
consumable
control system
conus
coolant
cordon sanitaire
COTS

deductible
degree of freedom
delivery in orbit
deploy
design driver
design lifetime
DRAM

Earth observation
EGSE
ELINT
environment (of space)
EPROM
experimental

firmware
forward contamination

Gaussian distribution
generation
Geographic Information System
   (GIS)
geolocation
GIS
glitch
graceful degradation
ground support equipment (GSE)
GSE

hardware
hi-rel
hyperbola
hyperboloid

in-orbit delivery
integration
intentional ignition
ITT

Laika
liability convention
lifetime

Mach number
maneuver
manoeuvre
manufacturer
margin

mass-limited
MET
MGSE
mil-spec
mission
mission elapsed time (MET)
module
Monte Carlo analysis
Moon Agreement
MOU
MTBF
MTTF
MTTR

NDT
negative margin
nominal
non-destructive testing (NDT)
non-operational
normal distribution

OEM
operational
operational lifetime
orthogonal axis
Outer Space Treaty
over-design

paper satellite
parabola
paraboloid
partial loss
passive
PDR
performance
PGSE
pitch
pitch axis
pixel
power-limited
pre-launch
pre-operational
prime contractor
prime system
probability distribution
processing

PROM
pyrotechnic

qualification
quarantine facility

radar
rad-hard
radiation environment monitor (REM)
radiation hardening
RAM
rat's nest
raw data
receiving facility
real time
redundancy
redundant
Registration Convention
reliability
REM
remote sensing
Rescue Agreement
residuals
retrofit
RFP
RFQ
roll
roll axis
ROM

sample return
SDI
sigma notation
sim
simulation
single point failure (SPF)
soft fail
software
software patch
SOW
Space Age
space colony
space debris
space environment
space exploration
space insurance

space law
space object
space-qualified
space science
space technology
space tourism
space treaty
spacecraft debris
spaceflight
spec
specification
SPF
START
statement of work (SOW)
strategic defense initiative (SDI)
subcontractor

subsystem
system

technology demonstrator
teleoperate
telepresence
threshold
TLO
total loss
tumbling
turnkey

vehicle

yaw
yaw axis